THE SCIENCE OF SELF

MAN, GOD, and the MATHEMATICAL LANGUAGE of NATURE

SUPREME UNDERSTANDING and C'BS ALIFE ALLAH

DEDICATION

SUPREME UNDERSTANDING

To the memory of David Walker, George Washington Williams, Duse Mohamed Ali, William Henry Ferris, Marcus Mosiah Garvey, Hubert Henry Harrison, Noble Drew Ali, Henry McNeal Turner, Wallace Fard Muhammad, Elijah Muhammad, Allah, Ivan Van Sertima, Cheikh Anta Diop, John Henrik Clarke, Asa Hilliard, and countless other ancestors who live on in this work, and in us.

C'BS ALIFE ALLAH

To Mrs. Janice and the Discovery Room at Martin Luther King School, Aunt Darlene for the subscription to 3-2-1 Contact, the Peabody Museum/Yale University summer camp, Mommy and Poppy, and the stack of science fiction books and comic books that made me want to know just how this world was put together and explore it for myself.

The Science of Self: Man, God, and the Mathematical Language of Nature. Copyright ©2012 by Supreme Design, LLC. All rights reserved. No part of this book may be reproduced in any form or by any electronic or mechanical means including information storage and retrieval systems without permission in writing from the publisher, except by a reviewer, who may quote brief passages in a review. Published by Supreme Design Publishing. PO Box 10887, Atlanta, GA 30310. First printing.

Although the author and publisher have made every effort to ensure the accuracy and completeness of information contained in this book, we assume no responsibility for errors, inaccuracies, omissions, or any inconsistency herein. Any perceived slights of people, places, or organizations are not intended to be malicious in nature.

Supreme Design Publishing books are printed on long-lasting acid-free paper. When it is available, we choose paper that has been manufactured by environmentally responsible practices. These may include using trees grown in sustainable forests, incorporating recycled paper, minimizing chlorine in bleaching, or recycling the energy produced at the paper mill.

Supreme Design Publishing is also a member of the Tree Neutral™ initiative, which works to offset paper consumption through tree planting.

ISBN: 978-1-935721-67-3 LCCN:

Wholesale Discounts. Special discounts (up to 55% off of retail) are available on quantity purchases. For details, visit our website, contact us by mail at the address above, Attention: Wholesale Orders, or email us at orders@supremedesignonline.com

Individual Sales. Supreme Design publications are available for retail purchase, or can be requested, at most bookstores. They can also be ordered directly from the Supreme Design Publishing website, at www.SupremeDesignOnline.com

Visit us on the web at www.ScienceOfSelf.com

TABLE OF CONTENTS

INTRODUCTION

BASIC INSTRUCTIONS

"The end of all explorations will be to come back to where we began and discover the place for the first time."
– T. S. Eliot, poet and playwright

Popular lore says that, sometime in the late 1800s, George Washington Carver rediscovered a unique shade of purple that had not been seen since Phoenician times. This dye, known as Tyrian purple, was such a unique formulation of pigments that we can't accurately represent it in the color pages of this text or online at our website. Until Carver rediscovered this formula, Tyrian purple had not been seen for over 2,000 years.[1]

George Washington Carver

Most records say Carver, one of the most important scientific minds of his era, was born into slavery in Missouri in January of 1864. Most of us learned about him for his work with peanuts. But Carver didn't simply invent peanut butter. Carver was a scientist, botanist, educator, and inventor. He invented over 500 plant-based technologies, ranging from synthesized rubber to industrial dyes.[2] He revolutionized southern agriculture, transforming its dependence on a single-crop economy, effectively "saving the South."[3] Though it may seem not to mean anything now, don't forget Carver and his rediscovery of that purple dye. We'll get back to why it's important.

WARNING

Now, fair warning: this book might give you a headache. It's not that we want to make your brain hurt. We didn't have a choice. We wanted this book to be clear and easy to read. We wanted to take difficult concepts and simplify them, while simultaneously removing some of the mysterious and fantastic elements used by other writers exploring similar themes; but this book is a textbook like no other. It covers everything from quantum physics to paleoanthropology to superorganism theory. Throughout this book, we explore African fractals, Middle Eastern architecture, Native American metaphysics, and East Asian oracles. This book connects the dots between intimidating topics like the self-creation of the first man and the spirals found in both our fingerprints and distant galaxies. So, unless you know it all already, or you're brain-dead, this book is likely to give you a headache at some point. Every time you learn something new or challenging, your brain gets another wrinkle – new synapses are formed between previously unconnected neurons. Your worldview may even change, which is probably the most painful part of the process, but it's also the most rewarding. We hope to change the way you see science, the way you see yourself and the way you see all life and matter…even IF it hurts.

This book is based primarily on scientific evidence. That is, while our language may be casual and even colloquial at times, we won't resort to fantastic claims about things we can't prove. We strived not to follow the paths of our peers and predecessors who've avoided scientific research in favor of myths, modern-day misconceptions, and urban legends. Yet this book still won't be well-received by the academic community. This book promotes many concepts that are considered unpopular in the Western academic community, such as the link between human melanin and the black matter found in interstellar Bok globules. Such a connection is well-documented and supported by many scientists, but for obvious reasons, it's not widely accepted. That's expected for a lot of what we'll say. However, that doesn't mean that we'll make the claim that melanin is a magical gift from the heavens above just because "it sounds good." Instead, we'll employ scientific inquiry, using both Western AND indigenous traditions, to determine the relationship between the two phenomena. Does that make sense so far? In laymen's terms, if we can't explain and validate it, then we haven't included it.

In fact, we're not going to push any beliefs on you. This book was written to promote scientific thinking. We want you to think critically about everything you read here. We want you to dig further and do your own research. We want to reintroduce science and history to everyone who fell asleep during 6th grade social studies or high school physics. Finally, we want to demonstrate how immensely rewarding it is to learn about reality…and hopefully, as a result, more people will become less immersed in the world of fantasy.

There are two primary authors involved in this project: C'BS Alife Allah and Supreme Understanding Allah. There are several others who contributed to (or peer-reviewed) this project who are or aren't Five Percenters, but we obviously are. We could've published the book under our "government names" and avoided some controversy, but we considered it important to highlight the fact that it was our involvement in the Nation of Gods and Earths (commonly known as the Five Percenters) that led us to critical thinking and scientific inquiry in the first place. We didn't receive "indoctrination" in our experiences with the culture. Instead, we were taught to question everything, including our own teachings. As a result, we have committed countless years, months, and days of our personal lives to studying something our schools wouldn't teach us no matter how many degrees we acquired: The who, what, when, where, why, and HOW of the Black man and woman.

"The opportunity of our time is to integrate science's understanding of the universe with more ancient intuitions concerning the meaning and destiny of the human."
– Brian Swimme, The Hidden Heart of the Cosmos

It's our goal to share this knowledge...and only that which can be literally considered "knowledge"...with anyone willing to reconsider how and why things came to be the way they are today. For more on the story of how this book came to be, see "Knowledge of Self" in the Appendix.

IS THIS BOOK FOR YOU?

- ❒ If you're interested in the history of how this Universe and all life in it came to be.
- ❒ If you want to know the history of all your ancestors, and how they came to become you.
- ❒ If you want to understand your place in the grand scheme of things.
- ❒ If you want to make sense out of the patterns you see in life and nature.
- ❒ If you're intrigued by science, history, anthropology, psychology, mathematics, quantum mechanics, etc.
- ❒ If you're looking for the real substance behind the esoteric ideas.
- ❒ If you read Knowledge of Self and you're looking to dig a little deeper.
- ❒ If you're skeptical about the metaphysical claims of others, and want to know what's real and what's not.
- ❒ If you want to read a gang of deep stuff to tell people so you can show off how brilliant you are.
- ❒ If you don't care about showing off, but you do need the equivalent of a teacher's guide or reference book to Black consciousness.

If you checked off any of those items, this book is for you.

Why this Book Exists

SUPREME UNDERSTANDING

"I was looking for myself and asking everyone except myself questions which I, and only I, could answer." – Ralph Ellison, Invisible Man, Prologue

Obviously, we wrote this book to counter all the lies that have been taught about Original people.* We wrote this book to introduce regular people to scientific inquiry, a process that can improve the outcome of everything we do in our lives, a process we may have shunned or ignored because we associated it with "they schools" and never saw the connection to our own reality. We also wrote this book to inspire a new generation of young anthropologists, physicists, mathematicians, psychologists, historians, geneticists, and archaeologists...people that can do a new wave of research that will REALLY construct a new paradigm (or would it be *the original?*).

THE SCIENCE OF SELF

"He who experiences the unity of life sees his own Self in all beings, and all beings in his own Self, and looks on everything with an impartial eye." – The Buddha

Now that we've covered the basic definition of science, let's talk about its meaning. To do so, let's start with the "science" behind the title of this book. In the symbolic language of Supreme Mathematics, each number (1, or knowledge, in this example) represents both a stage AND the process by which you arrive at that stage. So "knowledge" (or science, which literally means knowledge) means both the information and the process of acquiring or evaluating information. In that

* As we explained in *The Hood Health Handbook*, our usage of the term "Original" is a purposeful act of self-definition: "Throughout this book, "Original" will be used synonymously with Black or African-American, or as an umbrella term covering Black, Latino, Asian, Native American and other indigenous populations. While the use of the term "Original" to cover such a diverse population is not widely accepted, we argue that there is no other word that captures the deep commonality of people of color worldwide, without somehow being a subtle reference to a white or European standard. For example, "people of color" suggests that the norm is actually people without color, just as "minorities" suggests that Original people are somehow not in the majority worldwide. Similarly, "non-white" suggests that the standard is "white" and everything else is "other" and thus secondary. It is our argument that the world's first people, anthropologically known as indigenous people, may have been geographically separated and culturally distinct, but were never collectively oppressed, forcibly marginalized, and systemically set at odds with each other (creating the present need for a collective identity) until Europeans distinguished themselves as "white" (and thus "different") and went to work throughout the world, completely changing our paradigms of identity and ideology."

DID YOU KNOW?
An aboriginal tale about creation speaks on man's eternal search for Knowledge of Self:
One day...the gods decided to create the universe. They created the stars, the sun and the moon. They created the seas, the mountains, the flowers, and the clouds. Then they created human beings. At the end, they created Truth.
At this point, however, a problem arose: where should they hide Truth so that human beings would not find it right away? They wanted to prolong the adventure of the search.
"Let's put Truth on top of the highest mountain," said one of the gods. "Certainly it will be hard to find it there."
"Let's put it on the farthest star," said another.
"Let's hide it in the darkest and deepest of abysses."
"Let's conceal it on the secret side of the moon."
At the end, the wisest and most ancient god said, "No, we will hide Truth inside the very heart of human beings. In this way they will look for it all over the Universe, without being aware of having it inside of themselves all the time."[4]

sense, each degree of Supreme Mathematics is a closed system of its own, yet part of a larger system that is also cyclical (1-9, then back to 1). Much like how we as individuals are systems within ourselves (while also composed of a collection of systems that work together), as well as being part of a larger system where we must work together with other systems. And thus the concept of us "living mathematics" and us BEING a personification of "living mathematics." This also relates directly to the concept of the Original Man "being" Knowledge (and thus Science) personified. The Five Percent maxims, "I Study Life Around Me" and "In Self Lies All Mathematics" convey the idea that the "SELF" not only contains the "Science of Everything in LiFe," but that the self IS the science of everything in life.

On Ghostface Killah's *Ironman* album, Popa Wu explains:

> Some people don't have no direction, because they don't know the science of they self. See, the science of life is the science of you – all the elements that it took to create you. Cause everything in the universe - that created the universe - exists within you. You see what I'm saying?

This idea is simple, but not necessarily easy. It explains everything, yet it doesn't necessarily explain anything. The practice of "I Study Life and Matter" helps us move towards the reality of "I Self Lord Am Master." Without this process, we're very unlikely to become in tune with the nature of reality. After all, we've been quite far removed. This much should be obvious.

In nearly all my interactions with sensible people of different religions, cultures, and backgrounds, we have all agreed that we are perpetually trying to find a state of feeling "whole," complete, or balanced, through some means or another. Some of us fill the voids we feel within ourselves with alcohol or drugs, while others develop strange addictions and fixations. Some never find anything that works and become depressed with no coping mechanisms. Others cope and find hope in faith and religion, while still others do the same things in non-religious settings. Some realize that they've been "filling a void" with one unhealthy thing, stop that, and soon replace it with another unhealthy habit or fixation to fill that void.

"Do not feel lonely, the entire universe is inside you." – Rumi

The bottom line is that almost all of us are perpetually seeking to fill our voids and some of our means of doing so are unhealthier than others. We can measure how unhealthy they are by the ill effects they have on our lives and those of others. Even the "savior syndrome" can become unhealthy when people who sincerely want to "fix the world" become so driven to do so that they kill themselves in the process. I've honestly never met a person who was entirely complete. Anytime I think someone seems totally "whole," I later find a better hidden hole.

So why do we have these voids within us? My brother Born God observed, "In *The Celestine Prophecy,* the author speaks on how people seek to fill a void for energy, and this energy is the God within, but due to not understanding that the energy is within, they seek it without."

The *Mundaka Upanishad* of ancient India says as much:

> Within the lotus of the heart he dwells, where, like the spokes of a wheel in its hub, the nerves meet…This Self, who understands all, who knows all, and whose glory is manifest in the universe, lives within the lotus of the heart, the bright throne of Brahman…Self-luminous is that Being, and formless. He dwells within all and without all…The Self exists in man, within the lotus of the heart, and is the master of his life and of his body…The knot of the heart, which is ignorance, is loosed, all doubts are dissolved…[5]

And you can find 500 other quotes – from the Gnostic Gospels of Christ to the Sufi traditions of Rumi – all saying the same thing. But that brings us back to the process.

Everybody knows it's all within us. But what is the process by which we arrive at that understanding in a way that registers with our whole being. We can only say *The Science of Self* is our humble contribution to that process. We hope it can open doors for you, and – at the very least – prove to you that the universe and all nature are personified in your existence. That there is no need to seek outside of yourself for your salvation or your sense of completion. That the origin* of all

---

* In this text, we define "origin" as that which is first (in linear progression, or at the onset of a cycle), or that which is the source from which things are derived from.

things – positive and negative – is found in you. And that the answer to every question and every problem begins with you. This book is only a part of the process of changing the way we think and live, but we'd like to think it will open minds in a way that few other works have accomplished thus far. Please read it carefully and apply its lessons wisely.

"...the world is just as concrete, ornery, vile, and sublimely wonderful as before, only now I better understand my relation to it and it to me." – Ralph Ellison, Invisible Man, Chapter 3

How to Read this Book

SUPREME UNDERSTANDING AND C'BS ALIFE ALLAH

"Upon this first, and in one sense this sole, rule of reason, that in order to learn you must desire to learn, and in so desiring not be satisfied with what you already incline to think, there follows one corollary which itself deserves to be inscribed upon every wall of the city of philosophy: Do not block the way of inquiry." – Charles Saunders Pierce, "First Rule of Logic"

BE PREPARED

This is not a light read. This is heavy and deep. If you've never read a book that deals with science or history before, you have to get ready for this book. (And unfortunately, some of the stuff circulating as science and history in our communities is written for people who will believe anything.) We often suggest that people read our earlier texts, specifically *How to Hustle and Win* and *Knowledge of Self*, before tackling our other books. In this case, that's not a suggestion! Don't dive into this water unprepared! Not to insult your intelligence, but we've also included a breakdown of "How to Read a Book" in the Appendix, simply because it's an excellent guide to making sense of difficult literature.

Although this book isn't a comic book or a religious pamphlet, it's also not a long-winded, overly technical academic paper that is too boring to finish. It's written in plain English. Sometimes there's some intellectual, academic sounding content, and sometimes we write in conversational tone with some sarcasm in the mix. That's how we keep our books interesting yet heavy. FYI: Don't copy our style.

EMBRACE CRITICAL THINKING

It's important to be alert and get your brain actively engaged while reading this book. The average person just passively receives a stream of information without analyzing the information. Their brain usually only becomes active and starts analyzing when in classes that involve critical thinking (contrary to popular belief, not all classes utilize critical thinking and many classes are set up for you to succeed in passing only if you intake and regurgitate the information as presented). In order to appreciate the content of this book, you have to become familiar with the scientific methodology of critical thinking and understand its value in viewing the world around you. In doing so, the benefit is greater than how much you get out of this book – such a mindset significantly improves how much you get out of LIFE. And THAT is a major part of why this book exists.

EXTRACT THE LESSONS

In many sections of this text, we've actually written out the lessons that history and science are revealing to (or reminding) us. But not in every section. If this is a book promoting critical thinking, it would be backwards for us to do it for you every time. We want you to find where the information you're reading applies to you or to the world today. The key is to understand the science of analogy and symbolism. For example, when you read the section about a prehistoric meteorite that may have carried DNA into the depths of the Earth's oceans, where life was born on our planet, what does it make you think of? Picture that flying asteroid with its tail "impregnating" the Earth and it's clear.* Or you can look at the nature of quantum entanglement, where paired particles are able to respond to changes in one another at speeds faster than the speed of light. What can you extract from this as a lesson on brotherhood? Don't worry, this kind of thinking gets much easier as you get used to it. If you are a citizen of the Nation of Gods and Earths who knows the 120 Lessons, or a Muslim who knows the Supreme Wisdom, many of these parallels and analogies will be more readily apparent to you. In fact, some of this material will be much more accessible for those who know those lessons by heart, and who are thus already familiar with these ideas which aren't taught in any schools.

MAKE SURE YOU UNDERSTAND

Don't just read an entire paragraph and then say to yourself, "What did I just read?" Yes, that WILL happen. If you need to read a sentence

* An asteroid is what you call a meteorite when it's flying through space. If it burns up in the atmosphere (a "shooting star"), it's called a meteor. If it hits Earth, it's called a meteorite. Comits also fly through space, but are made of dirt and ice, unlike asteroids which are made of carbon, metal, and/or rock.

two or three times to "get it," do so. If you need to skip to another section, fold the corner of the page and come back to it when you feel ready. Just make sure you come back!

Also, you might want to read this book with a dictionary nearby. If you have a smartphone, there are plenty of free dictionary apps that can serve the same purpose: helping you understand an unknown word that you can't otherwise figure out with context clues from the rest of the sentence.

TRANSLATE IT INTO PLAIN ENGLISH

In many places, we've already translated difficult content into the plainest English possible. But in other places, we've intentionally kept the reading above an 8th grade level. We want to challenge you on all ends. So practice translating your learning into your own voice. It will increase your understanding as well as your confidence in these subjects. You might want to make a Science of Self journal and keep frequent notes. You should also practice talking to people about what you're reading, but try explaining the ideas in your own way. So keep the book on you, and watch the conversations it sparks!

MAKE THE CONNECTIONS

In many places in the text, we note cross-references in the footnotes. These provide suggestions where you can get background info on the topic just mentioned, or another angle with which to look at the topic. But there are literally hundreds of other spots in this text where one piece of information connects to another piece of information elsewhere in the book. Those are for you to find.

"He who asks a question is a fool for five minutes; he who does not ask a question remains a fool forever." – Chinese Proverb

DO YOUR OWN RESEARCH

We left a lot of questions unanswered. Our goal isn't to fill your head with facts, but to show you the process by which things came to be, while also showing you that this process must occur with you on a daily basis. A major part of this process is the initial stage: the acquiring of knowledge. You can speculate all you want, but it isn't knowledge until you've researched, investigated, and did some evaluation to make sure the information is sound.

We want you to engage in that process with the claims we're making (don't take our word for it!), but also to research further and look into topics we didn't explore in depth. There are some topics we intentionally left open to future research.

PURSUE YOUR INTERESTS

Speaking of future research, there's a reason why this book rotates around the sciences and scientific thinking. We need more scientists. By scientists, we include everyone from historians to military strategists to quantum physicists. Basically, we need people who can get busy in the world of "thinking" and not just for the sake of thinking, but for the sake of nation-building and progressive global change. So hopefully, the introductions to various sciences and the subsequent discussions of topics that rely on these sciences will compel some of our readers to pursue these paths in their adult life. Of course, it doesn't take becoming a rocket scientist to make use of the knowledge you find it this book. Trust me, you'll find ways to apply this stuff.

"Knowing a great deal is not the same as being smart; intelligence is not information alone but also judgment, the manner in which information is collected and used." – Carl Sagan

WHAT WE ASK IN RETURN

And what do we ask of you in return for this hell we've put ourselves through to produce this work? Above all, we ask that you be a student first. Open yourself up to the idea of learning and mastering a subject before feeling qualified to teach on it. Do you know how many people we consulted to assemble this book? Plenty, because we don't know it all. There's a saying that goes, "It's better to be a fool and remain silent than to open your mouth and remove all doubt." Some of us are out here embarrassing ourselves, trying to open people's eyes by teaching them stuff we can't prove or even explain…until someone with sense makes us look like a dummy. It's just like studying martial arts on YouTube, until that fateful day when you pick a fight with someone who really knows their art. So, to be absolutely clear, ignorance does not equal authority. That is, not knowing the evidence or methodology is a reason to STOP and STUDY…not a reason to argue louder.

With that said, we'd like you to share what you're learning in this book. Just without all the pomp and circumstance. You know, sharing knowledge…opening eyes…giving people something to think about…not bombarding folks with arrogance and extravagance. Sounds fair, doesn't it?

SPREAD THE WORD

I know we're saying you shouldn't beat everyone in the head as soon as you learn something, but we're really just talking about the WAY we spread knowledge. We can't be all snobby and condescending about it, or argumentative and crazy either. If you want people to learn anything, you've got to make it register and relate. The way you do that is make it real and relevant. So keep this book on you, show it to people, and let them read something you think they'd like, or talk about something interesting that you learned. You can even help spread the word by posting about this book on Facebook, Twitter, Tumblr, WSHH, MTO, your blog, etc. But, above all, let's get back to talking to each other in real life. As you'll see when you read this book, we've gotten FAR, far away from our nature...and that's when all the problems started. If you want to get serious, start a discussion group or a book club. You can even partner up with one other person and discuss some of these ideas. Trust me, having a peer to sort through ideas and problems with...makes ANYthing easier.

ONE LAST THING...

This is just the beginning. That is, this book is the first volume of a five-part series. This volume covers the nature of reality, the history of the physical Universe, the science of the Mind, the story of life, and the origins of the first Black man and woman. So anything after 100,000 BC is in a future volume. As much as you might be interested in ancient Black urban civilizations (they are definitely fascinating), trust us when we say you can't fully appreciate THAT story unless you are fully acquainted with THIS part of the story.

And just think about it, when have you EVER read a book on Black history that goes back before 100,000 BC?

What is Science?

THE BACKGROUND TO THIS BOOK

"When I'm asked about the relevance to Black people of what I do, I take that as an affront. It presupposes that Black people have never been involved in exploring the heavens, but this is not so. Ancient African empires -- Mali, Songhai, Egypt -- had scientists, astronomers. The fact is that space and its resources belong to all of us, not to any one group." – Dr. Mae Jemison, first Black female astronaut

For too many of us, we've been turned off by the idea of studying science (and sometimes all academic areas) for a number of reasons. For some of us, science was fun in elementary school but became boring and (seemingly) pointless by the time we hit high school. Of course, if you went to a "ghetto" elementary school, science probably wasn't fun even then. Just stuff to memorize so you could pass a test. For others, we've learned to associate anything coming from the Western (meaning "white") school of thought as repressive, corrupt, and full of lies...to the extent that we don't trust ANYthing that comes off as "the white man's science" or "the white man's medicine" or "the white man's politics." Let's be real, though. There's some validity to that viewpoint, because European anthropologists DID spend over 400 years calling indigenous Black and brown people subhuman "savages"...to the point that some of us grew up thinking "tribal" people really were "primitive"...when the reality was that most indigenous societies were safe, cooperative societies with advanced medical knowledge and egalitarian social standards. But this viewpoint – if applied to anything called science (or history) – becomes untenable because you have to throw out the baby with the bathwater.

"Learning without thought is labor lost; thought without learning is perilous." – Confucius

That is, although far too many European scientists spent the greater part of their careers lying about Black and brown people, it doesn't mean that Western scientists were entirely incapable of establishing anything truthful. The problem is, how do you tell what's true and what's not. Simple, you have to take ownership of the scientific method. You see, there's no such thing as the "white man's science." There's science and then there's "not science." Engaging in science is about pursuing the truth about reality* through testable means. And white folks didn't invent it.

WHAT YOU'LL LEARN

- ❒ The difference between belief and knowledge.
- ❒ What scientific thinking is all about and how Original people are the first to have used it.
- ❒ Which Original people gave the scientists of the West their methodology and earliest training.
- ❒ How YOU are science.
- ❒ Why "New Age" thought is not new to indigenous people.
- ❒ Why Europeans use "aliens" to explain everything Black and brown people have done in the past.
- ❒ The history of the Black and brown consciousness movements in America.
- ❒ How indigenous knowledge fueled the Haitian revolution.
- ❒ How the United States of America would not exist, were it not for the scientific knowledge of one enslaved African who most of us have never heard of.
- ❒ Who was the first Christian bishop to declare that God is Black.
- ❒ How the Five Percenters and Hip-Hop made it cool for people in the hood to talk like scientists.
- ❒ What is BIT Science and how it holds Original people back.
- ❒ How to tell real science from BIT Science and outright lies and scams.

BELIEF VS. KNOWLEDGE

C'BS ALIFE ALLAH

> "Do not believe in anything simply because you have heard it. Do not believe in anything simply because it is spoken and rumored by many. Do not believe in anything simply because it is found written in your religious books. Do not believe in anything merely on the authority of your teachers and elders. Do not believe in traditions because they

* Reality is defined in this text as "The field of thoughts, feelings and apparent sense impressions that organizes our experience into meaningful patterns; the paradigm or model that people create by talking to each other, or by communicating in any symbolism; the culture of a time and place; and/or the semantic environment." In other words, reality is what we perceive and agree on. Absolutely Reality, by contrast, is a metaphysical concept explored throughout this text and in Volume Three.

have been handed down for many generations. But after observation and analysis, when you find that anything agrees with reason and is conducive to the good and benefit of one and all, then accept it and live up to it." – Siddhartha Gautama, known as the Buddha

The catch phrase "the dumbing down of America" refers to the downward spiral in the education of the public. One result is that people have a poor understanding of the mechanics and language of science. Overall, this doesn't seem to affect the general public because only those who understand science are allowed to work with technology (the application of science), like all these great toys we depend on, from cars to cable. People don't have to understand the science behind these tools to use them, however – only to create them or improve them. If you're not concerned about either, you're good. But maybe not. There are far too many instances where our lack of scientific understanding can cripple or harm us. To understand why it is in our best interests to learn how to think scientifically, and NOT in the best interests of the people in power for them to teach us how to do so, we

Activity: Stop Everything and Do a Mantra

Before you read any further, stop whatever you're doing and say the following ancient Egyptian Mantra: **"Aum Pri Tee Zee Tu Fu."**

Say it ten times, out loud, and don't worry about who hears you. I'll explain why this is essential to your sense of self in just a minute. Don't read any further until you've done it.

Are you done? Did you make it to ten? Did you not do it at all? Was it because you're lazy? Or skeptical? Did you read ahead? (smart) Or did you realize right away that this "mantra" was total gibberish? Was it the fact that none of those words look anything like ancient Egyptian words? Or was it the fact the mantra, when read aloud, says "I'm pretty easy to fool." I really hope you didn't say it ten times. If you did, consider it a blessing that you're getting this lesson. Don't believe everything you hear, read, or even see with your own two eyes. People take advantage of those who do, and this is true whether you're in a church, a conscious group, or anywhere else. Always read the fine print. Always think twice. Always consider how something could be used to manipulate or exploit you, and make sure that naïveté doesn't cost you a ton of money, your health, your life, or your relationships. One of the elements that may have snared a few readers into trying this out was the element of trust. You wouldn't think that we would mislead you. And we wouldn't. But you should always think critically, no matter who it's coming from. After all it might not be corruption; it might just be an honest mistake. Still doesn't mean you should go down with the ship. Another snare was the use of a "buzz word." Buzz words like "Egyptian," "mantra," and other "New Age" words tend to "turn off" our BS detector and have us wide open for some BS that's ten times worse. How else do you think people could leave a church and then end up in an Egyptian alien sex cult? If you don't get anything else out of this book, we just want people to think critically about everything. Together with applying what we learn, scientific thinking provides the foundation for successful personal and community development.

> DID YOU KNOW?
> In the upcoming book, A Sucker Born Every Minute, Supreme Understanding discusses over 200 common scams, schemes, and rip-offs that rely on the willingness of people to believe extraordinary claims without extraordinary evidence. This text goes beyond financial fraud and discusses the type of deception that occurs in politics, socio-economics (such as the myth of capitalism), and the media.

need to explore the relationship between knowledge and belief.

Belief can give you a nice, warm fuzzy feeling, yet it isn't something that you should trust any aspect of your life with. For example, as youth, many of us were fed a belief in Santa Claus. Now that we're older, we know it was taught in order to encourage good behavior during the Winter season, and was then "verified" by the "miraculous" appearance of presents on Christmas Day. Belief is simply the acceptance of some statement or explanation without any requirement for proof. Typically, we'll grow out of "childhood" beliefs such as monsters under the bed, Santa Claus, or Tooth Fairies. Some don't and society labels them crazy. But what about the rest of us?

Consider this: The insidious truth is that these "childhood beliefs" are the training grounds for our later acceptance of adult beliefs (particularly in religion, but also in politics, economics, etc.). You are taught to have faith (read: "don't ask any questions") that a person can see you all of the time, know your entire moral record and will reward or punish you at some point. Now tell me if I'm talking about Santa Claus or an invisible deity! In the Christmas scenario, the presents (or lack thereof) are attributed to Santa when, in reality, the end result comes from your parents. In religious contexts, good "things" or bad "things" are attributed to an invisible deity when, in reality, they are simply the result of cause and effect in the natural world (and often directly the result of human agency).

According to conservative estimates 2.4 people get scammed each day. Not two adults and a small person, but an average of 2.4 people. This is because they choose to hinge their economic future on belief. It is the belief that some great "opportunity" will make them rich, and this isn't limited to pyramid schemes and such. If you take a casual look at the last century, you'll find no lack of people in religions and cults who were tricked out of their money, wives, and in some cases, even their lives. Yes, they were lied to, and there is certainly some victimization going on, yet the greater issue is that belief is like having sex without a condom. If that person has the cooties, you're going to get it also.

So you may be asking, "How can I not get cooties?" (unless, of course, you want cooties). Easy. You have to strap up with knowledge and information. Knowledge is your own personal awareness of the world, gained through direct interaction and experience. Knowledge deals with that which is detectable through your four senses. You can believe my car is red all that you want. In fact, I may have told you that it was red. You may have actually read a story that said my car was red. Yet the only way that you will actually know if it is red, is to see it with your own eyes and verify that it is my car. Now information can come from many sources: books, scientists, religions, etc. The key for you is to weigh the source because all sources are not created equal (check our section in the Appendix about doing just that).

As Supreme Scientist explained in his article "Knowledge Versus Belief" in *Knowledge of Self*, knowledge requires us to do more than simply "accept" a claim. It requires us to demand proof. It requires someone to back up whatever claim they are presenting. When a person is arguing for something they can't back up, they may start doing one of the following:

- ❒ Pushing 'group-think' instead of 'you-think' (otherwise known as "come on, everybody's doing it.")
- ❒ Saying that proof isn't needed (otherwise known as "putting their hands over their ears and saying 'la, la, la...I can't hear you.'")
- ❒ Throwing a tantrum (otherwise known as "throwing a tantrum")
- ❒ Downgrading science (otherwise known as "science schmience...yet I'm still gonna rely on science when I check the weather, take my car to the mechanic, or do anything where any sort of testing or verification is involved.")

If they do any of those things, then you know they just don't want to grow up. Belief is a part of many of our childhoods. For some of us, imagination, creativity, and beliefs went hand-in-hand. It was fun (sometimes).* But those who fully carry the believer's mindset into adulthood haven't fully completed their mental rites of passage. Are we saying that people who prefer belief over knowledge won't get anything out of this book? Just the opposite. We WANT you to see the beauty in thinking scientifically. As Dr. Neil DeGrasse Tyson has observed, the world of fantasy pales in comparison to the world of wonder that is really out there (and in us).

"If you don't understand how the world works, then everything is a mystery to you. If everything is magical and mysterious, then you really don't work on logic anymore. Then, everything is all about belief."— Dr. Joy Reidenberg, Anatomist

---

* We want to be clear that children CAN have healthy, fun-filled childhoods without believing in random nonsense. Teaching creativity and imagination don't HAVE to come with belief in supernatural behavior monitors or the like. Teaching our children to think rationally and ethically (of their own accord) are two of the most empowering things you can do to help them have a prosperous future.

DID YOU KNOW?

As noted in *How to Hustle and Win, Part Two*: "While Europeans were still living in caves, these Black men and women were building pyramids that folks still can't build today! In fact, the three pyramids of Giza are arranged in the exact same orientation as the stars in Orion's belt, and the Great pyramid lies in the exact center of the earth's landmass. The height of the Great pyramid is almost exactly one-billionth of the distance from the earth to the sun. The perimeter of the Great pyramid divided by two times its height equals Pi up to the fourteenth digit. In Europe, Pi was not calculated accurately to the 4th digit until the 6th century A.D. And these ancient Egyptians studied a system of symbolic mathematics as well, so our science is nothing new. These mathematical laws were seen as governing the structure and development of the universe. And who is the author of these mathematical laws? The same ones who'd eventually rediscover them. Us. In the ancient Egyptian "mystery schools," the students were taught that man was God on Earth, and trained in methods to unlock this great potential."

WHAT IS SCIENCE?

ROBERT BAILEY

"Science is organized knowledge. Wisdom is organized life." – Immanuel Kant

Before we get any deeper into this book, we obviously need a working definition of science. "Science" (from the Latin word *scientia,* meaning "knowledge") is both a noun and a verb. As a verb, it refers to the organized methodology of acquiring an objective body of information about different natural phenomena. This is important because, for one, it acknowledges there is a method to it; not everything qualifies as "real science," which is why there are labels such as pseudoscience to classify all the other stuff people try to pass off. Second, it must be kept as objective as possible. Peer-reviews, as well as the tests of replicability are required because some researchers (or their funders) are biased in one way or another.

As a noun, science refers to the orderly structure of knowledge gained through a strict methodology. Under the umbrella of science you have different categories such as natural, social, cognitive, formal and applied science, which each have their own specific focus. Natural sciences search for the laws that govern the natural world using empirical and scientific methods. Social sciences use the scientific method to study humans and our societies. The focus of cognitive science is the mind, what it does and how it works. Applied sciences are what you get when you apply science to the physical world. Formal sciences (for example, mathematics, logic, statistics, theoretical computer science, etc.) use formal systems to verify and generate knowledge.

"Science is knowledge and wisdom in tandem.
Observation and experiment. Study and experience." – J-Live

Science is not a modern discovery or practice, nor is modern science necessarily better than ancient science. There exists a tendency to downplay the accomplishments and intelligence of prehistoric societies. Yet, they often had highly sophisticated models of the natural world, especially when it came to astronomy and ecology, two of the "vital" studies of the prehistoric era. But while many of these models had practical applications, some of our ancient intellectual pursuits were purely leisure activities. Some of these scientific ideas spread across vast ranges, indicating long lines of communication in the prehistoric world (without the Internet!).

For example, the Lake Onega petroglyphs in Russia and the Nabta Playa megaliths of the Sahara share the same astronomical sophistication as Stonehenge. Also, the aborigines of Australia, North America and Siberia all called the Pleiades the "Seven Sisters." It's very unlikely that these similarities are just mere coincidences. These are obviously the results of scientific awareness being carried by three collaborating scientific communities, or perhaps even before those three communities separated – which could mean we were comfortable with astronomy over 50,000 years ago. But we didn't just pioneer astronomy and ecology. As we'll note when we describe all of the different branches of science, they often have their origins amongst indigenous societies, long before European "re-discovery."

The first scientific accomplishment of man must have been the development of the scientific method.

WHAT IS THE SCIENTIFIC METHOD?

ROBERT BAILEY

"If I have seen a little further, it is only because I stand upon the shoulder of giants." – Isaac Newton

In early indigenous cultures, knowledge was passed in oral tradition. Indigenous knowledge, sometimes called traditional knowledge or traditional ecological knowledge (TEK) is both a holistic understanding of the relationship between living things, as well as a way of life. In fact, science is but a small part of indigenous knowledge. It is learned by doing and experiencing and makes use of empirical data as well as what is considered supernatural. Nowadays, some natives have had some exposure to non-native values, institutions and patterns of thought. In addition, they don't just use what was been passed down to them, they live and experience life on their own, which further allows them to validate or discard traditional knowledge.

The most prevalent methodology today is the western Scientific method. This method typically follows this general outline: (1) Observation; (2) Hypothesis; (3) Experiment; (4) Findings; and (5) Conclusion.

The scientific method begins with observation of a phenomenon that you want to study. Basically, you see something you want to understand, so you come up with an idea for how or why it happens.

That is, you develop a hypothesis, which is simply an explanation for how something works. A null hypothesis is a hypothesis that a researcher is trying to disprove with an alternative explanation, called the research hypothesis. The research hypothesis is usually formed once an observation is made that the given explanation for something isn't always right.

The researcher then creates a controlled experiment to test the hypothesis. The experiment undergoes multiple trials to make sure that the first results are no accident and to check for consistency.

When done with the trials, the researcher compares the results with their hypothesis. The results are then usually published for others to read, review and replicate themselves. The scientific method has a long history with numerous contributors fine-tuning the process. Essential to the scientific method is that all data has to be empirical. Empiricism simply means that information, or knowledge, must be gained by experience, observation or experience. Basically, nothing superstitious, spooky or anything you have to just "accept" without any proof.

WHERE DOES THE SCIENTIFIC METHOD COME FROM?

It's now known as the western scientific method, but this approach, as well as empiricism in general, has its roots among Original people. In *Knowledge, Belief, and Witchcraft: Analytic Experiments in African Philosophy*, Barry Hallen and J.O. Sodipo describe how the Yoruba have a tradition of empiricism that distinguishes *ìmò* (knowledge) from *ìgbàgbó* (belief). They grade knowledge by varying degrees of *òótó* (truth or certainty), with the highest knowledge being what one has first-hand experience of. That is, when you lack first-hand experience, your *ìmò* comes with some degree of *ìgbàgbó*.[6]

"When an elder dies, a library burns." – Indigenous proverb

Yet this is only example. Most indigenous Black and brown traditions – which can date back 50,000 years or more – weren't well recorded until the time of ancient Egypt (where scientific inquiry quickly reached its apex in the ancient world), so we don't know how far back this African tradition goes. But ask any indigenous person about their scientific process and they'll tell you they've been doing it this way since the beginning of time. However, as we'll see later in this book, many indigenous people don't question the wisdom of their ancestors because they believe all the "scientific work" has been done thousands of years ago, and questioning their traditions, which continue to work for them, would be disrespectful (and unproductive). This perspective has its pros and cons, as you can imagine.*

We often hear that the first scientists were Greeks like Pythagoras, Anaximander, Galen, Hippocrates (the "father of medicine"), Herodotus (the "father of history"), Socrates (the "father of philosophy") and so on. But, as Martin Bernal explains in Black Athena, all of these individuals studied in Egypt. *Stolen Legacy* by George G.M James adds more to the story.

The Greeks also pulled from the knowledge of the Mesopotamians, who pioneered ancient astronomy, and developed the foundations for our modern calendar. But post-Hellenic Europe didn't do much with this Greek knowledge. In fact the "Dark Ages" of Europe was no misnomer. This was a time when – after the Goths sacked Rome – Europe was so "backwater" that they believed bathing was a sin and illnesses were caused by demons. Instead, the scientific knowledge of ancient Egypt, Greece, Mesopotamia, India, and China was consumed ravenously by Muslim scholars who traveled throughout Europe, Asia, and Africa as Islam spread its trade network (along with its scientific community). While scholars such as John Henrik Clarke have blamed Islam with destabilizing Africa (and others such as Ivan Van Sertima have credited Islam with re-stabilizing Africa after the Ptolemies wrecked Egypt), there's no one denying the role of Islam as a cultural transmitter. If nothing else, Muslim scholars preserved knowledge from throughout the ancient world and diffused it among cultures thousands of miles apart, all while improving their own understandings by comparing findings from these distant schools of thought.

During this process, numerous Muslim scholars helped to shape the "western" scientific method as we know it today. By 800 AD, Abu Jābir had introduced controlled experiments to the scientific process, eight centuries before Galileo. Al-Rahwi pioneered the peer-review process in his 9th century book *Ethics of the Physician,* to ensure that physicians met the standards of medical care. 10th century Egyptian polymath Ibn Al-Haytham developed modern scientific methodology. His process involved observing the world, stating a problem, hypothe-

---

* See "Mythology vs. History" in Volume Two for a more in-depth discussion.

sis-formulation, testing the hypothesis through controlled experimentation, analyzing the results, interpreting the data and making conclusions, then publishing the findings. Around the same time, Abū Rayhān al-Bīrūnī, understanding the possibility of mistakes and bias, emphasized the need for repeated experimentation. Meanwhile, Ibn Sina (also known as Avicenna) contributed heavily to dozens of hard sciences as well as metaphysics and philosophy. This period, between the 10th and 14th centuries, is known as the Islamic golden age of science.

When the Moors occupied Europe from 711 to 1492 AD, they introduced their knowledge to European scholars, effectively "saving" Europe from its Dark Ages and ushering in the European Renaissance. Although, in the early 1600s, Galileo was one of the first Westerners recognized for setting up formal experiments and explaining the results using mathematics, the rest of Europe was slow to accept his observations. In fact, he was tried by the Christian Inquisition, who found "vehemently suspect of heresy," forced to recant, and spent the rest of his life under house arrest!

You see, despite the influence of the Moors, Europeans weren't too interested in new ideas until the 16th and 17th centuries, when we have what's known as the Scientific Revolution, a time marked by the willingness to question previously held truths. After this period, "western science" took off and became what we know it as today. If only we all knew that "western" science was really "our" science redefined.

A perfect example can be seen in the fact that Galileo wasn't the first to propose that the Earth revolved around the Sun. A heliocentric worldview was promoted by the Greek Aristarchus of Samos (c. 270 BC), but it didn't catch on. As Joel Primack notes in *View from the Center of the Universe,* discoveries are typically attributed to the last – not first – person to make them.

How Lies Become the Truth

C'BS ALIFE ALLAH

THE WESTERN SCIENTIFIC TRADITION

The philosophical arm behind the Western Scientific Tradition (WST) declares that Europeans are the founders and greatest historical proponents of science. This is part of a larger "package deal" of indoctrination that says all good things come from Europe. Black and brown scholars, since at least the 1800s, have critiqued various mani-

> DID YOU KNOW?
> As Ivan Van Sertima notes in *Egypt: Child of Africa*, Aristarchus must have learned of heliocentrism during his studies at the University of Alexandria in Egypt. Sir Isaac Newton credited the Egyptians as the originators of the theory. This idea also existed among the Black people of India. Long before Aristarchus, Yajnavalkya, the 9th–8th century BC author of the *Shatapatha Brahmana*, recognized that the Earth is spherical and expounded a heliocentric concept. He wrote, "The sun strings these worlds - the earth, the planets, the atmosphere - to himself on a thread." Yajanavalkya also accurately measured the distances between the Sun and the Earth and the Earth and the Moon.

festations of this campaign, which finds its way into the standard telling of history, religion, sociology, psychology, and just about any other school of thought you can imagine.

The Sociology of Scientific Knowledge allows us to critically analyze how this same dynamic occurs in the sciences. Sociologists of scientific knowledge study the development of a scientific field and identify issues that could be interpreted any variety of ways. They examine whether the most favored interpretations are promoted because of scientific merit or because of political, corporate, racial, or economic factors. You can imagine that, quite often, the views that dominate, dominate for a reason…and the reason isn't better data.

The WST is a community. It is a world community and has such influence that in many cases the WST and the general scientific community are synonymous in practice or in the minds of the people. The WST community, according to the Sociology of Scientific Knowledge, is highly influenced and directed by the ideology of Western cultural imperialism (read: white supremacy), which itself is often synonymous with Western civilization. In effect, the WST community won't promote any scientific ideas that go against the accepted Western Paradigm. This is why the WST appears to be biased against some ideas that come from religion, but not others.

For example, civilization* is usually said to either have started in the Nile Valley amongst the Egyptians or in the Middle East amongst the Sumerians. Yet, although both areas are outside of the boundaries of Europe, these first civilizations are claimed to be Caucasian civilizations, despite all the evidence to the contrary. Any claimant to the throne of first civilization is either vehemently denied (no matter what the evidence) or it is "whitewashed" as a Caucasian civilization (no matter how far away it is from the epicenter of white societies, i.e. Europe). Why? Because the dominant racial ideology says "Black peo-

---

* In this text, "civilization" is defined as a culture whose value system provides for them a mechanism to execute collective goals that will outlast the present generation. For a full discussion of Black civilizations throughout the world, see Volume Two.

ple don't build civilizations. They're slaves and servants. Always have been, always will be." Anything oppositional to that paradigm is problematic for the WST.

Other examples can be found in mainstream scientific descriptions of "melanin" as a "waste product" (while, as Dr. Jewel Pookrum has noted, other segments of the scientific community were holding conferences to investigate its properties); the dismantling of the concept of race (for reasons we'll explain in Volume Two); and the grouping of indigenous cosmologies with religious belief rather than traditional interpretations of scientific observation.

NEW AGE IDENTITY THEFT

In this post-modern racial society, certain strategies of white supremacy are now outdated. They don't write all the books and studies anymore. We're living in an era where many of this country's disenfranchised groups are now represented in academia in strong numbers. In foreign countries as well, indigenous peoples have been let into the traditionally white male fraternity of science. As a result, Black and brown people are gaining "voices" with which they can speak on behalf of their people's scientific knowledge (which often rivals that of the West). In lieu of these transformations, new strategies and tactics have been undertaken in order to push forward the "white is right" agenda. One such tactic is essentially a version of "identity theft" that appropriates Original/indigenous knowledge under the guise of "New Age Thought."

The New Age is actually the Old Age. It is the theft of ancient storehouses of Indigenous knowledge traditions. It usually follows the model that some white person has been initiated into some secret ancient tradition or that this white person has deciphered some Indigenous book or script. Therefore this white person becomes the new "steward" and best knower of this ancient information through an implied "changing of hands." This removes it from its cultural context and recasts it in a white world view. Once they've done this with any school of thought, they don't ever have to reference the indigenous origins of this knowledge ever again. You'll see countless examples of this appropriate in Robert Bailey's introductions to the various sciences.

Another example can be seen in how the end count of the current Mayan calendar in 2012 has been appropriated by New Age enthusiasts as the "end of the world." This is not believed by any Mayans, but Europeans have made millions promoting this myth.

A more direct form of identity theft and appropriation involved straight-up hijacking an indigenous identity. Countless Europeans have become "honorary" medicine men, witch doctors, monks, swamis, yogis, and shamans to accomplish this goal. Vine Deloria Jr., a noted Native American writer, also noticed this trend:

> Beginning in the 1960, the federal census allowed people to self-identify their ethnic or racial background, and in the past three decades a startling jump in the Indian population has occurred. Where there were over a half-million Indians in the United States in 1960, in the last census the Cherokees alone totaled over 360,000, primarily the result of consciousness-raising efforts of New Age enthusiasts but nevertheless welcome as a politically significant figure. As whites get more familiar with Indian symbols and beliefs we can expect both the national figures and the Cherokee figures to skyrocket beyond belief by the year 2000. Indeed, today it's popular to be an Indian. Within a decade it may be a necessity. People are not going to want to take the blame for the sorry state of the nation, and claiming allegiance with the most helpless racial minority may well be the way to escape accusations.

Another tactic of the WST is to obscure the origin of artifacts that don't fit within the accepted time line. These artifacts reflect principles of science and technological advances that are considered "out of time" and/or place. The obscuring of the origin of these artifacts is done under the auspices of alien visitations. Whenever the Sphinx and the Great Pyramid aren't ascribed to a race of mythological, world-civilizing whites known as Atlantians, they're then ascribed to aliens. Either way, the role of the indigenous inhabitants of Egypt is removed from the equation. If they are added into the equation, it is only after they have been clearly stated to be of Caucasian stock. The "Chariots of the Gods" thesis took off in the 70's, but was suspiciously applied only to architecture or artifacts outside of Europe: the Nazca Lines of the Inca, the megalith site at Baalbeth (in Lebanon), and the metal spheres of West Transvaal, South Africa, for example. Yet, when things are found that are "out of space and time" in Europe (such as Stonehenge) they are never attributed to aliens. Think about that!

FYI, "BLACK" DON'T MAKE IT RIGHT

To be clear, we're not saying every claim that came from one of us was all good, either. Just because it came from Original People doesn't make it right. Despite all of the above, some things just ain't "deep" or "scientific." Also, just because your ancestors did something doesn't mean that it's automatically "right" or that it has a place within today's circumstances. Some adaptive strategies are just that, *adaptive*, not

DID YOU KNOW?
Shamans are individuals who guide indigenous societies with wisdom that is said to come from "another realm." Books like *Native Science: Natural Laws of Interdependence* by Dr. Gregory Cajete and *Blackfoot Physics: A Journey into the Native American Worldview* by theoretical physicist Dr. F. David Peat explore the indigenous shaman's forays into the world of quantum theory long before Heisenberg and Schroedinger ever stepped foot in a laboratory. Could the Shaman, in fact, be seeing life through the quantum lens? In Volume Three, we explore the world of the shaman, and what these men knew long before modern Western science emerged.

meant to be permanent. As much as we can identify and critique white religion, colonialism, genocide, and so on, we also must not fall into the trap of thinking that everything Original People have done is correct. In reality, there's a segment of what some Original People did in the past that formed the blueprint for the structure of white supremacy as we know it today!

There's something that permeates an aspect of Black research into history and origins. It's what's referred to as the "Black Jesus" syndrome. It is "painting" everything in the past as Black without critically weighing out the deeper constructs, events, and processes. The most important thing becomes proving that something is "Black" or "Indigenous" vs. discussing how that thing impacted the world. So for instance, some Black Christians will simply paint everyone in the Bible as Black, instead of really examining stories in the text as ancient manifestations of "Manifest Destiny" (like the conquest of Canaan by the Israelites). Which, as we know, certainly wasn't good for Original people in recent history, nor was it good then. So just because it is (or was) Black doesn't make it "right."

"No way of thinking or doing, however ancient, can be trusted without proof." – Henry David Thoreau

Another aspect of "Black" not automatically making things right is found in the way many of us look at Black scholarship as infallible. In the recent decade or so there has arisen a new "Black mythology." It references groundbreaking scholarly work done in the 70's and 80's by such men as Ivan Van Sertima, Chancellor Williams, Cheikh Anta Diop, Charles Finch, and so on, yet without embracing their methodology. So instead of embracing the method that produced their teachers, they wallow in a sea of restating what these pioneers have written *ad nauseum* (even when some of those findings have to be updated or revised thanks to more recent data). In that vacuum, psuedo-scholars have emerged with shiny trinkets in the form of bold claims and connections, but no methodology or data to support these claims.

This new era of anti-scientific thinking, which goes against the foundations of the Black scholarship that gave birth to it, has given rise to yet another bad creation: Social Network Scholarship, where the primary sources utilized are Wikipedia, Youtube, and personal websites and blogs. As noted above, when utilizing any source, one needs to be able to weigh the validity of that source. The dumbing down of American education has made it easier for anything to pass as "factual" these days. People aren't learning, and aren't interested in learning, but are "collecting" assortments of information (that may or may not be true) and thus allowing random beliefs to take center stage in their lives. It doesn't matter if that belief involves an astral, invisible god who makes us do what we choose to do, or alien intervention in the building of a Black civilization, or a secret global cabal who only recruits high school rappers, or the latest mind-control weapon in your cereal box. The dumbing down of America has made millions of people more susceptible to conspiracy propaganda (vs. analyzing historical events and trends), to being reactionary to their sense of helplessness (vs. proactive plans of engagement), and to being easily manipulated by those in power (vs. coming together as the collective majority to dismantle that framework). None of it is good for us.

The key is to embrace a new scientific narrative that puts Original people at the center stage. This has not been done before, at least not in a manner that encompasses the entire observable Universe, its history, and the place (and importance) of Original people in it. That is what we've set out to do with this work.

SO TELL ME AGAIN WHY LEARNING ABOUT SCIENCE IS IMPORTANT?

"The principal goal of education in the schools should be creating men and women who are capable of doing new things, not simply repeating what other generations have done; men and women who are creative, inventive and discoverers, who can be critical and verify, and not accept, everything they are offered." – Jean Piaget, influential child development theorist

These are some of the reasons why we feel that science is important:

Legacy and heritage: It's ours. We're the originators of science and technology. We need to re-embrace it to sow the seeds of a new inclusive scientific age.

Tell the Truth, Shame the Devil: As a companion to the above statement when someone takes your stuff you take it back and beat them or rather beat them at their own game. Western civilization may have been built off of our physical labor and scientific legacy yet its time to set the record straight.

"There's real poetry in the real world. Science is the poetry of reality." – Richard Dawkins

Wonder: What is beyond the night sky? What is the bubbly space between atoms? It allows you to retain your child-like sense of curiosity without social stigma or repercussions.

Elevation: Through science you can improve your own life and the world around you. It can heal, unify and stabilize. It has answers to many of the crisis in our communities. Through reclaiming it we place ourselves again on the world stage of human affairs.

"Nothing in this world is to be feared...only understood." – Marie Curie

Removes fear: It allows you to approach the unknown with maturity and a sense of responsibility. You no longer have to be afraid of monsters under the bed, men in the sky, and fire underground.

Art is strongest when it is inspired by science: As Kim Stanley Robinson has said:

> Art in our time is strongest when it is aware of science, includes science, is inspired by science, or is about science. On the linguistic level, the new words coined by scientists to describe their new discoveries form a giant growing lexicon that means English is simply bursting with new possibilities.

I want to bring you back to the man that, in 1941, *Time Magazine* called the "Black Leonardo," a nod to the Italian polymath, Leonardo da Vinci. I'm talking about George Washington Carver. Though he is primarily remembered for the many products he developed from peanuts, that's only part of the story. The real story is that he wanted poor farmers to grow alternative crops, both as a food source and as a source of other products to improve their quality of life. He developed alternatives to the single-crop cotton industry, focusing on hardy food crops like peanuts, soybeans, and sweet potatoes. His research into those areas improved nutrition among poor farming families. Some of the products he developed from these plants included (yet weren't limited to) lubricants, cosmetics, paints, plastics, gasoline, and nitroglycerin.

While inventing all these products, Carver also developed various shades of dye which either had been lost to antiquity or were brand new. Within this rainbow of dyes was a purple which was similar to Tyrian purple, the famous Phoenician purple or indigo. Why is all of this important? Because a former slave's will, through the path of science, transformed the world around him.

He didn't just inject crafts of functionality into the landscape. He didn't just improve quality of life or health outcomes. He brought beauty into the world. All of this was done by embracing science. Now what isn't magical about that? And who doesn't want to bring a little more magic into the world? And while we're thinking magic, let's just ask ourselves, "What was it that emerged in Carver's consciousness that had him reproducing a purple dye the entire world had not seen since the North African Phoenicians – also master scientists – produced that same dye 3,000 years ago?"[7]

DID YOU KNOW?
There are now many popular books detailing the intersections of Buddhism and modern science. Notable titles include *Buddhism and Science* by Donald S. Lopez Jr., *The Quantum and the Lotus* by Matthieu Ricard and Trinh Xuan Thuan, *Science and Mysticism* and *Piercing the Veil* by Richard H. Jones, *The Universe in a Single Atom* by The Dalai Lama XIV, *Hidden Dimensions* and *Choosing Reality* by B. Alan Wallace, and *The Evolving Mind* by Robin Cooper.

EVERYTHING ISN'T A RELIGION

C'BS ALIFE ALLAH

A religion is any systematic approach to living that involves beliefs about one's origins, one's place in the world, or a responsibility to live and act in the world in particular ways. Religion is often equated with faith and belief in a higher power or truth, but it's more properly defined as the habits that express that faith, and reinforces it in day-to-day living. Thus, one can share the philosophy of a religion, believing in its higher truth, without manifesting that faith religiously.*

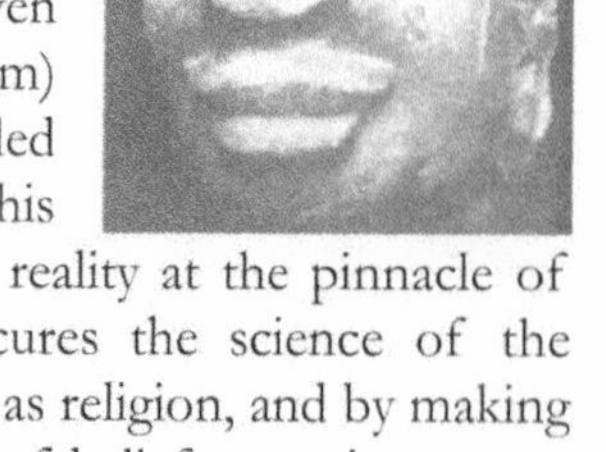

The Western Scientific Tradition (WST) imposes a hierarchal view upon the world. At the top of the pyramid is the modern WST. At the bottom of this pyramid are the traditions of the indigenous world. One way that it demonizes other world traditions is by labeling everything outside of WST a religion. Even traditions that are clearly agnostic (Buddhism) or philosophical (Confucianism) are labeled religions in the average Western textbook. This is done to place European perspectives on reality at the pinnacle of human development. This effectively obscures the science of the world's indigenous populations by labeling it as religion, and by making it seem as if all of our traditions are systems of belief, not science.

* For most of us, our religion is merely the culture we were raised into it, complete with a set of practices, values, and ideas. This is not the same as "religious thinking" which is the refusal to think critically about one's beliefs or ideas, and the desire to hold onto one's beliefs even in light of new knowledge. Many traditions that are called religions have good values and ideas, but most of them require religious thinking. Typically, when people want to get away from religious thinking, they grow distant from the religion or religious community that requires this type of disposition.

On the other end, you have indigenous paradigms that haven't been renamed as religions by the WST, but aren't widely known or promoted by WST academia. Here are just three examples:

- In the empirical tradition of the Yoruba, there are three degrees of knowledge: knowing-that, knowing-how, and knowledge-by-acquaintance, followed by three varieties of belief: believing-that, believing a person, and believing-in.[8]
- The *Carvaka* of ancient India were adherents of *Lokayata*, a materialist doctrine rooted in skepticism (skepticism means simply "to question"). It developed and solidified during the Mauryan period in India where skepticism of faith-based appeals and other unproven claims predominated.

  "Large skepticism leads to large understanding. Small skepticism leads to small understanding. No skepticism leads to no understanding." – Xi Zhi

- *Kaozheng* ("evidentiary research") in China connected the rationalistic sages of the past with WST, an indigenous response to Jesuit priests attempting to use science as a shield for religious conversion.

Throughout this book, we'll discuss dozens of other examples of indigenous science, many of them forming the root of later Western traditions.

REWRITING HISTORY

SUPREME UNDERSTANDING

Within the past century, Black and brown authors have begun reclaiming both the methods of scientific inquiry along with the body of available data to reconstruct the "truth" about Black and brown people. Since the earliest "Afrocentric" books were published in the 19th century, we've been on a quest to undo the lies. And this campaign has extended far beyond the reaches of published literature. Since Hubert Henry Harrison was standing on a soapbox with J.A. Rogers in 1917, telling Harlemites to shun religion and get into the science of themselves, we've had brothers on the street corner pushing knowledge like stolen goods.

Of course, the oral tradition of teaching Black consciousness began long before that, with enslaved Africans and repressed Native Americans passing on teachings of their people's science and history "from mouth to ear," just as their ancestors had done for thousands of years. In 1805, Shawnee Prophet Tenskwatawa chided his people to return to the ways of their fathers, declaring:

> In the Beginning, we were full of this shining power, strong because we were pure. We moved silently through the woods… That was our

> state of true happiness. We did not have to beg for anything. Thus were we created. Thus we lived for a long time, proud and happy. We had never eaten pig meat, nor tasted the poison called whiskey…nor hunted and fought with loud guns, nor ever had diseases which soured our blood or rotted our organs. We were pure, so we were strong and happy.[9]

Tenskwatawa, or the Prophet, as he was known, went on to demand that his people give up the white man's liquor, his "filthy swine," the teachings of his "Jesus missionaries," and all other European customs.[10] With the skill of a preacher, he illustrated how the Indians' use of, and dependence upon, the white man's ways was a setback to traditional Native American society and had caused them to lose their ability to take care of themselves.[11]

Tenskwatawa, the Prophet

Tenskwatawa's message was felt by many who heard him and his followers grew in number. Thousands accepted him as a spiritual leader when, in 1806, he correctly predicted a solar eclipse. Culling together indigenous science, culture, and history, his speeches resonated with Shawnee and non-Shawnee alike. His following spread among the Senecas, Wyandottes, Ottawas and Shawnees. In one of the most significant resistance movements of the 19th century, Tenskwatawa brought the ideology, while his older brother Tecumseh developed the military strategy. With their headquarters in Prophet's Town, Tecumseh and the Prophet together built a "Pan-Indian" confederation of Native American Nations that numbered in the thousands.

Similar awareness campaigns occurred on the slave plantation, but most were crushed with brute force. Keep in mind that it was indigenous knowledge that fueled the Haitian Revolution. Knowing this, whites drove "knowledgeable" Blacks into hiding or punished them into submission. Yet some of this early knowledge survived among the "runaway" communities of Maroons. (Check out *Black Rebellion* for some of that history). But when white people were able to make USE of our knowledge, they didn't complain. In fact, if it weren't for an enslaved African named Onesimus, who taught whites the science of vaccination (based on a practice well-known among his people in Africa),[12] General George Washington's forces wouldn't have survived the smallpox outbreak that decimated the British forces, allowing for the decisive victories that launched what is now the United States.[13] Yes,

DID YOU KNOW?
There are African traditions that parallel the speech given by Tenskwatawa. The Akka (or Akwa) pygmies of West Africa have a chant recorded by Christian anthropologist Henri Trilles as saying, "When he lived with us, – Him, giving his orders, and us obeying Him – we were powerful and strong, we were the Masters." Later scholars have critiqued how European missionaries heavily influenced the accounts of the indigenous traditions they recorded, but the concept of being filled with a "shining power" is unquestionably an Original tradition among Africans and other indigenous people.[14]

read that again. If it weren't for Onesimus (whose "original" name we may never know), there wouldn't be a United States of nothin'.

But no matter which way you chop that cherry tree, the bottom line is that we've worked HARD to preserve our ancestral knowledge. Yet at a time when most Blacks in America were illiterate, the Great Migration to the North (1910-1970) further unraveled some of the cultural links that were already being forcibly dismantled in the South. Once a new wave of Black migrant workers from all over the South congealed in the Northern states, you had a mass of people looking for direction but not finding any that pointed back at themselves.

By the late 1700s, educated brothers like Prince Hall (the founder of the Prince Hall Masons) and Richard Allen (founder of the African Methodist Episcopal Church) were teaching Black people to study their past and respect their heritage. They hinted that Blacks occupied a special place in the universe, but left much unsaid. In 1827, John Brown Russwurm and Samuel Cornish published the first edition of *Freedom's Journal*, the first Black newspaper. *Freedom's Journal* would soon feature some of the earliest published commentary on ancient Black history. In 1829, David Walker published his infamous *Appeal*, which spoke of the "God of the Blacks" in opposition with "heathen" whites "acting like devils," and the idea that "your full glory and happiness, as well as all other coloured people under Heaven, shall never be fully consummated, but with the *entire emancipation of your enslaved brethren all over the world.*" Walker also knew of the importance of the ancient past, and the lessons it held for the present:

> When we take a retrospective view of the arts and sciences – the wise legislators – the Pyramids, and other magnificent buildings – the turning of the channel of the river Nile, by the sons of Africa or of Ham, among whom learning originated, and was carried thence into Greece, where it was improved upon and refined. Thence among the Romans, and all over the then enlightened parts of the world, and it has been enlightening the dark and benighted minds of men from then, down to this day. I say, when I view retrospectively, the renown of that once mighty people, the children of our great progenitor I am indeed cheered. Yea further, when I view that mighty son of Africa,

> HANNIBAL, one of the greatest generals of antiquity, who defeated and cut off so many thousands of the white Romans or murderers, and who carried his victorious arms, to the very gate of Rome, and I give it as my candid opinion, that had Carthage been well united and had given him good support, he would have carried that cruel and barbarous city by storm. But they were dis-united, as the coloured people are now, in the United States of America, the reason our natural enemies are enabled to keep their feet on our throats.

In 1883, George Washington Williams published his *History of the Negro Race in America*, which included chapters dedicated to the history of African civilization and the unity of the Black people of Asia, Africa, and the Americas. By 1890, prominent AME Bishop Henry McNeal Turner went ahead and put it all the way out there, proclaiming that God was Black:

> We have as much right biblically and otherwise to believe that God is a Negroe, as you buckra or white people have to believe that God is a fine looking, symmetrical and ornamented white man. For the bulk of you and all the fool Negroes of the country believe that God is white-skinned, blue eyed, straight-haired, projected nosed, compressed lipped and finely robed white gentleman, sitting upon a throne somewhere in the heavens.

In the following years, the emergence of Ethiopist traditions in the Black church turned the attention of Black Christian back toward their African roots, and eventually towards Black conceptions of Christ. This development gave way to Pan-African movements like that of Marcus Garvey in the early 1900s. Garvey taught over 2,000,000 followers that God was Black and that Black was beautiful. UNIA Chairman William Henry Ferris's phenomenal two-volume series *The African Abroad*, published in 1913, dug deep into everything from the idea of man as God (and God as man) to evidence that Blacks built most of the world's ancient civilizations. Meanwhile, Hubert Henry Harrison campaigned against religious thinking and rallied for Black liberation on Harlem street corners.[15]

The African Abroad (2012)

The UNIA's widespread influence helped facilitate the rise of Noble Drew Ali's Moorish Science Temple in the 1920s. The Moorish Science Temple, in turn, with its emphasis on the "science" of the Asiatic* Blackman and his "hidden" history, opened the doors of Black American consciousness to Islam. When Wallace Fard Muhammad introduced the theology of what was then known as the Allah Temple of Islam to Detroit in 1934, it was like nothing the people had heard before…yet still not entirely unfamiliar.

When Fard began teaching about prehistoric Black scientists, the Black Gods of the East, and the process by which the atom became man himself, it was a revolutionary curriculum. Yet – despite the public renown of Elijah Muhammad or even Malcolm X – the "high science" of the Nation of Islam remained "hidden" within the halls of NOI temples, sometimes unknown even to the rank-and-file members. In 1964, one of these members, an FOI named Clarence 13X not only took the NOI's "secret teachings" (the *Supreme Wisdom*) out of the temple, but left the NOI to share this revolutionary curriculum with Harlem's street youth. These street youth became known as the Five Percenters, who – after Allah's assassination – embraced a "free culture" approach with no formal leadership, a model whose closest precursors were the indigenous cultures of Native American and African societies.

The Five Percenters took the NOI's curriculum and minimized the religious emphasis of its teachings, emphasizing the analysis and investigation of everything found within the Supreme Wisdom lessons (which they renamed "120 Degrees") and added the study of Supreme Mathematics (a symbolic system of numerical laws) and Supreme Alphabet (a related system of principles) as decoding tools. Through often intense exchanges that would have put Socrates and his peers to shame, growing circles of teenage Five Percenters developed a new framework through which they could understand themselves and their relationships with the universe in which they lived, and pass these teachings on to ever-increasing numbers of people in the streets. Through their presence in urban communities across the Northeast, Five Percenters both stimulated the development of the burgeoning Black Consciousness movement and helped the latest developments

* Another term referencing a collective global identity for Original People. It is derived from *Asia* which, although Latinized, is the oldest placename still in usage for that part of the planet (older than Africa, Europe, etc.). *Asia* appears to come from the Black Akkadian word *Asu* which means to "to go outside" or "to ascend," referring to the direction of the sun at sunrise. As such, this refers to viewing the eastern horizon and thus the whole planet. Essentially, it is one of the first "global" concepts. As we'll see in Volume Two, "Asiatic" also emphasizes the continuity of Black civilizations across the Red Sea (similar to the term introduced by Dr. Wesley Muhammad, "Afrabia").

within the movement reach laypeople in ghettos across America.

Five Percenters in 1980s Brooklyn (renamed Medina), courtesy of photographer Jamel Shabazz

When Hip Hop emerged in the late 70s, it gave Five Percent ideology a vehicle to reach untold millions via rappers "dropping science" in their songs. These brothers and sisters effectively made it cool for the hood to read heavy books and talk like scientists. In an era when there were ciphers on every street corner, Five Percenters introduced people to books by Dr. Richard King, Ivan Van Sertima, Cheikh Anta Diop, Frances Cress Welsing, Dr. Yosef Ben Jochannan, Chancellor Williams, and John G. Jackson. And this "street corner campaign" led to the production of more literature, written to satisfy a growing demand for science and history among Black audiences. The work that was produced between the 70s and mid 90s was nothing short of mind-blowing. Even some white folks had gotten in on the act of telling Black folks the truth about themselves. This field of literature, known collectively as "Afrocentric studies" and "Black Consciousness literature" became increasingly well-documented and evidence-based and by the late 90s, the picture was getting much clearer.

But with the explosive growth of the Internet and the dawn of the Information Age, the picture started getting cloudy again. What happened? Evidence-based scientific inquiry in the Black and brown "consciousness" communities gradually became replaced by a growing menace: BIT science.

How the Truth Can Become Lies

SUPREME UNDERSTANDING

DOES MELANIN HAVE MAGIC PROPERTIES?

Have you ever heard that melanin is the foundation of the physical universe? That EVERYthing is made of melanin? That melanin has metaphysical properties? Well, claims like this aren't "lies" in the same sense of how Western scholarship has lied to us. Instead, they're examples of what we call BIT science.

WHAT'S BIT SCIENCE?

That is, somebody took a bit of something and a bit of something there, and they put things together to form a fantastic claim, not knowing that you just can't put things together like that. Some of the "experts" who promote BIT science are intentionally deceiving their followers for personal gain, but many just don't know better because they're not acquainted with the methodology of the fields from which they're pulling their "bits" of information. We're not mad at honest, well-intentioned people attempting to empower the masses with information, as long as they're teaching them how to think and research for themselves. If that's not the case, the teaching promotes powerlessness, not empowerment. But we're not here to knock the hustle and tell people to stop teaching about science and history. We're just here to show you how it's done, because there are a lot of people out here doing it wrong. As for the people telling lies for personal profit, this book just might put an end to all that. Once you've read this book, you'll be able to tell what's real, what's a lie, and what's BIT science.

There are thousands of examples of BIT science circulating through the hood right now, but let's focus on some of the above claims about melanin (all of which can be found on various YouTube videos, DVD lectures, and published books). BIT stands for "Based In Truth." That is, there IS some truth behind the claim that melanin is the foundation of the physical universe,* because the essential element of melanin, carbon, is found throughout the universe AND carbon is the foundation of all living organisms that we know of. BUT carbon is not melanin. Melanin is an organic compound with plenty of other ingredients. And there are other elements that are much more fundamental

* Throughout this book, the terms "universe," "visible universe," and "physical universe," refer to everything that exists that is observable, quantifiable and qualifiable.

to the physical universe than even carbon, like hydrogen and helium (from which all the other elements came). Do you see where we're going with this?

When people say that melanin has metaphysical properties, there's a "bit" of truth behind it, because melanin clearly is a conductor of electromagnetic energy and has many other exciting properties (as we'll see in this book), but unless you define EXACTLY what you mean by "metaphysical," we have no way of verifying the truth of your claim (hence the reason why we use working definitions in this book, to be clear about what we mean and what we don't mean when we say something). And if people were giving you the WHOLE truth, rather than bits and pieces thrown together and dressed up...they'd tell it that way. They'd even acknowledge that there are some questions they simply can't answer because they don't understand all the information, or there simply isn't enough data available to come to a conclusion. But who's honest enough to do that? Maybe not the YouTube experts, but we are.

By BIT science, we're referring to connections that are overly simplistic and unsupported by evidence; claims that sound exciting, but which nobody can prove; and ideas that can't be pursued to their logical ends (e.g. if "man was always here [in human form]" then how did man survive before the planet cooled down and formed a crust for him to stand on?). We're also talking about scientific and historical arguments that employ the common logical fallacies that negate the validity or credibility of ANY argument. A "fallacy" is basically an argument that might sound good, but is logically flawed. Common fallacies include:

- ❒ Ad Hominem (attacking the person making the argument instead of the points made in the argument),
- ❒ Straw Man (setting up a stupid version of someone's claim to attack that easy-to-destroy misrepresentation of their claim),
- ❒ False Analogy (comparing different situations as if they are the same), and
- ❒ Faulty Cause and Effect (assuming that because things happened with some relation, that one must have caused the other).

There are so many more examples of "not science" that we've included a "Guide to Critical Thinking" in the Appendix, which goes into much more depth.

With all that said, let's get one thing clear. People can't "believe in science," so you can't dismiss scientific claims as if they're merely opinions or beliefs. True scientific claims are based on evidence and testing. If you agree with a scientist's findings (because you understand their methods and how they arrived at their conclusions), then you are not simply "believing what they said." However, you can take what a scientist said without understanding ANY of it, and in fact be "believing what they said." On the other hand, if you don't understand any of it but still disagree a priori (I can't always get my head around a priori), perhaps due to pre-conceived notions or beliefs that you don't want undermined by contradictory evidence – without any scientific or evidence based reasons to disagree, once again, you are merely "believing" that they are wrong.

Are we saying everything that's supposed to be scientific is correct? Of course not. All scientific research is not made equal. As we noted in *The Hood Health Handbook*, some studies are funded and publicized over others because they promote the agenda of a specific lobbying group, corporation, or government agency. Those studies tend to be biased in subtle ways that are designed to give the funders the results they want to see. This is why scientists do "literature reviews" where they examine the entire body of studies published on a topic and see what the majority of studies found. Another consideration is where does the information you're reading come from? Is it:

- A primary source (which means it's the original evidence or study),
- A secondary source (where the author is compiling and sometimes interpreting primary sources), or
- A tertiary source (where the author is a third party who has possibly never even looked at the primary sources on the subject and is simply rehashing the work of secondary researchers, who may or may not have been accurate)?

To be all the way real, within the Black consciousness community today, we're not even seeing tertiary sources anymore. Much of the "information" we've been getting comes from lecturers who mix up bits and pieces of tertiary sources in the most "exciting" way possible. So it's been four or more degrees of separation from the primary source, and most of this "information" is nearly impossible to verify against the original sources. So it's like playing the Telephone game. By the time we get the info, it's been through so many transmitters that we don't know what the hell the original data was anymore.

So one of the things we'll do in this book is (wherever we make any claim) cite the ORIGINAL research on the topic, or a solid, reputable secondary source that provides an easy read on otherwise difficult information. Because even primary and secondary sources can be biased or straight up inaccurate. For example, many of the earliest "anthropologists" to publish studies on the cultural mythology* of various African societies were actually Christian missionaries, and, as a result, these accounts suffer from slanted reporting and indigenous traditions were made to sound more "Biblical" than they really were. Unfortunately, this not only happened throughout the world but the influence of these Europeans was so pervasive that many indigenous societies have gradually altered their own traditions. In effect, some of their original knowledge has been lost. But there are methods to reconstruct a lot of what's been lost, and we've put them to use throughout this book. It's been painstaking work to sift through the information and filter out the bias of the authors, to read between the lines, to look for what's been left out, and to dig for all the information that's no longer being reprinted.

We can't afford to give you anything less. As we noted earlier, science HAS been used against us. In fact, science continues to be used against us. That's why this book exists. What's unique about this text is – although we'll be revisiting some of the distortions still accepted by mainstream scientists and exposing the truth that others have covered up – we'll be using the existing data to do so. No made up stuff. No illogical conclusions. No claims you can't look up on your own. No "sources" that are merely "hearsay" from people that don't do primary research themselves. No "proof" in scriptures and myths that can't be corroborated by other solid evidence. We won't do that to you. You deserve better. And so, throughout this book, you'll see us writing in the casual, conversational parlance – slang and all – that you're used to seeing in books from from Supreme Design Publishing, and then switching gears to the type of heavy content that you'll have to read three times to understand. We didn't want to make it all "easy reading" because we want you to be able to think critically about information when it's in the original sources, the kind of papers and books that the scholars read, rather than just some third or fourth party, watered down interpretation of those sources.

THE KISS METHOD OF CLAIM EVALUATION

SUPREME UNDERSTANDING

To REALLY make things easy, I'm going to end this introduction with

* In this text, we define "mythology" as a narrative device that played a multifunctional role in many pre-industrial civilizations. A myth can preserve history, propagate religion, and/or, through symbolism and metaphor, preserve and promote scientific technique and method. For a full discussion of mythology, see Volume Three.

a set of questions I've developed, which you can ask yourself about an extravagant claim. It's called the KISS (Knowledge is Supremely Simple) Method of Claim Evaluation:

- ❑ First things first: Am I really high, drunk, or schizophrenic right now? If so, let's revisit this claim later.
- ❑ Next, what exactly is this claim saying? What are the specifics?
- ❑ What is the evidence provided? Extraordinary claims require extraordinary evidence. How significant is this claim and how detailed is the evidence provided?
- ❑ Does this claim contradict things I already know? Do I simply WANT to believe (or ignore) this claim, and am - as a result - not thinking critically about it?
- ❑ Who is the source? Do they have anything to gain? Are they trustworthy? What else have they produced that makes them trustworthy?
- ❑ Are there pictures that are so colorful, etheric, or cartoonish that it seems they're trying to wow me with the visuals?
- ❑ Where else can I find more about this claim? Are most of the sources just similar versions of the same set of information?
- ❑ Does a Google search only turn up blogs and message boards, but no books or journal articles? Are most of the sites I find either selling something or full of other "impossible" claims? Have I checked this claim on skeptic/debunking/rumor-correcting sites?
- ❑ Is there an independent review of this information anywhere?
- ❑ Is there a logical linear explanation of the cause and effect, or other processes, involved in this claim? Or is it just speculation, theory, or straight-up "this is what it is" with no real details?
- ❑ Do I know the background information required to critically analyze this claim? Do I know the meanings of all the words used in this claim?
- ❑ Finally, do I care about looking like a dumb dummy when I try to tell other people about this claim and they laugh because it's a rumor or fabrication that was debunked years ago?

Bok Globules in IC 2944 (Lambda Cen Nebula)

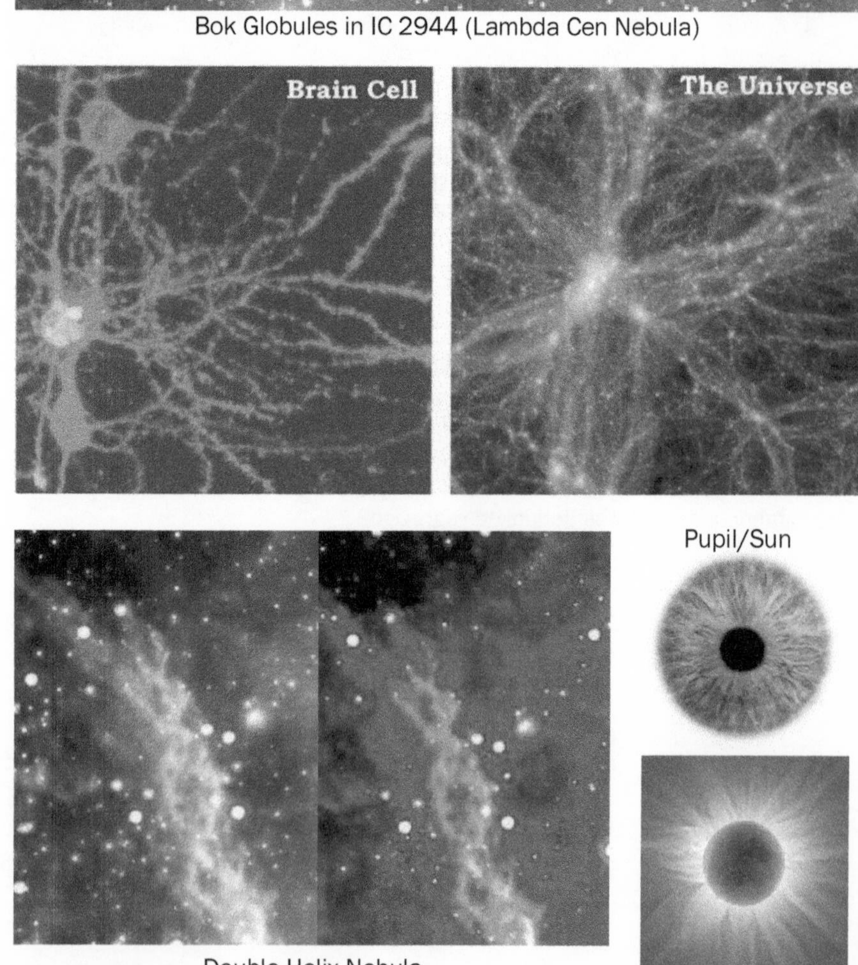

Pupil/Sun

Double Helix Nebula

The Laws of Nature

THE WAY EVERYTHING WORKS

> *"The mathematical phenomenon always develops out of simple arithmetic, so useful in everyday life, out of numbers, those weapons of the gods: the gods are there, behind the wall, at play with numbers." – Le Corbusier, The Modular*

Look all around you. In fact, look around with that new lens on life we discussed, the lens of scientific inquiry. This doesn't mean you have to analyze and break everything down its smallest components. In fact, science allows you multiple vantage points (or perspectives) with which to look at things, so you can take the Western route and analyze or the Indigenous route and synthesize (seeing the bigger picture). Or you can do both, by looking at the parts, the whole, and then the relationship between the two. But the most important part of any of these processes is simply to LOOK.

Equipped with this news lens, what you'll see when you look around might surprise you. You may notice patterns you once ignored. In fact, you may begin seeing patterns everywhere. You're not going crazy like John Nash in *A Beautiful Mind.* There really ARE patterns everywhere. So, now that we've covered how to think (or more importantly, how NOT to think) about the science of life, we're going to begin our excursion into the universe of self with a study of the laws that create the patterns and structure we see all around us. What you'll learn by the end of this chapter is that these laws are everywhere, they've been here since the "beginning," and these laws are all about you.

You may not see that aspect of it just by looking around you, but studying life and matter are ways in which you see manifestations of the principles we'll explain to you here. Some of this content draws on quantum mechanics and other concepts we won't be able to fully explore until future chapters, but bear with us. We're taking you somewhere.

WHAT YOU'LL LEARN

- ❒ How and why the patterns and processes of nature are mathematical.
- ❒ How understanding nature's mathematical patterns can help you make predictions about the future.
- ❒ How mathematics is something you live and witness all around you every day.
- ❒ Which ancient civilizations were using complex geometric patterns and mathematical knowledge over 70,000 years ago.
- ❒ Why mathematics could be hard-wired into our brains in such a way that people who don't know "school math" still know the mathematics of life.
- ❒ Why you could be wearing your ancestors on your ears.
- ❒ Why the Universe is full of self-replicating patterns, from the level of galaxies to seashells and snowflakes.
- ❒ What these laws tell us about their origin.

AN INTRODUCTION TO MATHEMATICS

ROBERT BAILEY

"Go down deep enough into anything and you will find mathematics." – Dean Schlicter

Much more than high school algebra or money problems you need a calculator for; mathematics is essentially the science and study of patterns. Mathematicians study patterns in quantity, structure, space, change (i.e. arithmetic, algebra, geometry and analysis), shape, motion, behavior, and so on.[16] As Keith Devlin notes in *Mathematics: The Science of Patterns: The Search for Order in Life, Mind and the Universe*, these patterns can be real or imagined, visual or mental, static or dynamic, qualitative or quantitative. After observing patterns, statements are formed about abstract concepts, then proven true or false through deductive reasoning; those proven true become theorems. Mathematics as a subject can be divided into three major divisions: foundations and philosophy (studies the essence of mathematics), pure mathematics (like algebra, arithmetic, geometry and analysis), and applied mathematics (focuses on the mathematical methods of other sciences and practicality). What good is studying patterns? As Galileo put it, "The great book of nature can only be read only by those who understand and who know the language in which it was written. And this language is mathematics." As Napoleon Hill (author of *Think and Grow Rich*) and countless others can attest to, when you can identify patterns, you can find out how people who are doing what you want to do, got to where you want to be. Recognizing patterns increases our ability to do anything from making predictions to working more effectively.

"Now we can see what makes mathematics unique. Only in mathematics is there no significant correction – only extension." – Carl B. Boyer

So when and how did it begin? The origins of mathematics are said to lie in the concepts of number, measurement, magnitude and form. In *Ethnomathematics and its Place in the History and Pedagogy of Mathematics*, D'Amrosio proposes that mathematics have been culturally derived since people first walked the Earth.[17] In *The Mathematical Ways of an Aboriginal People: The Northern Ute*, Jim Barta and Tod Shockey explain how in many cultures throughout the world, there may not be one specific word for mathematics; instead, it is something you live and is witnessed through its applications. As Fabian Jenks, Ute Tribal elder states:

> Math is a part of our lifestyle too but we don't have it in the way that the White man has it, say...math like times tables, and addition or subtraction. We have those things also in our traditional ways of beading, our raising of our horses and cattle, or building fences, putting up teepees, etc.

Barta and Shockey add:

Ishango Bone

> Ute elders described uses for the basic math operations of adding, (combining) subtracting (taking away), and dividing (sharing or splitting up). Repeatedly adding similar quantities when such amounts were counted took the place of multiplication. Division and fractional quantities were aspects of sharing items such as food and calculations were rather intuitive. One might get "a half" of the bounty and another might get another portion of what was left. Equal shares did not always mean everyone got the same amount but rather got what they and their family needed. "One-half" was described as "in the middle or center."[18]

In addition to mathematical artifacts such as the Ishango bone and Lebombo bone (about 37,000 years old), paleontologists have even found ochre rock art decorated with complex geometric patterns (suggesting we were able to think abstractly older than previously thought)

ACTIVITY: ALL ROADS LEAD TO 123

"Start with any number that is a string of digits – say, 9,288,759 – and count the number of even digits, the number of odd digits, and the total number of digits it contains. These are 3 (three evens), 4 (four odds), and 7 (seven is the total number of digits), respectively. Use these digits to form the next string or number, 347. If you repeat the process with 347, you get 1, 2, 3. If you repeat with 123, you get 123 again. The number 123, with respect to this process and universe of numbers, is a mathematical black hole."[19]

which they date to more than 70,000 years ago in South Africa.[20] Ancient Mesopotamia, Egypt, China, Greece, India, the Incas, the Mayans, and many other cultures have contributed to modern understandings and applications of mathematics.*

THE UNIVERSAL LAWS

C'BS ALIFE ALLAH & SUPREME UNDERSTANDING

"When God wrote the universe it was in the language of mathematics." – Galileo Galilei

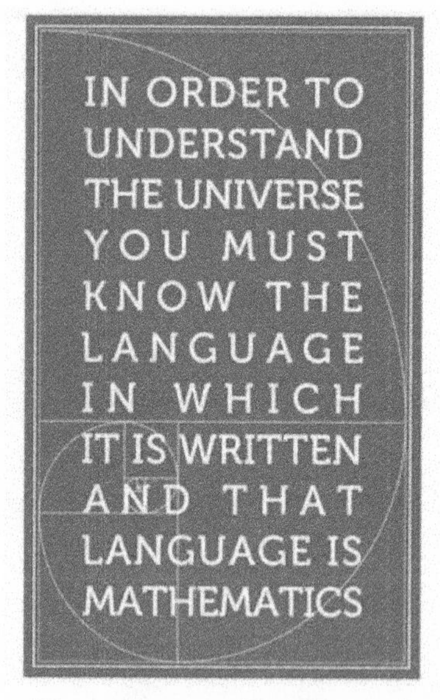

Mathematics is the language of the universe. While every important scientist from Imhotep to Einstein has acknowledged this fact, it doesn't take an advanced degree to understand why. It's simple really. The language of mathematics is one that is the same no matter whether you're in China or Chicago, or if you're on Pluto. In fact, the laws of abstract mathematics (and this is what Western scientists have said) can be used accurately describe real phenomena anywhere in the visible Universe.

Mathematics is a device used for logical measurement. Mathematic deals with the relationship between things that can be expressed through the language of arithmetic numerical relationships. Through mathematics, we can reach new conclusions in the study of structure, quantity, change, spatial relations, order, etc. Patterns can be detected through mathematical models which can be used as models and guides for phenomena in real life. So mathematics becomes a means to gain great insight and a method for prediction. Physics deals with how matter and energy move through space and time. The word itself comes from the Greek word meaning nature. Mathematics is the language in which physics is written. So by learning the language of nature, we gain a greater understanding of the universe. So mathematics is literally the (scientific) law (description of what something does) of the universe. Everything falls within the parameters of the laws of mathematics. Everything is

* For the origin of written numerals, see "The Origin of Numbers" in Volume Two.

built up (lives) and broken down (expires) via physics. So everything adheres to the script of this language. Just like any other language, you have to learn its fundamentals so that you can understand what's going on. By learning the language of mathematics, you can see the script as it unfolds in nature around you. This is the structure of the universe.

Except, there's a limitation to this. Scientists have realized that the laws that govern most of the physical universe, described in Einstein's Theory* of Classic Relativity, cannot be applied to the happenings of the quantum universe, that is, the world that exists at the subatomic level. There is where quantum physics replaces classical mechanics. Classical mechanics can work for anything from pieces of dust to the movement of entire galaxies, but it just doesn't govern quarks and neutrinos and stuff. So scientists, since the earliest scientific observations, have cited the need for a "Theory of Everything."

In 1974, scientists developed the idea of a Grand Unified Theory (GUT) that could be crucial step towards a Theory of Everything.† Perhaps, however, the GUT is not something we have to dig for in our brains, but something we know in our GUT. I'd argue that it is, because the mathematical laws of nature are certainly within you (they govern your very biology), but we really would much rather argue the point through scientific reason than make an argument from emotion. There are too many charlatans and frauds taking advantage of sincere people because we've grown so accustomed to going with our feelings rather than analyzing and investigating claims more thoroughly. The goal of this book is to detox you from that, so let's talk about the science of mathematics and see what it reveals.

LIVING MATHEMATICS

C'BS ALIFE ALLAH & SUPREME UNDERSTANDING

We know everything in nature grows, develops, and even decomposes, according to a pretty general set of mathematical laws. For example, the law of reproduction, also known as multiplying. Or the relation-

* As we discuss later in this book, a scientific theory is not a "guess" or belief. It is an explanation that has been strongly supported, strongly substantiated, and well documented through repeated observations. The theory is an umbrella that brings together all the facts so that an explanation that fits all of the observations can be used to make predictions about future behavior. A theory is not a law. A scientific law describes things. A scientific theory explains them.

† A GUT would need to cover everything from classical physics and quantum mechanics, but not gravity. A TOE would have to cover everything, including gravity.

ship between predator and prey, which boils down to simple subtraction. And there's clearly an additive factor involved in the emergence of groups, colonies, herds, and other communal networks of living organism. Even the process by which we go from single-celled organisms to complex creatures requires our cells to divide. This is the basic math of life. There are much more complex examples. For example, as Gary Zukav explains in *The Dancing Wu-Li Masters*, your very DNA is a mathematical algorithm.

But DNA is just a part of the code to program the system. The real work is in the system processing the code. This is a theme you'll find anywhere we talk about math and systems. The math is the math – it's unalterable – but the way it's programmed into something, or the ability of the system to follow the code…that's what determines success or failure. This is especially true for a code like DNA, which is not so much a complete set of instructions as it is a program or algorithm that must be read by a system.[21] This is because even mathematical laws or codes require context. This goes back to the hardware/software entanglement Paul Davies describes in *The Fifth Miracle*. So what's the hardware that processes this software? In a broad sense, the universe. In a specific sense, man.

IS THE HUMAN BRAIN A SUPERCOMPUTER?

With all this talk about the mathematics in nature, you may be wondering about this "computer" we have in our heads, and whether we were born with math on our minds. As Paul Davis once asked, "Just why is the world structured in such a way that we can describe its basic principles using 'do-able' mathematics? How was this mathematical ability evolved in humans?"[22] Did we "make math" to fit the world we live in, or was it the other way around?

Albert Einstein once said in connection with his celebrated mathematical insights, "Words and language...do not seem to play any part in my thought processes." A French scientist, Stanislas Dehaene, used this quote as support for his claim that human brains possess a "number sense" that is independent of language and symbols, including even the numerals we use in arithmetic. The numerals, says Dehaene, are needed *only* in "exact arithmetic," which is a cultural invention and unrelated to the "number sense." Exact arithmetic, in fact, is an activity of our left brain where language is processed. Our general number sense, though, is sited elsewhere; the parietal lobe, to be specific. Dehaene's experiments with babies demonstrate that, even before they can speak or do exact arithmetic, they can do "approximate arithme-

DID YOU KNOW?
We don't even need number words or numerals to know mathematics. European anthropologists used the lack of these symbols to argue that indigenous people lacked number-sense. But research among indigenous cultures has proved them wrong time after time. A recent study of children from two Aboriginal groups in Australia's Northern Territory found that, though the communities had no words or gestures for numbers above two, they were still able to copy and perform number-related tasks. The findings showed that Aboriginal children did better than European children, and suggest that Aboriginal people possess an "innate system for recognising and representing numerosities." The researchers also voiced concerns that many indigenous children may have difficulties learning math the Western way because of cultural differences.[23]

tic." That is, they could distinguish between these two sequences of tones: "beep-beep, beep-beep, beep-beep" and "beep-beep, beep-beep, beep-beep-beep." This brings us back to rhythm, a mathematical pattern found within our very pulse and heartbeat.*

As another example, the Munduruсu people are an isolated indigenous people (anthropologists say "tribe") living deep in the Amazon jungle. Their language hardly has any words for specific numbers or geometric shapes. Yet, researchers found that they scored just as well as American and French school children on geometry tests, actually scoring higher in an area dealing with non-traditional Euclidian geometry. In this case, it appears that the traditional Euclidian teachings actually got in the way of more appropriate intuitive thinking. This led researchers to the conclusion that geometry must be hardwired into the brain.[24]

In Botswana, Hilda Lea found that Bushmen have their own "informal" mathematical traditions "suitable for their traditional way of life, and their highly developed spatial abilities are very necessary for survival in their harsh environment." Lea describes a binary number system (one, two, two-one, two-two, two-two-one etc.), measurement, time reckoning, classification, tracking and other mathematical ideas in technology and craft. "Bushmen have the oldest pattern of life found in the world today," she notes, adding that, "A hunting and gathering community does not have need of counting precise measurement" but requires very special skills of a more visual sort. In other words, they have to be able to see the math, not write it down.[25]

And this suggests that mathematical understanding, sometimes known as number sense, is "hardwired" in a specific part of the mammal brain.[26] Considering the connections between our brains and representatives of our simplest ancestors, we might wonder if mathematics is

* See "The Origin of Music" in Volume Two.

DID YOU KNOW?
Sharks use math to hunt. A study by the Marine Biological Association of the United Kingdom confirms that marine animals follow fractal patterns of behavior as well. Their study found that, in the featureless expanse of the oceans, a shark forages for food along a complex mathematical pattern called a Lévy flight, a type of fractal. In fact, a shark's movements more closely conform to that pattern when food is scarce, suggesting that stricter conformity to the mathematical law ensures better outcomes.[27]

wired into us from day one.

THE MATH IN LIVING ORGANISMS

And by day one, I'm talking about single-celled organisms. After all, as we note elsewhere, those single-celled organisms were no dummies. SOMETHING is processing inside that slime mold. It's the math of nature. It's in everything living. Mathematics may be such an integral part of the structure of consciousness that even plants, insects, and animals "know mathematics." And we're not talking about monkeys learning how to count. Recent studies have found that slime mold are able to solve geometric puzzles, sharks hunt using fractal patterns, bees employ complex geometry to organize their hives, bacteria can solve Sudoku-type puzzles, and cicadas use the science of prime numbers and divisibility to avoid emerging from their underground habitats in the same years as their predators, who spread during even years.[28]

There are countless other examples of an ingrained, genetic, chemical-coded mathematical pattern of organization that appears in the living world. In fact, mathematics is what gives the leopards its spots and the zebra its stripes. Scientists have used a unique set of differential equations to explain everything from the patterns of the wings of butterflies to the colored patterns of exotic fish. The same basic equation explains all of the patterns (which are mostly melanin, by the way).[29]

In *The Algorithmic Beauty of Seashells*, Hans Meinhardt also describes "natural" equations behind the patterns on seashells. The equations are straightforward for the simple patterns, but there are intricate patterns that require modelers to imagine traveling waves of excitation, signals that travel faster than chemical diffusion, and long-range synchrony employing a "global control element" that can't possible occur on the cellular level, and reminiscent of quantum phenomena. Russian chemist Ilya Prigogine described such "self-organizing" phenomena as "the spontaneous creation of order,"[30] hinting at the morphogenetic field later described by Rupert Sheldrake and the Mind described in this work.[31]

DID YOU KNOW?
Some scientists contend that genes don't actually carry the 'blueprint' for the construction of the whole organism at all; they merely code for the production of proteins. Just as honeybee specialization occurs at the hive level (not the genetic level of each honeybee) our genes don't carry instructions for how the proteins are molded into tissues, organs, and complex living organisms, nor do they explain instinctual behavior, consciousness, or other important aspects of living things. So, if it's not in the genes, where is the true blueprint? The blueprint is within the Self.

BACK TO THE MATH IN MAN'S MIND

Based on what we know about computer software, we know that just about any program or algorithm can be coded entirely in a language of 0s and 1s. This binary language can produce any sort of data, but there's a possibility our brains process things on a level above and beyond that of computers.

Consider the ability of chess geniuses like Maurice Ashley, Baraka Shabazz, Justus Williams, Joshua Colas and James Black Jr., who can calculate the outcomes of millions of possible moves in order to select the best strategies.* In one famous example, chess master G.K. Kasparov was able to compete with a supercomputer named Deep Blue, which could evaluate 20 billion moves and countermoves every turn. IBM's A.J. Hoane, Jr., remarked that chess geniuses like Kasparov "are doing some mysterious computation we can't figure out." Perhaps the math our brains do is less mechanical and more organic.[32] It's possible that our brain is more than a binary processer, and has some mathematical capacity that overarches all mathematical computations.

In his groundbreaking book on consciousness, *The Emperor's New Mind*, theoretical physicist Roger Penrose argued that the brain had the ability to go beyond what could be achieved by axioms or formal systems. That is, the brain had some additional function that was not based on algorithms (a system of calculations), whereas a computer is driven solely by algorithms. Penrose asserted that the brain could perform functions that no computer could perform. He called this type of

* Bobby Fischer became famous because he was young, but some of the youngest chessmasters of recent years have been Black. See Daa'im Shabazz's site www.thechessdrum.net for more examples of Black chessgrandmasters and experts.

processing non-computable.[33] If folks like Penrose and Bohm are correct, it would help answer many questions about the connections between math, the Mind, man, and reality.

So is the human brain a supercomputer? Philosophy professor John R. Searle says the compute is only the most recent in a long line of mechanistic analogies for the brain:

> Because we don't understand the brain very well we're constantly tempted to use the latest technology as a model for trying to understand it. In my childhood we were always assured that the brain was a telephone switchboard. (What else could it be?) And I was amused to see that Sherrington, the great British neuroscientist, thought that the brain worked like a telegraph system. Freud often compared the brain to hydraulic and electromagnetic systems. Leibniz compared it to a mill, and now, obviously, the metaphor is the digital computer.

DID YOU KNOW?

Mathematics may be hardwired into our very biology. You know that the sperm makes a journey to meet the egg to form a baby. The egg releases attractant chemicals that manipulate the concentration of calcium inside the sperm. This affects how fast the sperm is going to wag its tail and swim. What scientists discovered, however, was that it wasn't the concentration level that was affecting the sperm's speed, it was the change in the concentration. So the sperm are calculating derivative levels of calcium. The sperm is performing calculus.[34]

To answer the question, if by "supercomputer" we mean a processor that exceeds the capabilities of most computers, perhaps on some levels the brain is like that. But, all things considered, the human brain is even greater than that. Because the brain accesses the Mind, which no computer can do, the brain has access to "non-computable" functions that distinguish it entirely from any mechanistic device.

So let's recap. There's clearly some programming to the way things work. It may exist at a quantum level, or it may be sub-quantum. What is clear is that it's beneath the surface. We, as humans, are particularly unique because are able – unlike most other creatures – to reflect back on this programming and consider what lay beneath the surface of the classical world's material veneer.[*] Part of this reflective process involves considering what truly constitutes the Self.

As Dr. Denise Martin notes in her illuminating introduction to Pan African Metaphysical Epistemology, indigenous African traditions have taught this reality long before Western philosophers considered it:

> Generally speaking, African understandings of person are multifac-

[*] Hence, one Five Percenter etymology of the word "wnderstanding" is to see the knowledge under where we stand, a process described in metaphor as seeing through muddy water to what is below the surface, and in the maxim, "Understanding is seeing things clearly for what they really are, not what they appear to be."

> eted and contain several intrinsic characteristics: (1) a person is made up of numerous components, (2) a person has an active moral component; and that (3) the components are synchronized between the physical and metaphysical bodies. This complexity described in the Bântu-Kôngo word for person, muntu is a "set of concrete social relationships…a system of systems; the pattern of patterns in being." Muntu is "n'kingu a n'kingu" a principle of principles; such that muntu is able "to produce materially or technologically other mechanical systems." Therefore, a muntu is distinguished from other beings by intelligence and a unique human quality.[35]

Self-awareness comes with its downside, however. In being perhaps the only organism capable of studying our own mathematics, we also have the free will to deviate from the dictates of nature. An individual organism's deviation from its mathematical programming only spells disaster for the survival potential or that one organism…but when groups – entire populations even – begin going "against the grain" of nature, it spells disaster for the entire species. In extreme cases, the systemic abuses of a prevalent species can be so significant that they can have a profound effect on the entire environment and can – through a number of chain reactions – spell disaster for the planet itself.

This is what we've seen several times throughout the history of man. As life emerged from single-celled inception all the way to dominion over the earth and seas, there are times when this dominion put him in competition with the rest of the world (nature in most cases), and breakdowns follow. This build-destroy process facilitates the survival of some and the destruction of others. Sometimes entire populations of man have gone extinct. Other populations successfully adapted to the demands of the changing environment and thrived.

Today we, as Homo sapiens, are both the Original man and the descendants of those who survived everything that came before us. As history continues, there will be new threats to our existence and new groups who have survived while those who failed did not. Sometimes, nature has to give way to us (such as with the extinctions of thousands of species that occur when the environment has shifted to conditions that were ultimately more conducive to the emergence of man) and other times, nature has eliminated millions of men so that nature itself wouldn't self-destruct. It's the code we wrote, and it's a system that designed to prevent abuse. We built a fail-safe mechanism into the laws of nature that prevent us from going beyond the dominion of God to the abuses of a tyrant.

I know, I probably lost some of you when I argued that WE wrote this math into the system. But we did, long before the physical universe was constructed based on those same mathematical constructs. This is where quantum mechanics meets classical physics, and it's all governed by the same laws. We'll explain. Just keep reading.

The Mathematics of the Universe

C'BS ALIFE ALLAH

"The Tao gives birth to the One; The One gives birth to the Two; The Two gives birth to the Three; The Three gives birth to every living thing. All things are held in Yin, and Carry Yang: And they are held togther in the Chi of teeming energy." – Lao Tzu, Tao Te Ching, Chapter 42

THE HARMONY OF THE SPHERES

Observing the movement of the sun, moon, planets, and stars above their heads, the Ancients developed an idea known as "the harmony of the spheres," referring to the way each object followed a discernable path and didn't invade the path of another. These short term and long term observations had direct impacts on the growth and development of civilization. They became the root of agricultural science as the seasons were mapped out via the journey of the Earth around the Sun and the Moon around the Earth. They also became the root of astronomical science as the precession of the Earth around the pole star (and its own axis) was insituted. These observations guided the development of navigational science as well.

With astronomical constants, the first short and long term calendars were constructed. Certain syncroneous events between the heavens and the Earth also made a lasting impression on measurements regarding human life and activity (like noticing that the time between menstruation corresponded with the the time between full moons). Even our music was mathematical, coordinated to the cycles, patterns, and harmonies we observed in the natural and celestial worlds. How long ago? As we'll explore in Volume Two, this knowledge didn't emerge among the scientists of our ancient urban cities. It was well-known long before that, with evidence for all of the above over 30,000 years ago.

CONSTRUCTING THE UNIVERSE

"I want to know how God created this world. I am not interested in this or that phenomenon, in the spectrum of this or that element. I want to know His thoughts; the rest are details." – Albert Einstein

Our present day universe started with the expansion of the universe from the singularity.* The singularity is the point in space/time where

* The idea of a "start" point is relative, however, as there is actually no "beginning"

all matter/energy in the universe (in actuality it WAS the universe) was infinitly dense and had zero volume. Mathematically this is expressed as the first dimension as just a dot.* In various traditions this singularity is referenced as the point from which all creation emanated.† As Michael S. Schneider explains in *A Beginner's Guide to Constructing the Universe: Mathematical Archetypes of Nature, Art, and Science: A Voyage from 1 to 10*, cultures throughout the world have used mathematical language (either in numerals or through geometry) to represent not only this initial stage, but all evolution that followed, up to the the mathematical connection between God, man, and the universe.

Indeed, meditations on the mathematical constants within the universe and man has been a hallmark of many ancient civilizations. Though there are many traditions, one of the best known is the Islamic tradition of Arabesque (which synthesized elements from Persia, China, Arabia, Spain, North Africa, India, etc). Islamic decree prohibited to-depictions of animals or humans. Such "creation" was only supposed to be done by Allah. Thus, they developed the "Arabesque" style of artwork by creating decorations based on rhythmic linear patterns of scrolling and interlacing foliage, tendrils, script, or lines.[36]

The complex geometry embedded within Arabesque was developed via the artists' connection with Islamic mathematicians. Arabesque artists also made subtle references to infinity (by having patterns extend beyond the frame of the piece). These works became meditations on the universe because, instead of representing the "crude forms" detectable to the naked eye, the art unvieled the "true reality" of pure mathematics in a manner that remained veiled to those who didn't speak the language (mathematics). Similar traditions can be found among other Original people, who used complex geometry in their artwork to represent the same ideals. Examples include the intricate cross designs of the Coptic Ethiopians, the mandalas of Tibetan Buddhists, and even the hair-braiding patterns used by the Yoruba people in West Africa.

UNIVERSAL CONSTANTS

"Nature is almost always describable by simple formulas not because we have invented mathematics to do so but because of some hidden mathematical aspect of nature itself."
– Clifford Pickover, The Loom of God

nor "ending" to the universe. There are just beginning and endings to cycles. This is further explored in the section dealing with the grand expansion vs. the big bang.

* The second dimension would be a line and the third dimension would be "depth."

† Such as the *bindu* tradition within Hinduism

Constants are measurements that, according to all the available data, remain consistent no matter what equation you put them in. This means that they hold true whether they are utilized to measure something under a rock, in a hill, or on the dark side of the moon. There are two types of universal constants, physical and mathematical. Physical constants refer to phenomena that are constant in nature and consistent in time while a mathematical constant refers to a number that has been found to have some significant relevance in more theoretical equations.

There are a host of physical constants within the universe. These constants actually give form to our universe.[37] There are "universal" constants, electromagnetic constants, atomic-nuclear constants and physio-chemical constants.[38] Two of the most notable are the speed of light and the gravitational constant. The speed of light is approximately 186,000 miles per second. According to Einstein's Theory of Relativity, this is the maximum speed that anything (energy, matter and information) can travel. The gravitational constant refers to the relationship between the mass of two objects and the gravitational force between them. It's found in both Sir Isaac Newton's Law of Gravitation and Einstein's Theory of Relativity.

There are also a host of mathematical constants. Most of them probably aren't familiar (or relevant) to the average person.[39] One you may know about is Pi. In Euclidean geometry, Pi refers to the ratio between the diameter and circumference of any circle. Though it's most well-known for that function, it's also utilized in other fields, like physics (e.g., in the Heisenberg uncertainty principle and Coulomb's inverse-square law).

In addition to Pi, there's also *Phi.* Trudy Garland, in her book *Fascinating Fibonaccis: Mystery of Magic in Numbers*, not only explains the Fibonacci sequence found in so many of nature's spirals, but documents how ancient African mathematicians were cognizant of this transcendental function, known by the Greek letter *Phi* (1.618), and later known as the "golden ratio" in Europe during its emergence from the Dark Ages.[40] You'll learn of other mathematical constants when we discuss the "cosmic coincidences" and in the following section on fractals.

THE CYCLICAL UNIVERSE

Why do most people think that the Universe had a definite starting point and that a "God" created it? It's not because of our scientific knowledge, but because of our religious upbringings. And let's be specific, we're talking about the story of creation in the Book of Genesis. But in many Original cultures time doesn't just start like that. It runs in cycles. The Mayan calendar (No it doesn't "end" in 2012),* various East Indian calendars, and many ancient Near Eastern calendars are cyclic. Various Native Nation languages in the Americas don't have time tenses in their language. And conceptions of time among Australian Aborigines and many African societies are so non-linear, they can best be explained by quantum mechanics.

That is, scientists have proposed that many kinds of sub-atomic particles can move forwards and backwards in time, or like photons, not exist in time at all. In the cosmology of the Australian aborigines, the realm of creation, so-called Dreamtime, is not bound to classical physics and thus not time-limited. Therefore, creation is continuous. As one Aborigine has explained:

> My soul comes from a storyplace of my ancestors. It will go back there. It has always been there and always will be. Dreaming stories and songs are the way we continue creation, which is still happening now – not in some "dreamtime," as it was mistranslated fifty or so years ago. It was, and it is. The rainbow snake is still pushing up the mountains. This is a circular thing, and has no time. In this way, entropy through western linear concepts of time is thwarted, and my people and land endure.[41]

Among many Aboriginal cultures, kinship systems spiral, so that grandfather and grandson is the same word, as is past and future tense.

A QUICK NOTE ON GOOD AND EVIL

There's no such thing as good or bad. There is just what we "see" as good or bad. It's entirely subjective. A catastrophe for one person can be an opportunity for another. And people who lack worldly attachments tend to lack the suffering that comes along with those things (such as loss, or fear of loss) as well. But that doesn't mean you can be a horrible human being and everything's going to be okay. It means there is a way the world works, and our ideas about what's "good" are in line with that nature. This is why we say being "righteous" means doing what will provide the greatest good, not necessarily what others think is "right" or "wrong." If it harms more than it helps, it's most likely not a "good" thing. Some harmful acts may, however, work out in our favor in the end, but unless you're a master at prediction, you're better off doing what you know is good. And doing good starts with thinking good. As the *Tejabindu Upanishad* notes: "Brahman [the Self/God] cannot be realized by those who are subject to fear, greed and anger. Brahman cannot be realized by those who are subject to the pride of name and fame. Or to the vanity of scholarship. Brahman cannot be realized by those who are enmeshed in life's duality." This means you can't function in tune with your nature until you let go of unhealthy thinking. For more on this, see Volume Three.

DID YOU KNOW?
American physicist John Wheeler describes the universe as a gigantic feed-back loop: "The Universe starts small at the big bang, grows in size, gives rise to life and observers and observing equipment. The observing equipment, in turn, through the elementary quantum processes that terminate on it, takes part in giving tangible "reality" to events that occurred long before there was any life anywhere."[43] In other words, the man comes into being in the universe, but man – through his observations of the universe – brings the universe into real existence. George Greenstein is more direct, saying "the universe brought forth life in order to exist...the very cosmos does not exist unless observed."[44]

The word for soon and recent is often the same. Stories and sentences are often circular, with repetition used to arrive back at the same point where you started. Cycles of birth and rebirth go on through sacred places, which are referred to simultaneously as times and places (as in time-space in western physics). John Mbiti has shown that African conceptions of time are very similar.[42] *The World of Quantum Culture*, edited by Manuel J. Caro and John W. Murphy, shows that indigenous American cultures also held circular concepts of time and space.

Scientists are only recently beginning to appreciate the non-linear thinking that Original people have employed since time immemorial to describe the past, understand the present, and predict the future. They're beginning to realize that indigenous people have known about chaos theory, complexity theory, non-linear systems dynamics, and perhaps even the history of the Universe. I'll explain.

A cyclic view allows you to see things that are parallel, such as sequence patterns, triggers and reconstructions, catchment, and circularity. Looking at life or history this way, you will start to see clusters of information emerge, revealing the way the patterns work. You'll realize that it isn't about things starting from a rudimentary point and always becoming more advanced or better. Nor is it about things "declining" from some "golden age" in the past. Everything fluctuates according to a rhythm that is never-ending, so there's always a beginning after an end, and always an end before every beginning. In reality, there's no beginning or ending at all.

This notion of "cyclic" time can help explain the way the Universe works. Universe literally means "that which returns to itself" (from *uni-* meaning "one" and *-verse* meaning "travel"), and that's what scientists have proposed as an explanation for where our Universe came from and where it is heading. In other words, even if the Universe (after several billion years) reaches its maximum point of entropy and expansion, ultimately "phasing out" into pure energy, just keep in mind that pure energy is how this Universe began!

This is the explanation offered by theorists like Roger Penrose. In his 2010 book *Cycles of Time: An Extraordinary New View of the Universe*, Penrose proposes the Conformal Cyclic Cosmology, which says that the Big Bang happens *after* the end of the universe, starting a new one. The expanding universe ends because, if everything keeps going as it has, eventually all matter collapses into black holes, and all black holes evaporate into energy. So the universe ends up being nothing but energy at very high entropy. Penrose says that when there is no mass in the universe, time no longer applies. This makes this point timeless. This is because there is only energy left, with gravitational waves left behind in the material universe's wake. These waves translate to density variations in the next Big Bang. Which, with no space/time to define whether it was in the past or future, was OUR Big Bang.[45] Thus, what happened before the Big Bang is happening now! Again, no beginning, no ending.

Penrose himself has noted that his cosmology is "a bit more like Hindu philosophy" than anything in the Judeo-Christian traditions, but he has no particular religious leanings. Traditional Indian cosmological concepts say that the universe is endlessly destroyed and created again, with many Hindus believing each cycle represents a day in the life of the creator God Brahman.[46]

So don't worry about the end of time. You're there at the beginning, there at the ending, and there when it all begins again.

AFRICAN FRACTALS

SUPREME UNDERSTANDING

THE SCALES OF EXISTENCE

Throughout nature, we find innumerable reminders of the mathematical nature of the way things work. The symbolic language of Supreme Mathematics is a useful tool to identify some of these patterns, but you don't have to know those laws to see them in action. One of the core principles of our numerical system is that, once we travel from 1 to 9, we cycle back to 1 at a higher scale by adding a cipher, or zero, and then starting back at 1 when we say 11. And this pattern continues at every scale of 10, meaning that we're dealing with the same type of cycle even when we go from 99,999 to 100,000 and then start back at 1 with 100,001. We're just at a much higher scale.

In terms of size, scales can be quantified by the powers of ten, using the centimeter as the base unit. This is for our purposes, as a size scale is not a physical entity but just a setting on our intellectual zoom lens. People are at the 10^2 cm range (100-200 cm in height), and mountains (which are measured not in meters, but kilometers) are at the 10^5 range (100,000 cm). As you can see, the exponent shows you how many zeros come after the 1. A negative exponent would show how many zeros come BEFORE the 1, as for a cell in your body which is in the 10^{-3} range (0.001 cm). In *View from the Center of the Universe*, Primack and Abrams use largest and smalles scales to demonstrate that we are at the center of the universe.

One thing that's interesting about the various scales of existence is that we find the same patterns and processes replicated at each scale. That is, we find aggregation (coming together) on the scale of subatomic particles, bacteria, animal herds, and even galaxies.

WHAT IS A FRACTAL?

In geometry, the idea of a design property that replicates at every scale is known as a fractal. According to mathematician Benoit Mandelbrot, a fractal is a rough or fragmented geometric shape that can be subdivided in parts, each of which is (at least approximately) a reduced/size copy of the whole.[47]

This is a property called self-similarity. In layman's terms, a fractal is a pattern that is the same no matter what scale you're looking at. At a purely mathematical, theoretical level, this iteration (repetition of the pattern) continues infinitely, but organic models of fractal mathematics can be found throughout nature. There are two relevant images on the cover of this book: one is a tree, whose limbs have grown in an organically fractal pattern, and another that is the pure mathematical

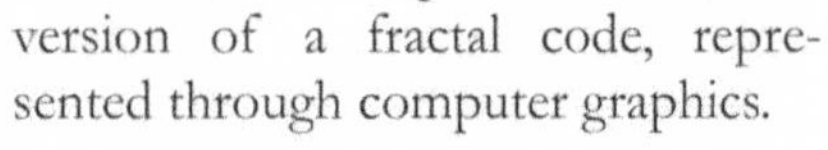

version of a fractal code, represented through computer graphics.

FRACTALS IN NATURE

Tree, lightning, or blood vessel?

"If the cosmos were suddenly frozen, and all movement ceased, a survey of its structure would not reveal a random distribution of parts. Simple geometrical patterns, for example, would be found in profusion – from the spirals of galaxies to the hexagon shapes of snow crystals. Set the clockwork going, and its parts move rhythmically to laws that can be expressed by equations of surprising simplicity. And there is no

> logical or a priori reason why these things should be so." – Martin Gardner, "Order and Surprise"

And fractals are not just "abstract" mathematics. They're found everywhere in nature, from the way a fern (one of the oldest plants on the planet) is shaped, with its branches following the same design as the leaves on the branches (and those leaves themselves having the same internal structure), to the branching of a riverbed. In fact, at every scale of existence, we find fractals.

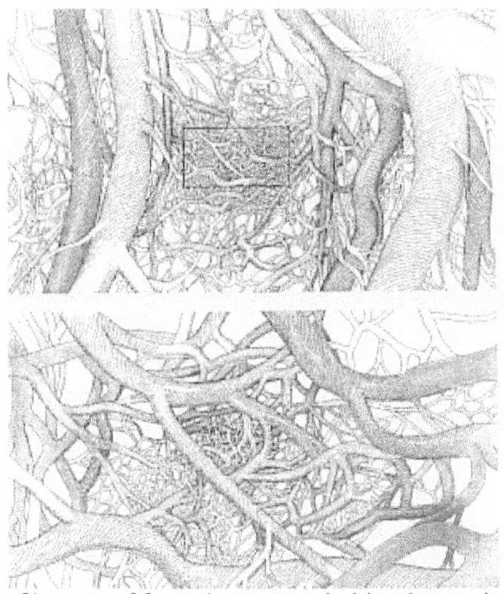
Closeup of fractal structure in blood vessels

You see, nature has a tendency to recycle patterns. In fact, the cyclical patterning of nature is the reason why fractals exist. These patterns serve as reminders to us that there is both massive universe beyond us, and similar pattern to be found on very fingertip. They are human-scale reminders about the mathematical "coincidences" of the universe, coincidences we may not otherwise consider since the abundance of them occur on scales too large or too small to observe with the physical eye. Such "iterations" also occur in behavior patterns, in the development of physical features, or even in the rhythms of light or sound. There's a reason why river networks, plant root networks, and lightning strikes reproduce the structure of blood vessels and pulmonary vessels. It's the same reason why a fully undilated pupil resembles a fully eclipsed Sun. Or why our body's cells resemble the structure of the Earth. Perhaps it even has something to do with the 360 waves on your head resembling the ocean waves.

Behavior patterns also progress fractally. Scientists can now use technology to identify everything from the fractal nature of a human heart beat to the splitting up of an ant migration to an urban growth pattern. Self-similarity at various scales epitomizes the ancient saying, "As above so below." It also emphasizes a corollary of that saying which is "As within, so without." In essense, fractals reveal the universe in man, or rather the universe as man.

At the molecular level, we know that molecules are organized in geometric formations. Such formations become multiplied in elements with crystalline structures that reproduce the geometric forms within them. On an organic level, we know that – despite the vast differences in their appearances – our body's organs are composed of cells which themselves are composed of similiarly-functioning organelles. Now, this is where I should note that, on a purely mathematics level, fractals are perfect and continue infinitely. In nature, however, "organic" fractals demonstrate the properties of self-similiarity, but don't need to go on forever to make their point.

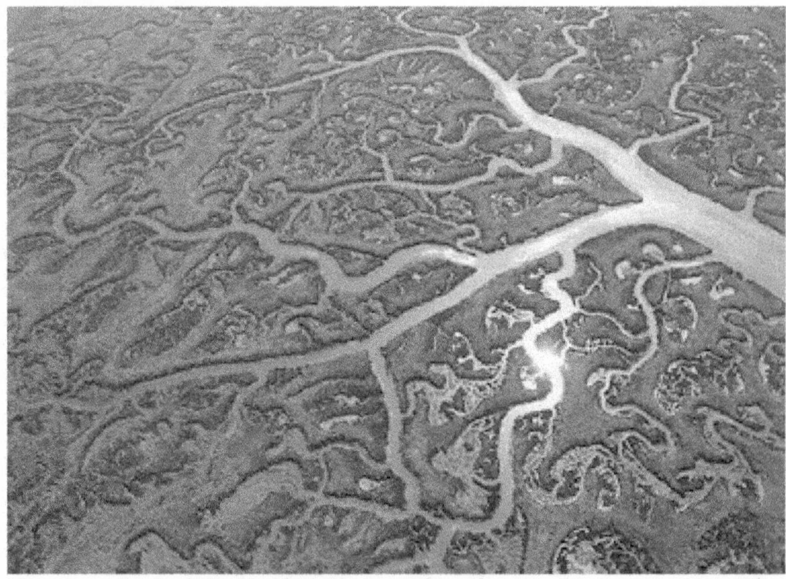

On a geological level, fractals are found just about anywhere the Earth's surface is cracked from the heat of the sun (as in deserts), rippled from the movement of its tectonic plates (as in mountain ranges), or split into an increasingly complex system of waterways (as in the mouth of a river). There are even fractal patterns to the increasing complexity of layers below the Earth's surface.

And why stop at our home planet? We know that we find mathematical structure embedded into even the earliest stages of the universe. One of the discoveries of the Keck 10-meter telescope in Hawaii was the appearance of substantial amounts of structure early in the history of the universe, far earlier than expected. Why (and how) was there so much structure at this early point? According to James Glanz, it's as if the structure was already written.[48] We also have evidence that the structure of universe is fractal in nature.[49] Of course, you don't need a microscope or a telescope to see the connections. Look at your fingertips and you'll see spirals the replicate the various spirals of galaxies throughout the universe. "Fingerprints of the Gods" for real.

But when it comes to the mathematical patterns that reproduce across various scales of the Universe, my favorite example is the connection between the structure of the brain cell and the Universe itself. Just look at the image (in the insert). If seeing a galactic spiral on your fingertip didn't speak to you, perhaps seeing the universe in your brain cell will.

According to Rupert Sheldrake's theory of morphic resonance, the mere existence of a pattern or structure (at any point in the past) makes it more likely for the same structure to be duplicated elsewhere in the universe or at a later point in time! So what came first, your brain cell or the universe? Wait 'til you read "The Cyclical Universe" for insight into THAT question.

> "There is an idea – strange, haunting, evocative – one of the most exquisite conjectures in science or religion. It is entirely undemonstrated; it may never be proved. But it stirs the blood. There is, we are told, an infinite hierarchy of universes, so an elementary particle, such as an electron, in our universe would, if penetrated, reveal itself to be an entire closed universe. Within it, organized into the local equivalent of galaxies and smaller structures, are an immense number of other, much tinier elementary particles, which are themselves universe at the next level, and so on forever – an infinite downward regression, universes within universe, endlessly. And upward as well...This is the only religious idea I know that surpasses the endless number of infinitely old cycling universes in Hindu cosmology." – Carl Sagan, astronomer

AFRICAN FRACTALS

Let's get back to the human scale. Where else on the Earth – besides the riverbeds, mountains, desert sands, and ocean floors – do we find fractals? In the African village. According to Ron Eglash, Senior Lecturer in Comparative Studies at Ohio State University:

> In Europe and America, we often see cities laid out in a grid pattern of straight streets and right-angle corners. In contrast, traditional African settlements tend to use fractal structures—circles of circles of circular dwellings, rectangular walls enclosing ever-smaller rectangles, and streets in which broad avenues branch down to tiny footpaths with striking geometric repetition. These indigenous fractals are not limited to architecture; their recursive patterns echo throughout many disparate African designs and knowledge systems.[50]

And these fractal patterns aren't found anywhere else in the world.

And Eglash looked! He could only find such modes of organization among the indigenous people of Africa. Eglash's book, *African Fractals* investigates fractals in African architecture, traditional hairstyling, textiles, sculpture, painting, carving, metalwork, religion, games, practical craft, quantitative techniques, and symbolic systems. He also examines the political and social implications of the existence of African fractal geometry.

Fractal structure in African village of Kotoko

Eglash theorizes that some of these instances, like the way the village was organized, wasn't "purposefully" designed as a fractal, but that they emerged "naturally" as if the builders had such systems built into their culture on an unconscious level.

Looking at the "human fractal" makes us wonder how all these inanimate objects developed fractal properties. Is it the law of aggregation? Is it the way a particular chemical's matrixes and latices work? Or is it more organic than that? Is there a life to this math? Is there a consciousness, or intelligence, to it?

WHY NO TWO SNOWFLAKES ARE ALIKE

We've all heard that no two snowflakes are alike, because they're all so intricately complex. But if that's so, how do they grow so symmetrically in six different directions at the same time? Could the same non-local "control element" that governs the complex patterns of seashells be what orders the radial growth of a snowflake? Scientists still can't explain thr process by which this happens.[51] Perhaps it's something organic again. After, all snowflakes aren't entirely inorganic. They're actually just ice crystallized around bacteria from the sea. Yup, there's life in there. Perhaps we're getting somewhere.

BIOGENIC MINERALS

The triangular pattern on Cymbiola innexa suggests the presence of a "global control element" that turns the pigment-secreting cells on and off in the correct order – like a computer-controlled loom! Google the "Sierpinski Triangle" fractal and make the connection.[52]

Let's talk about "life" for a minute. Animal, vegetable, mineral. Which one is nonliving? Mineral, right? Maybe not. While people typically think of minerals as having been formed by "inorganic" purely chemical processes in the Earth, it turns out that many minerals are actually "biogenic," meaning they're formed from living organisms.

In addition to many diamonds; common biogenic minerals include travertine and opal (deposited through algal action); pyrite and marcasite (from bacterial sulfate reduction); quartz (bacterial breakdown of oil produces organic complexes that dissolve, transport, and precipitate quartz); magnetite (living cells synthesize isometric crystals of magnetite.); hydroxylapatite (crystals manufactured by mitochondria); apatite (found in bones and teeth); and the aragonite, calcite, and

A QUICK NOTE ON MAGIC CRYSTALS

New Age "healers" and other esoteric thinkers have a special love for another crystalline mineral, quartz. Because quartz displays some unique qualities in the way it resonates a frequency it is connected to, crystals of quartz have become a "must have" item in many New Age toolkits. To be fair, quartz crystals have been collected by ancient people in the past. But we're not sure how they used them, outside of being revered for their looks. What we know now is that quartz crystals have a piezoelectric effect, meaning they acquire a charge when compressed or twisted, allowing them to act as a transducer between electrical and mechanical oscillations. As a result, quartz crystals are used in watches and radio transmitters. But there's no evidence for a how a handheld crystal could have any meaningful effect on a human body. Yet, when New Age authors began touting the "miraculous benefits" of crystal therapy in the 1970s, the price of rock crystals more than tripled. Many of these authors just "happened" to be selling these crystals, which were previously cheap or easy to find. A handheld piece of quartz will now run you about $30-80. But quartz is actually the most common mineral on the planet. It is a component of a huge variety of rock types, and comes in an extensive range of colors and varieties. The most well-known form? Sand. Yes, most sand is quartz. But nobody's going to buy a jar of sand for $50.

DID YOU KNOW?
According to a study published in the Journal of Sedimentary Research, diatoms (microscopic organisms) played a foundational role in the development of sedimentary quartz formations.55 Another study focused on quartz silt from black shales in the eastern USA, dating back to the Late Devonian period (about 370 million years ago). The authors concluded that common sedimentary quartz does not originate from the crust of the Earth: "Instead, it appears to have precipitated early in diagenesis in algal cysts and other pore spaces, with silica derived from the dissolution of opaline skeletons of planktonic organisms, such as radiolaria and diatoms. Transformation of early diatoms into in situ quartz silt might explain the time gap between the earliest fossil occurrences of diatoms about 120 Myr ago and molecular evidence for a much earlier appearance between 266 or even 500 Myr ago."56 In other words, quartz provides us evidence that diatoms were in existence over 300 million years before the earliest fossil evidence. This tells us a lot about the shortcomings in dating the origin of a species based on its fossil record, but also reveals an interesting origin for the most abundant mineral on Earth.

fluorite in the vestibular systems of vertebrates.[53]

In other words, many of the "crystalline" minerals are formed from living organisms. Naturally, we've got to talk about the biggest star out of that bunch: diamonds.

DIAMONDS ARE [US] FOREVER?

The main constituent of diamonds is carbon, but even chemically "pure" carbon is technically contaminated by light carbon (or C12), an isotope used by living organisms. The high C12 content of some diamonds suggests they have an organic origin. These jewels may have started out as the prehistoric single-celled organisms that lived around hydrothermal vents, which heat and pressure converted into "eclogitic" diamonds. That diamond on your ear might be the product of a billion-year-old single-celled organism![54] Or it could have come from carbon that came from the breakdown of some plant or animal that died in the Carboniferous period, 350 million years ago. Companies like Lifegem can now produce diamonds from the ashes of your deceased loved ones, but perhaps ALL diamonds come from one of our ancestors? (FYI, this does not mean go buy more diamonds)

WHAT'S MY POINT?

There's life behind these mathematics. We're not saying that every natural process is guided by little tiny organisms, or that subatomic particles only aggregate because something living told them to. What I'm suggesting is simply that MUCH of nature is more "organic" than we think, and the baseline tendence of organic nature is to self-replicate. This clearly occurs in reproduction (sex), and its related processes (cell division, etc.), as well as the exponential growth that follows ("Be fruitful and multiply" remember?). But it also occurs in the fractal organization of such things as seashells, ferns, blood cells, snowflakes, and acacia trees. But beyond the ecological scale, there's no life floating around in space, right?* Perhaps then, there is an "organizing principle" (or "intelligence") that guides the formation of less organic process like the formation of riverbeds and the structure of the universe. As Dr. Bruce Lipton writes in *Spontaneous Evolution*:

> Nature is a dynamic system, founded on iterated processes and chaos mathematics, and subject to sensitivity. The fact that fractal geometry is the specific mathematics to model such a chaotic system supports that nature should be fractal, but it does not necessarily provide a reason as to why. However, there is another compelling reason, based strictly on mathematics, that suggests why the observed parallels between fractal geometry and the structure of nature are more than coincidence.[57]

What's the reason? Because we wrote the Universe in the language of mathemtics. Because fractals and other mathematical principles where patterns reproduce across scales (like the Fibonacci sequence) allow for a great deal of information to be encoded in a very simple algorithm, it seems that (a) the universe could be encoded by very simple laws that operate across every scale and in every field, and (b) man is capable of grasping these laws quite easily, because they are "hardwired" into his brain and biology, and carried by his consciousness.

Returning to our discussion of mathematics, it becomes clear that the mathematical laws that govern life and reality are not merely hardwired into man's brain, they are the fundamental structure of his very consciousness. Effectively, man living by nature is a personification of these laws; "living mathematics." As we'll see in greater depth in the following chapter, this mathematical structure determined the course of the visible universe's development and structure, a concept Clifford A. Pickover calls the "Loom of God."[58] Five Percenters synthesize these ideas by deconstructing the word "Allah" as "All Law" and connecting both concepts to the Original man. The notoriously non-esoteric astrophysicist Carl Sagan says as much in his work *Broca's Brain*, when he writes:

> Some people think of God...busily tallying the fall of every sparrow. Others – for example, Baruch Spinoza and Albert Einstein – considered God to be essentially the sum total of the physical laws which describe the universe.[59]

This brings us to the Science of the Mind, which, as you'll see, is in-

* Or is there? See "The Story of Life."

deed the Science of Self.

THE RAINMAKERS

SUPREME UNDERSTANDING

MATHEMATICAL SYSTEMS FOR DIVINATION

In the Ishango region of Zaïre (now called Congo) near Lake Edward, archaeologists discovered a bone tool engraved with notches and a sharp piece of quartz affixed to one end, perhaps for further engraving or writing. The three rows of notches led scientists to assume at first that it was a tally stick, but others have suggested that the groupings of notches indicate a mathematical understanding that goes beyond counting. The Ishango bone tool is at least 20,000 years old.[60] On the tool are three rows of notches demonstrating some of the mathematical knowledge among the people of this region.* Jean de Heinzelin, professor emeritus of Ghent University and an authority in African archeology, compared prehistoric Ishango harpoon heads to those found in northern Sudan and ancient Egypt and proposed a link between Ishango mathematics and the origins of ancient Egyptian mathematical knowledge. Regarding the Ishango bone, de Heinzelin reported:

> It is possible to trace the influence of the Ishango technique on other African peoples by examining harpoon points at other sites. From central Africa the style seems to have spread northward. At Khartoum near the upper Nile there is a site that was occupied considerably later than Ishango. The harpoon points found there show a diversity of styles. Some have the notches that seem to have been invented first at Ishango. Near Khartoum, at Es Shanheinab, is a Neolithic site that contains harpoon points bearing the imprint of Ishango ancestry. From there the Ishango technique moved westward, but a secondary branch went northward from Khartoum along the Nile Valley to Nagada in Egypt...The first example of a well worked out mathematical table dates from the dynastic period in Egypt. There are some clues, however, that suggest the existence of cruder systems in predynastic times. Because the Egyptian number system was a basis and a prereq-

* The number of notches on either side of the central column may indicate more than simple counting. The numbers on both the left and right column are all odd numbers (9, 11, 13, 17, 19 and 21). The numbers in the left column are all of the prime numbers between 10 and 20 (which form a prime quadruplet), while those in the right consist of 10+1, 10−1, 20+1 and 20−1. The numbers on each side add up to 60, with the numbers in the central column adding up to 48. Both of these numbers are multiples of 12, suggesting an understanding of multiplication and division.

> uisite of classical Greece, and thus for many of the developments in science that followed, it is even possible that the modern world owes one of its greatest debts to the people who lived at Ishango. Whether or not this is the case, it is remarkable that the oldest clue to the use of a number system by man dates back to central Africa of the Mesolithic period. No excavations in Europe have turned up such a hint.[61]

Alexander Marshack concluded that the Ishango bone may actually represent a lunar calendar,[62] and Claudia Zaslavsky went further to suggest that the creator of the tool was a woman, tracking the lunar phase in relation to the menstrual cycle.[63] Similar paleolithic era "tally sticks" have been found, including present day Swaziland (the Lebombo bone, dated 37,000 years old and also possibly a lunar calendar[64]) and Czech Republic (the Wolf bone, approximately 30,000 years old[65]). Such "calendar sticks" are still used by Bushmen in modern day Namibia. Yet this hasn't prevented the spread of modern misconceptions about the mathematical abilities of such indigenous people. For example, another indigenous population, the Australian aboriginals, were once thought not to have a way to count beyond two or three. However, Alfred Howitt, who studied the peoples of southeastern Australia, disproved this in the late nineteenth century, although the myth continues in circulation today. The Australian Aboriginal counting system was used to send messages on message sticks to neighboring clans to alert them of, or invite them to, corroborees, set-fights, and ball games. Numbers could clarify the day the meeting was to be held (in a number of "moons") and where (the number of camps' distance away). The messenger would have a verbal message to go along with the message stick.[66] Similar forms of enumeration and recording utilized planks of wood, although obviously these would not survive as long as fossilized bone. Also related, but similarly unlikely to survive from a Paleolithic origin are recording systems using knotted cords, a system found among indigenous people throughout the world,[67] including the ancient Indians of present-day Peru (a system later used by the Inca, who called them quipus, or "talking knots"),[68] the ancient Persians, and the ancient Chinese. According to one Chinese text:

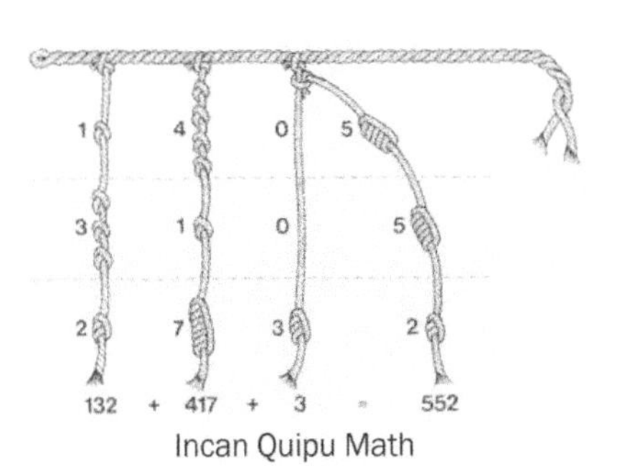

Incan Quipu Math

> In Early Antiquity, knotted cords were used to govern with. Later, our saints replaced them with written characters and tallies. In the ancient

> past, during the time of Rang Cheng, Xuan Yuan, Fu Xi, and Shen Nong, people tied knots to communicate. For a major matter, use string to tie a big knot; for a minor matter, tie a small knot. The number of knots corresponds to the number of matters to be dealt with.[69]

Both these systems, tally sticks and knotted cords, eventually became widespread in Europe, lasting even after the introduction of modern numerals (again from Original people) made them unnecessary.[70] In fact, the Catholic rosary is based on the knotted cord system. Yet when colonizing Europeans would later rediscover these intricate systems among Original people, they'd label them primitive.

Another system of divination involved the tossing of two-sided objects. In China, this became the *I Ching* system of divination, which today is done with coins, but was once done with cowry shells. In West Africa, this system occurs as Ifa. Both are binary systems that rely on mathematical laws of probability to "divine" (or predict) possible outcomes in scenarios that have multiple possibilities. Several other indigenous people have also used a binary system involved a two-sided stick. But such systems aren't simple. The binary system of *I Ching* has 64 hexagrams as possible results, each with multiple possible interpretations. In the past, only an adept shaman could reliably explain the results.* This ties us back to the mathematical language of the universe, where gender and polarity are binary indicators of the dual nature of reality itself. By having some way to measure the binary "biases" of a random event, our ancestors sought to see which way the future was headed.

Other cultures used objects with multiple faces. In Volume Two, we'll talk about the ancient six-sided number cubes found in India, Egypt, and Mesopotamia. Our people have been playing dice for thousands of years. But before dice, binary was all we needed. This way, or that way. And we could make sense of the entire universe from that.

This leads us to a final form of divination. Scapulimancy – a form of divination done by "reading" the cracks formed by heating caribou shoulder-blades – flourished among the Algonquin people of North America. The practice may have diffused from the "oracle-bones" of Shang Dynasty China, which were themselves preceded by the ancient Akkadians of Mesopotamia, who used sheep shoulder blades. But this

---

* See "The Origins of Religion" and "The Science of Shamanism" in Volume Two.

DID YOU KNOW?
God just might play dice. In 1926, Einstein was critiquing Niels Bohr, who proposed that there was no way to have total knowledge of any quantum-level event, only its probability. Einstein said God "does not throw dice." In the 1960s, Allah (the Father of the NGE), a known dice-player, allegedly said that he could not only predict the outcome of his rolls, but could intentionally affect them. His favorite combination? Six and one. Since then, scientists have confirmed Bohr's quantum mechanics theories. And some scientists have confirmed that Allah may have been telling the truth as well. A 1991 review of five decades years worth of studies on the "psychokinetic control of dice" (covering 148 studies involving more than 2 million dice throws by 2,569 subjects) provided "evidence of a genuine relationship between mental intention and the fall of dice." The most successful target? Six.[71]

practice may date back to the heating of stone tools – which left similar fracture patterns – over a million years ago. What's so special about fracturing? Consider the root word. Fracturing is fractal. This takes us back, once again, to the mathematical nature of the universe. Our ancestors felt they could read the direction of an emerging fractal and see what direction to take themselves. It was our means of decoding the complex variables of chaos theory to see which way the wind was really blowing. Can we do it now? I'd say our brains may be a bit too polluted by religious thinking to truly see through the matrix into the world of pure mathematics. We need a detox first!

But before the advent of organized religion, where science and sense are often smothered by rituals and doctrine, we may have been better equipped to accurately predict and plan for the future. I've heard that Elijah Muhammad taught that "circumstance times fact equals prophecy." This simple algorithm describes the process of considering all the variables involved, and the many ways those variables can interact, in order to deduce the likeliest outcomes of those interactions over time. Is it unrealistic to think that the scientists of 15,000 years ago could predict the future we're living now? Considering that white folks can become quite successful using the same science to get rich investing in stock options and hedge funds, I don't think it's impossible. I know white folks rely on massive bodies of data (and, to some extent, swarm intelligence) just to decide which areas to buy up in the hood, in anticipation of future gentrification.

So how far into the future is it realistic to predict? Here's where it gets deep. When you consider the fractal nature of the splits that can occur in any timeline of events, it certainly is quite unpredictable, as Chaos theory explains. But the form of these trends generalizes according to certain patterns that are predetermined. This is because the mathematical laws of the universe were pre-coded into the physical universe and the mind that organizes all of the life and matter in this universe. Those among us who were most in tune with this sub-quantum consciousness, who anthropologists now call Shamans, were the scientists who could see what really was (as opposed to what it appeared to be), and what was coming. When human societies became more hierarchically structured and less communal, the role of the shaman gave way to holy men, master teachers, mystery schools, and ultimately, priesthoods and organized religion. We'll see how exactly that happened in Volume Three.

AN INTRODUCTION TO PHYSICS

ROBERT BAILEY

Physics is a natural science that studies the laws and rules of matter and energy, as well as how they interact. Physics is present in every action around you. Energy is made evident in different forms such as motion, light, electricity, radiation and gravity. Matter exists on scales from the largest to those that you need the best microscopes to see. All of the above has an effect on something else. Physicists study those effects to figure out what forces are involved, how they operate, and what causes things to be the way they are and to work the way they do. From its name, you'd guess that physics was all about the physical world, but keep in mind that many things that aren't visible to your eyes are still a part of the physical universe. In short, physics studies the "mechanical" workings of the reality we live in, and attempt to describe in the simplest, most accurate language available: mathematics.

According to Aristotle, the philosopher Thales of Miletus invented the founding idea of science (that the seemingly infinite complexity of the world could be explained by a small number of hypotheses) and introduced it to Greece after studying in Egypt. There, Thales had learned about matter, geometry, and even some geophysics.

Geophysics is just of the many different areas of physics, such as astrophysics, cosmology, quantum physics, theoretical physics, atomic physics, and nuclear physics. Astrophysics and cosmology studies matter on its more massive scale by seeking to understand the birth of stars, origins of our universe and things of that nature. Geophysics focuses more on the Earth from an electromagnetic, seismic and radioactive standpoint. Atomic and nuclear physics look to make sense of the atom.

Theoretical physics uses mathematical explanations for natural events, but often develops theories that can't be tested empirically. Most of these studies fall under the laws of Newtonian Mechanics, which earned its name from Sir Isaac Newton, who is said to have discovered the law of gravity after watching an apple fall from a tree. Newton was no joke, however. His observations were on point regarding hundreds of physical phenomena. It seemed that his laws covered everything under the Sun (and the Sun too!). But in recent years, once scientists began to look at things on a subatomic level, they realized that the laws of Newtonian Mechanics (or Classical Physics) no longer applied! This is when Quantum Physics was developed to describe the subatomic world, where discrete, indivisible units of energy called "quanta" operate according to a very different set of rules. For other insights, see "An Outline of Africa's Role in the History of Physics," in *Blacks in Science: Ancient and Modern*, edited by Ivan Van Sertima.

KEY CONCEPTS IN PHYSICS

There are many important concepts in physics, so we're going to focus on some of the key concepts you'll see throughout this book.

The Laws of Motion: You may be familiar with Newton's laws, which describe how objects in motion stay in motion or stay at rest unless acted on by a force, and that for each action, there is an equal and opposite reaction.

The Universe is Made of Matter and Energy: At the smallest level, matter is made of elementary particles which have mass and charge. On a large scale, matter ranges from everyday objects to vast galaxy super-clusters. Energy has many different forms.

Waves Carry Energy: Energy propagates through materials and space by means of various types of waves, for example, sound waves in air, seismic waves through the earth, electromagnetic waves, including light that may travel through materials or empty space.

The Composition of Matter: Over the past 3,000 years, scientists have found smaller and smaller "building blocks" of matter. Scientists are now way past the idea that everything is simply made of protons, neutron, and electrons. There are currently three groups of sub-atomic particles:

- There are quarks which come in six flavors (types): up, down, strange, charm, top and bottom.
- There are leptons which also come in six flavors: electron, muon, tau, electron neutrino, muon neutrino, and tau neutrino.
- And finally there are the bosons: gluons, W bosons, Z Bosons,

photons, and the theoretical graviton.*

The first two groups are the sub-atomic particles that form matter. The third group is mostly related to forces of energy (or particles without mass) like light and gravity.

The Law of Conservation of Energy: Energy cannot be created nor destroyed, it only changes form. Thus, $e=mc^2$, the formula for the conversion of energy into mass and vice versa. This also means that overall they remain unchanged by an interaction or transformation.

The Theory of Relativity: There are many different elements to Einstein's theory, but the basic idea is that his General Theory of Relativity confirms Newton's theory, but also expands on our understanding of some of the key principles.

The Laws of Thermodynamics: These laws describe the way energy and matter work. Of particular importance is the second law, which explains how order and disorder come to be. It states that any system gradually breaks down because of entropy, or disorder. The only way

A QUICK NOTE ON METAPHYSICS

Commenting on the advances of Western knowledge, Senegalese scientist Cheikh Anta Diop observed, "Man is a metaphysical being," and to take man's desire to "find himself" through the reductionism of Western scientific knowledge would be a catastrophe. Man would "cease being himself," Diop wrote in Civilization or Barbarism. We know that indigenous people embrace multiple "ways of knowing" and that metaphysical perspectives are the hallmarks of traditional African, Asian, and Aboriginal culture worldwide. But understand this: We are in an age of cons, frauds, charlatans, and schemers. We have been exploited and manipulated to no end, in great part due to our willingness to believe those who intend to take advantage of us. In this era, we must revisit our roots to reclaim a true sense of self, before all this deception began. The root word of metaphysics is physics, meaning a study of the natural world. As we've explained, Western science doesn't own that study. We originated this science, through our observations of the way the world works, long before Newton and Pythagoras put the calculations on paper. Metaphysics literally means "after physics" or "beyond physics," which means one must first understand the basic workings of the world before embarking on such a potentially precarious study. Why is it precarious? Because if you don't have the foundational knowledge, people can tell you ANYthing and call it metaphysical knowledge. After all, people feel there's no way to verify such knowledge, and thus won't provide you any proof of their claims. This is dangerous territory. This is why, in this volume, we are providing you with the FOUNDATION. In Volume Two, we show you what we built with this knowledge (civilization), and in Volume Three, we get into the metaphysical perspectives that our ancestors developed "after physics." In Volume Four, we show you how these perspectives were used against us, and in Volume Five, we show you how the resurgence of such perspectives can be again used to our benefit – but only when there is a foundation of critical thinking (this volume) firmly in place.

to create order is by exporting disorder (or negentropy) outside the system, which creates heat energy.

The Four Forces that Hold it All Together: All interactions originate in four fundamental forces of nature. The force of gravity acts between all bodies and depends on their masses. The electromagnetic force acts between charged particles or between magnetic poles and is responsible for electric and magnetic fields and electric currents. The strong and weak nuclear forces operate between protons and neutrons in the nuclei of atoms, holding them together and sometimes resulting in radioactive decay.

There are many other laws, theories, key concepts and terms you will come across, such as self-organization, space/time, and nuclear reactions. Wherever possible, we've defined them when they occur for the first time. If we haven't defined something, definitions (and examples) are easy to find online or in a dictionary. For example, Wikipedia has a great body of credible information of self-organization throughout the universe.

AN INTRODUCTION TO QUANTUM MECHANICS

ROBERT BAILEY & SUPREME UNDERSTANDING

"Quantum theory is bizarre. In order to try and understand it we need to forget everything we know about cause and effect, reality, certainty, and much else besides. This is a different world; it has its own rules, rules of probability that make no sense in our everyday world" – Richard Feynman

Quantum mechanics (also referred to as quantum physics or quantum theory) is the study of physics at the level of individual atoms and subatomic particles like electrons, photons, protons, and neutrinos. As Richard Feynman put it, the behavior and movement of particles on this level is unlike anything else we witness in everyday life. This type of behavior required a whole new explanation, a new theory – hence the name quantum theory – to explain the quantum realm.

In our everyday world, phenomena are classified into two types by scientists: particles and waves. In general, think of a particle as a "thing," while a wave is more like a movement or motion of energy, not something you can sit down somewhere or put in a box. So an atom would be a particle, and light would be a wave, right? Not always. It's easy to distinguish between these two concepts in the world we can see, but on the quantum level, waves will sometimes behave as particles and particles as waves. Scientists can't even predict where any particle will actually be at any given point in time, only its probabilities of being there. It's physically impossible to simultaneously know the position and momentum of a particle – this is known as Heisenburg's uncertainty principle.

Earlier, we noted that 99.9% of the physical world is really empty space. Quantum mechanics has shown us that even the 0.01% of reality remaining may not be so real after all. Quantum mechanics is such a "weird science" that Einstein went to his grave attempting to explain some way around these quirky findings. When scientists demonstrated that two entangled particles could instantly "communicate" with one another (even at great distances) to cause one particle to immediately change if the other one was changed – at a speed faster than light – Einstein called it "spooky action at a distance."

But quantum mechanics isn't spooky. It's been verified in literally thousands of experiments. Scientists still can't explain all the hows and whys of the quantum realm's workings, but they know it's real. Or at least as "real" as the quantum world can be. For an introduction to some of the most important concepts in quantum mechanics, see our quick overview in the Appendix of this book.

As James Kakalios has said, one of the greatest things about quantum mechanics is that we can use it correctly and productively even if we're confused by it. We have quantum science to thank for technological advances such as the transistor and laser, without which we wouldn't have Blu-rays or PCs.

"I am now convinced that theoretical physics is actually philosophy. It has revolutionized fundamental concepts, e.g., abut space and time (relativity), about causality (quantum theory), and about substance and matter (atomistics). It has taught us new methods of thinking (complimentarity), which are applicable far beyond physics." – Max Born

But there's another use for quantum mechanics. It can show us the mathematical nature of the universe, and of reality itself, in a way Classical Mechanics never could. Studying quantum mechanics is truly like Alice peering into the looking glass, and ultimately peeking behind the curtain to realize that it's ALL an illusion, and the mind – the sub-quantum mind, or simply "consciousness" – is both the source, the observer, the maintainer, and the final destination of all reality. This will become increasingly clear as you read the next chapter.

REVIEW

So let's recap. We've seen that Life and Matter = Language and Mathematics. We know that DNA, or Genetic Code is a sort of software, written in instructional language composed of words, sentences, and punctuation, with a built-in error-checking feature! Non-coding DNA, for now, is like an undeciphered language. In addition to the A,

C, T, and G proteins that DNA uses to encode its instructions, there are X and Y chromosomes, and mtDNA and y-DNA haplogroups that range from A to Z. These may appear to be arbitrary names, but they're no accident. The fact that DNA is both mathematical and linguistic is widely accepted and has worked its way into many common conventions among geneticists. When Noam Chomsky proposed the idea of a "universal grammar" that was innate to man, scientists wondered exactly how far back the roots of language go. We'll explore the origins of spoken language in Volume Two, but we know that a language does not have to be spoken.

There are primates who communicate with each other using ultrasonic tones that no other animal can hear.[72] Nature photographer Kjell Sandved has found that every letter of our modern alphabet, including many characters from ancient scripts, can be found on butterfly wings.[73] We know that such patterns are a form of communication between, but on a deeper level, could the genes that encode such patterns be pulling from the same morphogenetic reservoir of "universal language" that we do when we construct our written and spoken languages? Perhaps the "universal language" emerges from a "universal math."

Math is woven into the very fabric of life. This is why indigenous cultures who lack the words for numbers still possess a profound number sense. There is math hardwired into our brains, and it comes from the same conscious patterning that structures the universe mathematically. This is why the folding proteins of our genes (the hardware of DNA's software) form Platonic solids.[74] In fact, geometry is woven so deeply into the physical universe, that molecules combine based on their geometric shapes (like puzzle pieces).[75]

According to quantum electrodynamics, particles are simply "excitations" in a field, along a wavelength of mathematical probability that is not narrowed down to a single data point until it is consciously observed. In other words, it's all mathematical waves of probability, with nothing "real" unless we make it real. In even simpler terms, this universe is simply math. Our brains do the math and we see what we see. But, like Neo sees in *The Matrix*, beneath the surface it's all math.

We can see this in the Fibonacci spiral found in everything from a Nautilus shell to a distant nebula, in the fractals found at every scale of the universe, and in the patterns of organization that can describe the developmental stages of anything we see. Throughout nature, we see addition, multiplication, division, subtraction, and every other mathematical operation in process. Again, it is from this "universal math" that there emerges the "universal language."

In Physicist Michio Kaku's interpretation of string theory, the universe exists in dimensions of 10 and 26:

> The heterotic string consists of a closed string with two types of vibrations, clockwise and counterclockwise, which are treated differently. The clockwise vibrations live in a ten-dimension space. The counterclockwise live in a twenty-six-dimensional space,* of which sixteen dimensions have been compactified [to correspond to the 10 clockwise dimensions].[76]

In the tradition of the Five Percenters (also known as the Nation of Gods and Earths), the keys used to "unlock" the meanings of the universe are known as the Supreme Mathematics and Supreme Alphabet. The Supreme Mathematics consists of ten numerals (1-9, plus 0), each of which represents a quality, ideal, or process. These aren't merely principles, but laws of reality. The 26 letters of the alphabet, however, are used to represent about 30 principles, ranging from the nature of man and woman, to governance, to ethical concerns. The 26 letters of the Supreme Alphabet can be compacted to correspond to the number values of Supreme Mathematics. If string theory is one day recognized as the Theory of Everything.

Yet the 5% weren't the first to develop and study a symbolic system of numbers and letters to explain the workings of life and the universe. In their previous incarnations, the Supreme Mathematics and Supreme Alphabet were known to the Jews as the Ten Sephiroth of the Kabbalah and the letter-number system of Gematria, to the Greeks as the Pythagorean Numbers, to the Arabs as the Abjad numerals (a system where each letter of Arabic alphabet has a numerical value, from where they derive the ilm-e-jafar (Science of Cipher), and ilm-e-huroof (Science of Alphabet Letters), and to various other cultures across the world. Most of this comes from Original people in some derived or direct form. The Kabbalah, for example, was formulated in Spain under Moorish rule, and Pythagoras studied in Egypt. The 5% weren't the first to propose this way of revealing the world's workings through letters and numbers, but the 5%'s system is perhaps the simplest of any such system. So, despite the 5,000 years of history over which the Nile Valley ideographs transformed into the present day English alphabet of 26 letters, and the 5,000 years of history over which the Indus Valley symbols became our 10 present day numerals,

---

* How do physicists explain all the extra dimensions in such theories? Most theories propose that the dimensions are either not perceivable, or they are embedded within the perceivable dimensions (length, width, depth, time).

some cosmic coincidence, some "string on our finger," must have left us with a set of letters and numbers that would ultimately help us – particularly those of who are young, Black, oppressed, and not privy to discussions of quantum mechanics – see ourselves as the authors and geometers of the universe and life itself.

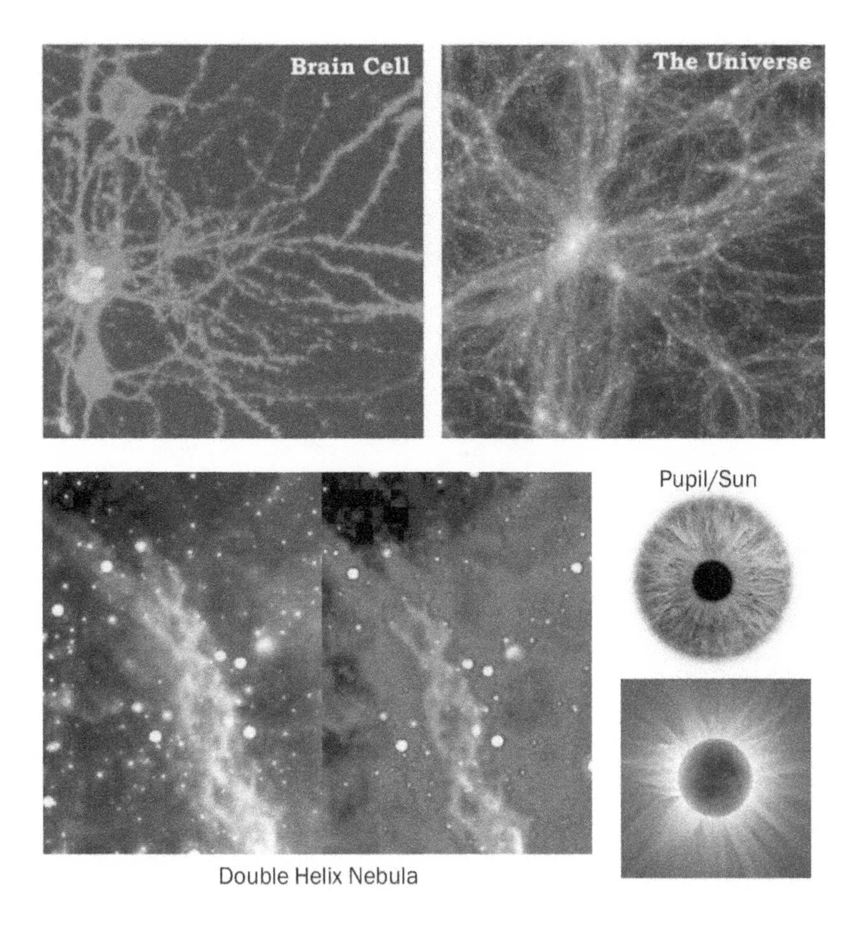

Double Helix Nebula

KNOW THY SELF

OUR PLACE IN THE UNIVERSE

"The self-existent Lord pierced the senses to turn outward. Thus we look to the world outside and see not the Self within us. A sage withdrew his senses from the world of change and, seeking immortality, looked within and beheld the deathless self." – Katha Upanishad

Can you see the patterns yet? Can you see the structure of reality? Can you see through the layers of material into the substance below the surface? If so, you know what is at the heart of all reality. In the precursor to this book, *Knowledge of Self: A Collection of Wisdom on the Science of Everything in Life*, Supreme Understanding wrote about the conscious intelligence running every system in our bodies:

> This "mind that you can't see," this innate intelligence keeps your heart beating and your lungs breathing without you even having to think about it. How else could all those processes go on inside of you without a glitch? But there's nothing spooky about it. This intelligence is definitely within you. It's even within your control. You simply don't know about it. You see, there are actually five "stages" (or frequencies) to the mind: (1) Conscious, (2) Subconscious, (3) Superconscious, (4) Magnetic Conscious, and (5) Infinite Conscious.
>
> Most people are only aware of the base stage – their conscious thoughts. They are reminded of their subconscious thoughts when they dream at night or have fantasies, but even then they don't realize that their mind is functioning on more than one level. A computer works the same way. There are processes that happen on the desktop screen (like the program I used to type this book), and processes that happen in the background (like the computer managing its memory so it doesn't slow down while I type).
>
> If you were to start investigating your subconscious thoughts, you could learn a lot about yourself. You could learn "what makes you tick" as they say. You could understand what you're REALLY thinking about or why you REALLY did something you did.
>
> It's not that hard to do: It begins with looking past the basics, and asking questions like "Why?" about everything. If you ask yourself what everything means, even the little things you do, you'll begin to see clearly in no time. Unfortunately, most of us function solely on the

> conscious level: satisfying our most basic needs (food, clothing, shelter, sleep, and sex). We don't know why people (including us) do what they do, nor do we try to learn and find out. I'm hoping you're not like that, because I'm about to take you deep into the rabbit hole.[77]

Consider *The Science of Self* a headfirst dive into the rabbit hole we opened up in *Knowledge of Self*. In this chapter, we'll show you how there's more to your mind than what you sense on a conscious level, more to your self than a merger of mind and body, and how the same consciousness that governs your body also governs all reality. When you're done, you will understand what is meant in the ancient Indian scripture, the *Kena Upanishad*, where it says:

> That which cannot be expressed by speech, but by which speech is expressed – That alone know as Brahman and not that which people here worship.
>
> That which cannot be apprehended by the mind, but by which, they say, the mind is apprehended – That alone know as Brahman and not that which people here worship.
>
> That which cannot be perceived by the eye, but by which the eye is perceived – That alone know as Brahman and not that which people here worship.
>
> That which cannot he heard by the ear, but by which the hearing is perceived – That alone know as Brahman and not that which people here worship.
>
> That which cannot be smelt by the breath, but by which the breath smells an object – That alone know as Brahman and not that which people here worship.

WHAT YOU'LL LEARN

- ❒ Why man is mind and mind is man.
- ❒ How your mind is (connected to) everything in the universe
- ❒ The differences between the brain and the mind.
- ❒ How the Mind structures reality at all scales of existence.
- ❒ How EVERYTHING is a form of energy.
- ❒ How the third eye is actually a real organ in our body.
- ❒ Why no one ever really dies (yet there's no life after 'death' either).
- ❒ The science behind ideas like Chi, Feng Shui, Karma, Levitation, ESP, Telekinesis, the Aura, etc.

WHAT IS THE MIND?

C'BS ALIFE ALLAH

"All that we are is the result of all that we have thought. It is founded on thought. It is based on thought." – Buddha, the Dhammapada

Are we mere products of the universe? Are we the sum of extensive evolution that extends beyond biological evolution into chemical and anatomical evolution? Are we the original seed potential out of which the universe unfolded? Are we the product of the environment or are we the producers of the environment?

Some are under the illusion that science doesn't seek to answer the above questions. They think that only religion has been set up to contemplate man's role in the universe. Science doesn't just seek to understand the underlying structure of reality; it also works at answering how we are connected to the structure of the universe on the quantum and astronomical levels. By recognizing that things reproduce after their own kind on the biological level, science isolates the principle of reproduction. Reproduction thus becomes not just a principle that is seen on the biological level, it extends into the chemical, elemental, planetary and other realms of existence. Seeing that there is a principle that is shared throughout existence propels science to look for other connections between these levels. This structure or resonant relationship between different levels of the universe can be referred to as the mind.

MAN MEANS MIND

The word "Man" comes from the Sanskrit word *manus* which means "Mind." Of course, Man meaning mind is not limited to Sanskrit and its relative languages. In the language of the Awabakal aborigines of Australia, the word for "self," *koti* comes from the word for thought or thinking, *kota.*[78] Among the Bantu of West Africa, *muntu* is the Self, and muntu means mind. In fact, *Bantu* is simply the plural of *muntu.*

The mind, as a noun, does not exist as a "thing." The mind is the order, structure, and organization that energy and matter takes. Order means "arrangement" and it also means "condition of a community which is under the rule of law," from its early etymology. Patterns clearly evident from galaxies to atoms show there is a "rule of law" or "order" in the universe. Though the word law is utilized in a sociological context, one should become familiar with the scientific context. A scientific law is a description of an observed phenomenon (while a scientific theory is an explanation of an observed phenomenon). So if energy and matter are the building blocks, then the mind is the order in which you arrange the blocks. Some people have a hard time seeing the mind as not being a "tangible thing." They confuse the mind with the brain. The mind exists independent of the brain. It is like heat. You can't touch "heat." You can only touch those things whose atoms have been accelerated to generate "heat." What is the length of despair? What is the humidity of peace? They don't have a context because things like love, peace and happiness come from order or the relationships that things share with each other. Order is not less "real," however. It is just not a thing that you can measure by a yard stick rather it is a quality of organization.

This "implicate order" is often referred to in some circles as the "non-local" principle. In some Chinese philosophies it is referred to as the "no-mind" (*wu shin*)[79] or simply the Tao.[80] In quantum physics it is coined by Dr. David Bohm as the "mind-like" aspect of the Hidden Variable, which he says underlies the holographic structure that allows information of any part of the universe to be equally present in another part of the universe.[81] I'll explain.

WHAT IS THE STRUCTURE OF REALITY?

In a paper titled "Space-Time Transients and Unusual Events," Persinger and Lafreniere write:

> We, as a species, exist in a world in which exist a myriad of data points (events or actions, verbs not nouns). Upon these matrices of points we superimpose a structure (models or maps, static things; nouns not verbs) and the world makes sense to us. The pattern of the structure originates within our biological and sociological properties (brain hardware and software).

Testing for some elements of this quantum level structure came about with Bell's Theorem in 1964.[*] In summary, Bell's Theorem postulated a device which emitted a ray of photons which went to two different devices separated by space. The measurement of the light at each device showed that they were mathematically complementary down to the same property. It's as if each group of photons "knew" what measurement was being carried out. Bell's Theorem was actually tested in 1974 and it acted in the exact manner that Bell had postulated. Let me make it clear what this implies. Whether they are separated by an inch or a million miles, subatomic particles can communicate instantly. Bohm explains this is because the Universe is not as real as we think – it is in fact like a hologram that gains its illusion of depth from our

[*] Not to be confused with the infamous Bell Curve, a racist piece of pseudo-science, written by people who had neither extensively studied in that field nor compared their findings with fellow scholars before publishing, and which has since thoroughly critiqued for flaws in its premise that Blacks have lower IQs than whites. For beginners, Black babies have higher IQs at age 4, suggesting that society impacts the development of IQ. Looking deeper, IQ is a culturally biased measure that fails to recognize aspects of mental ability besides those valued by Western culture.

conscious observation. As Michael Talbot explains in *The Holographic Universe*:

> Bohm believes the reason subatomic particles are able to remain in contact with one another regardless of the distance separating them is not because they are sending some sort of mysterious signal back and forth, but because their separateness is an illusion. He argues that at some deeper level of reality such particles are not individual entities, but are actually extensions of the same fundamental something.[82]

Even Einstein said that some notions of non-local quantum mechanics appear "spooky." Yet this is what the data (through repeatable experiments) has established and there is no conflict within the scientific community about the data. The only thing being discussed is what the data actually implies. In general, there are three lenses through which to filter this data. They were all determined by David Bohm yet given playful names by the author Robert Anton Wilson. Wilson has labeled these three modes as the Philosophical Monist alternative, the Science Fiction alternative and the Neo-Kantian alternative. He stressed that they are not the only three choices, yet, in general, most views fall within these three slots.

"For a parallel to the lesson of atomic theory... [we must turn] to those kinds of epistemological problems with which already thinkers like the Buddha and Lao Tzu have been confronted, when trying to harmonize our position as spectators and actors in the great drama of existence."
– Niels Bohr, co-developer of Quantum Mechanics

THE PHILOSOPHICAL MONIST ALTERNATIVE

The Philosophical Monist alternative affirms what is alluded to in Buddhist and Taoist texts. It also correlates with modern theories such as Rupert Sheldrake's concept of the morphogenetic field[83] or Carl Jung's concept of synchronicity.[84] It implies that all is one. It also resonates with the "Grand Expansion" (or "Big Bang") theory, in that if all particles were together at that first moment of this current epoch, then they are still non-locally "connected." In the words of Dr. Bohm, "everything in the universe is in a kind of total rapport, so that whatever happens is related to everything else."

THE SCIENCE FICTION ALTERNATIVE

The Science Fiction Model states that energy can either be enfolded (implicate) or unfolded (explicate). That is, objects in the universe have an underlying program of activity. This is also known as the FTL or "Faster Than Light" Model. In this model, information appears, or is traveling faster than light because it is "programmed" into energy and matter on the quantum level. In the words of Dr. Bohm, "It may mean there is some kind of information that can travel faster than the speed of light."

THE NEO-KANTIAN ALTERNATIVE

The third model, the Neo-Kantian alternative, is summarized clearly in Dr. Bohm's statement "Or it may mean that our concepts of space and time have to be modified in some way that we now don't understand." For the record, Dr. Bohm strongly supports the Science Fiction model.

So within this series, that "implicate order" where it is viewed through any of the above lens, is referred to us as the mind. It also is related to acausality (having no apparent cause) on the mesoscopic level (macro- being the astronomical level, meso- being the biological level, and micro- being the quantum level). This is the experience of two or more events that are apparently causally unrelated or unlikely to occur together by chance and that are observed to occur together in a meaningful manner. The concept does not question, or compete with, the notion of causality. Instead it maintains that, just as events may be grouped by cause, they may also be grouped by meaning. A grouping of events by meaning need not have an explanation in terms of cause and effect.

WHERE IS THE MIND?

This is my brother Supreme Scientist's response to this question:

> Where is "$e=mc^2$"? Where is "Wednesday"? Where is "Uncle Sam"? Where is "information"? Where is "security"? Where is "an anthem"? Where is "digestion," "consciousness" or "metabolism"? Are these things real? What are they composed of?
>
> Better yet...Where exactly is the program that is controlling your computer at, right at this moment? Is it in the base of your computer? Or can you "see" it in your monitor, too? Is it in your printer (when you print)? Or in your mouse? Or your speakers (when you listen to music)? Or is it "in" the cable wires and cords connecting those above mentioned things?
>
> Better yet...At what point did the computer program (Windows 7, Microsoft Excel, etc.) stop being "in" the brain of the person who invented it and "in" the brain of the person who developed the program or developed a blueprint to manufacture it? And, at what point did it cease being "in" that small jump drive they loaded "into" the usb port and downloaded "into" your computer?
>
> Answer: The Mind, like a computer program, IS ORGANIZATION ITSELF. If someone ever discovers a definite shape, size, position, momentum or orientation for Mathematics, let me know.
>
> What does it do? Nothing. It is only a DESCRIPTION of how people, places and things can be ordered on their most fundamental level. The more one discovers which phenomena employs, or fail to employ these basic building blocks of physics, one can maximize their success

> in all their under and overtaking.
>
> The Mind is not a "thing" that causes life and matter. It is a DESCRIPTION of something in life and matter that IS ORGANIZED. Asking what does it do is tantamount to asking, "what does '$e=mc^2$' do?" or Einstein's theory on Special Relativity do? It is not an "entity" out there "causing" things to happen. That's like saying that Newton's laws of motion CAUSE motion.

"AS ABOVE SO BELOW"

In *Prometheus Rising*, Robert Anton Wilson writes:

> Consider the human brain a kind of bio-computer—an electro-colloidal computer, as distinct from the electronic or solid-state computers which exist outside our heads. The brain appears to be made up of matter in electro-colloidal suspension (protoplasm). Colloids are pulled together, toward a condition of gel, by their surface tensions. This is because surface tensions pull all glue-like substances together. Colloids are also, conversely, pushed apart, toward a condition of sol, by their electrical charges. This is because their electrical charges are similar, and similar electrical charges always repel each other. In the equilibrium between gel and sol, the colloidal suspension maintains its continuity and life continues. Move the suspension too far toward gel, or too far toward sol, and life ends.
>
> Any chemical that gets into the brain, changes the gel-sol balance, and "consciousness" is accordingly influenced. Thus, potatoes are, like LSD, "psychedelic" – in a milder way.
>
> The changes in consciousness when one moves from a vegetarian diet to an omnivorous diet, or vice versa, are also "psychedelic."[85]

A QUICK NOTE ON BEING "DEEP"

"Higher consciousness" is not when you can judge people who ain't where you at. Higher consciousness is when you understand why everyone is where they are, including why you aren't where you could be. As we explained in the beginning of this book – despite how honorable our quests for knowledge and perfection may seem – sometimes we are merely trying to fill a void within our sense of Self. We're just trying something different. Now, there's nothing wrong with "finding yourself" in "deep" knowledge, but if this process involves alienating yourself from everyone else, there's nothing indigenous, African, traditional, healthy, or holistic about that. We shouldn't aspire to be above others (talking over their heads), or so deep that people can't see us at ground level. If we end up doing so, we should be honest and consider if our "personal journey" is more about feeding our own internal demons than the work of the gods. The purpose of "higher knowledge" is not simply to get "high" off the knowledge, but to bring this knowledge back down to Earth so the rest of us can benefit. And yes, we know it's hard, because many just don't get it. But there was a time when you didn't get it either, and some ideas (and methods of delivery) worked better for you than others. Remember that, and instead of becoming frustrated with the sick, work hard to become a better healer. And yes, some of us (including you?) don't know that we're sick!

The brain is not just the organ that interfaces with this non-local structure known as the mind. It is actually built according to the specifications of the mind, replicating its structure. For example, there's plenty of research demonstrating that the brain has "holographic" functioning, similar to the holographic universe theory that says every piece of the universe entails the structure of the greater whole. You can read about both in Michael Talbot's *Holographic Universe.*

In addition to a wealth of research showing that the brain's "thoughts" are non-local, there havea also been many experiments showing that consciousness cannot be explained strictly via biology.[86] In his book *The Universe in a Single Atom*, the Dalai Lama – who is known for his interest in reconciling science and spirituality (particularly the non-theistic dimensions found in Buddhism) – observed: "I do not think current neuroscience has any real explanation of consciousness itself."[87] In *The Undiscovered Mind*, science journalist John Horgan agrees: "Mind-scientists and philosophers cannot even agree on what consciousness is, let alone how it should be explained."[88]

There is also the holonomic brain theory, originated by psychologist Karl Pribram and initially developed in collaboration with physicist David Bohm. Pribram was originally struck by the similarity of the hologram idea and Bohm's idea of the implicate order in physics, and contacted him for collaboration. In particular, the fact that information about an image point is distributed throughout the hologram, such that each piece of the hologram contains some information about the entire image, seemed suggestive to Pribram about how the brain could encode memories. So what we are coming to realize more and more is that it's not a matter of which came first, the chicken or the egg, the seed or the tree. It is more of the reality that the chicken IS the egg and the seed IS the tree.

The mind is the structure of that electronic plasma we call thought in the brain. Plasma is charged electrons. The structure of the neurons of the brain is the mind's utilization of physical laws which are mathematics. Thoughts don't exist other than carried on by signals-chemical, electromagnetic, gravitational, force, etc.

The Original man is the ultimate architect in the manner in which matter and energy are organized by the process of identification, observation, measurements, etc. The mind determines angular momentum, so forth or else they are not known (quantum physics). Therefore the position or momentum of a particle is not certain until measurement. The Original Man is the most high, meaning the one drawing matter and energy up into an organized form. On its own tends to entropy and marinate into a medium of mediocre equilibrium. This means that it tends to behave like stagnant water.

This deals with Mentalism, i.e. everything that is mental is physical and everything that is physical is mental, or that the universe is in fact all mental. This means is that order exists, even before an object materializes and forms a kinetic energy from a potential energy existing from some geometric configuration of elements in this or a previous form of space time.

The photons in your brain existed physically before your own personal brain, yet they are still your thoughts that compose of your intelligence. All of the material in the universe came from radiation (thought). This is how the universe is the Black Man body, the same way the Universal Family is the body of the Black Man. Black means "dominant" and mind means "order." The reality of conservation of energy advances the principle of correspondence. You can't pollute your environment (Universe) without ultimately polluting your own biological body.

> "God is the highest extent of the mind. And energy and matter is the medium through which he expresses his ideas. And the black physical body of the Original man is the Supreme medium. In other words, the amassment of elements...which comprise the physical body of the Original man, is the only vehicle in the universe through which (in all its essence) the great mind in the universe, in any and all ramifications and manifestations, manifested in the character or nature of the Original man." – Infinite Al'Jamaar U-Allah, The Black Family (a 5% periodical), 1981[90]

So again you can see that the original man isn't just "related" to the order, structure, mind. He IS the order, structure, mind. This order and structure on a fundamental level is the instructions of the universe. The instructions are everywhere. How do we see that the Original Man, the Highest Form of Living Instructions, The Instructor, and the teacher is the Supreme Being? Because he was the only One that was able to modify the sequence of the instructions on the fly on a local or global level, before other people made from the

DID YOU KNOW?
It's estimated there are 85 billion cells in the human brain that are not neurons, according to a 2009 study by Brazilian neuroscientists. Neurons make up less than 50 percent of all brain cells. Researchers have only recently begun to catalog the functions of the remaining cells, known as glia. They may insulating material for neurons; digest and clean up dead cells in the brain; help form and maintain the connection points between neurons and other cells; or some other yet-undetermined functions.[89] Stuart Hameroff, in collaboration with Roger Penrose, proposed that the glia are where consciousness interfaces with the brain through quantum-level activity.

Original Man were on the scene. How? By observation, development, engineering, cultivation, fortification, standardization, organization, building and reproducing. Our bodies are the physical "command centers" of the mind. The instructions last forever and are transformed and carried out forever through communication and radiation.

"One must know oneself. If this does not serve to discover truth, it at least serves as a rule of life, and there is nothing better." – Blaise Pascal, mathematician

The original man was the first to recognize his integral part to the whole (universe) and use it. The consciousness that organisms carried evolved and continued to expand to reach a state of cognizance necessary to perform its duties to bring forth the next phase of existence. When the conscious outgrows a particular vehicle it brings into existence something better or at least something that can adapt to its surroundings. Everything played a role in developing the infrastructure and when it's no longer needed it's composted into another form to serve another purpose. Mind is the central structure that tethers all of reality.

The instructions do not call for the biological organism to emerge that rejects vital radiation, making it mentally and physically weaker and not being able to sustain the organization of the compounds within its form. Therefore, when the biological organism reemerges, it's going to be black (complete), to absorb the vital radiation. Thus, the direct connection of the Original man as a manifestation of the instruction, or rather he is the biological development of the meso incarnation of the instructions (which he is, not just a product of).

The individual isn't greater than the whole, nor is the whole greater than the sum of all of its parts. Every cell, atom, and molecule vibrates relative to the purpose it is in existence for. Thus, when the original man recognizes the properties of the aforementioned he can add on by modifying their sequence as well as study them to discover new potentialities in relation to his ideas or what is known.

The mind is summation of all the fundamental order(s) that exists. Whenever a thought is manifested, it is built and constructed from the very simple rules (fundamental order(s)) from the mind into a more complex expression in order to effect a change in the Universe. It's a change dependent system in a series of cause and effects.

In *Prometheus Rising*, Wilson continues:

> Every computer consists of two aspects, known as hardware and software. (Software here includes information). The hardware in a solid-state computer is concrete and localized, consisting of central processing unit, display, keyboard, external disk drive, CD-ROM,

> floppies, etc.—all the parts you can drag into Radio Shack for repair if the computer is malfunctioning. The software consists of programs that can exist in many forms, including the totally abstract. A program can be "in" the computer in the sense that it is recorded in the CPU or on a disk which is hitched up to the computer. A program can also exist on a piece of paper, if I invented it myself, or in a manual, if it is a standard program; in these cases, it is not "in" the computer but can be put "in" at any time. But a program can be even more tenuous than that; it can exist only in my head, if I have never written it down, or if I have used it once and erased it. The hardware is more "real" than the software in that you can always locate it in space-time—if it's not in the bedroom, somebody must have moved it to the study, etc. On the other hand, the software is more "real" in the sense that you can smash the hardware back to dust ("kill" the computer) and the software still exists, and can "materialize" or "manifest" again in a different computer. In speaking of the human brain as an electro-colloidal bio-computer, we all know where the hardware is: it is inside the human skull. The software, however, seems to be anywhere and everywhere.[91]

IS THE MIND CONSCIOUS?

Yes and no. The mind as the structure, implicate order of the universe is not conscious, yet, the Original Man is the conscious manifestation of the mind which through atomical, chemical and biological evolution, built upon its preceding elements to develop itself. Consciousness is the property of complex, highly organized systems. It functions and is exhibited by a system. The Original Man is the penultimate conscious manifestation of this system which we refer to as the "mind."

Yet, even most physicists will acknowledge that the universe on the quantum level does many "conscious like" things. This is the "unfolding information" of Bohm, electrons and other subatomic particles having "feelings" or affinities, the naming of the properties of certain subatomic particles as "flavors." This makes perfect sense in that even if the mind isn't conscious, it is the fundamental structure from which consciousness arose. Man is the mind which is mentality and mentality is mathematical order.

WHAT IS CONSCIOUSNESS?

Both science (through psychology and biology) and philosophy have sought to tackle the notion of consciousness which can be defined as the awareness between the mind and the world in which it interacts. There are a host of systems which seek to explain the levels, types, and variety of "consciousnesses." Even at this point in time there has not been a general consensus on many points. We will offer some general elements.

The part of consciousness that most of us are familiar with is the ego. This is your personality which you have been nurturing since birth. It is what makes you an individual. No one else shares your ego. In addition you have your own personal subconscious which has been referenced both by Freud and Jung (the shadow).[92] These two aspects of consciousness are pretty well known by the public at large and people are familiar with them. They are the aspects of consciousness that deal with individuality. So of course in Western society (which is hyper-individualistic) they have received the most attention. They both develop from the "data stream" of information which you are bombarded with from birth. They both actually develop to form part of your "operating software." The "program" has basically four subsections. There is your genetic imperatives (the portion which is biologically hardwired into your genes), imprints (hard wired programs which the brain is genetically designed to accept only at certain points in its development. These points are known, in ethology (scientific study of animal behavior), as times of imprint vulnerability), conditioning (similar to imprints. They are programs built onto the imprints. They are easier to change then imprints with some reverse conditioning) and "learning" (basically softer conditioning). The purpose of your "program" is to develop a tether point in time and space. Without your conscious (ego) and subconscious you would have no sense of center, individuality or personhood in the midst of the constant stream of "data." The human brain, being an animal brain behaves in a similar fashion, as an electro-colloidal computer. It doesn't act as a solid-state computer. Following the same laws which govern animal brains and electro-colloidal computers "programs" enter into the brain as electro-chemical bonds. This happens in discrete quantum stages as mentioned above. There are two other aspects that people may not be familiar with. They are the collective unconscious and the collective consciousness.

The collective unconsciousness is often referred to as genetic memory. It is proposed to be a part of the unconscious mind, expressed in humanity and all life forms with nervous systems, and describes how the structure of the psyche autonomously organizes experience (DNA-RNA brain feedback). It is also referred to as phylogenetic unconsciousness, the neurogenetic archives, the morphogentic field, long memory, aliyavjana (treasury mind) or even reflected somewhat in new age thought as "the Akashic records." In the Eight Circuit Model of Consciousness of Timothy Leary (further expounded on by Robert

Anton Wilson) it is referred to as the Neurogenetic or Morphogenetic Circuit (Buddha–Monad "Mind").[93]

Though these concepts don't all neatly overlap and there are some aspects of each one that may be questionable, they all point towards the same concept. The most important thing to take away from this concept is that there is a collective genetic base to your consciousness that you share with everyone, and which passes on to your children.

There is yet another structure. It, too, is the Mind, but it is not the "personal" mind nor is it "consciousness." It is referred to by many names also such as the "no mind," non-local consciousness, the Buddha mind, non local/quantum circuit. In the Eight Circuit Model of Consciousness, it is referred to as the Psychoatomic or Quantum Non-local Circuit (Overmind). By operating off of the implicit structure of energy and matter (the Mind), another further layer is developed which can be referred to as the collective consciousness. This is not the Mind as we have defined earlier, yet it reflects many of the elements of the mind such as order, structure, etc. It is reflected in such concepts as collective consciousness, the noosphere, global brain, groupthink, peer pressure, etc.

These are all manifestations of consciousness that are made manifest through the vehicle of the human body. There are even other hypothesized notions of consciousness such as the Neurosomatic Circuit (Zen–Yoga Mind–body Connection) which is defined as the consciousness of the body or the Neuroelectric or Metaprogramming Circuit (Psionic Electronic-Interface Earth Grid Mind) which is defined as the nervous system becoming aware of itself.[94] Though consciousness may be seen as a byproduct of a biological vehicle, the mathematical structures which are inherent within consciousness predate the biological organism(s). The mind is the blueprint upon which the brain and consciousness are constructed.

An Introduction to Psychology

ROBERT BAILEY

Psychology is the study of the mind, behavior, mental states and processes. Psychologists utilize research from a variety of different fields to arrive at their findings. At times they might employ empirical and deductive methods, while making use of symbolic interpretation and inductive techniques when appropriate. From the infant stage and continuing throughout the course of our lives, we use psychology for different purposes. Knowing how our mind works and why we do the things we do allows us to accept greater responsibility and control over our lives. We might use psychology to help us relieve stress, make better decisions, pull a number, court someone, and ultimately lead us to finding a desirable companion.

In his book *The Story of Psychology*, historian Morton Hunt credits an experiment performed in seventh century BC by Psamtik I, King of Egypt, to be the first recorded psychological experiment. However, many cultures around the world outside of, as well as before Egypt, contemplated the ways and complexities of the mind. They just never called it "the mind"! They referred to the conscious intelligence that direct's man's activities by a variety of names, some of them represent different "stages" or aspects of man's consciousness. This is why some cultures distinguish between man's "soul," his "spirit" and "God" himself. Some West African cultures have seven or more such "stages" of self. But all this refers to some aspect of consciousness. And throughout man's history on this planet, we have known that this reality is the work of this same consciousness. According to Dr. Edward Bruce Bynum, Director of Behavioral Medicine, University of Massachusetts Health Services, Amherst:

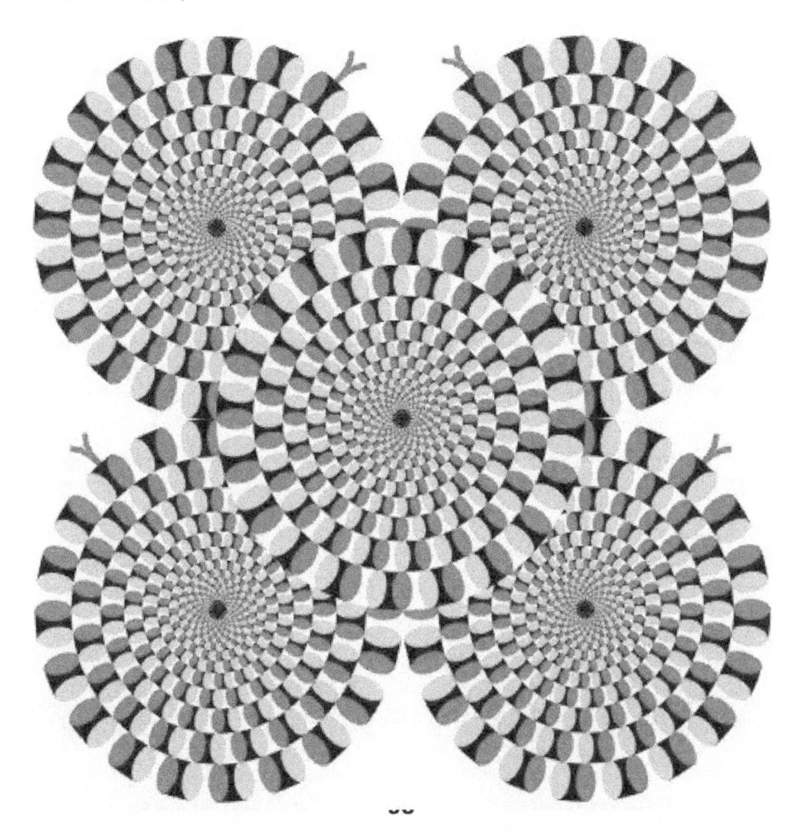

> The workings of the dynamic consciousness on dreams and symptoms production were known to the Kemetic Egyptians of 3000 BCE and that they consciously employed it in their clinical and religious work. They referred to it as the all Black Underworld of the Amenta and the Primeval Waters of Nun. They literally invented biological psychiatry and were aware of the dynamics of the neuromelanin nerve tract and light Later, some West African civilizations, especially the Yoruba, formalized a variant of the dynamic unconscious in philosophy and medicine, centuries before Freud."[95]

So we know that psychology has been around for a while, way before the accepted time of the early Greeks, though it may not have always been recognized as such. Psychology was originally a branch of philosophy and didn't break off into its own academic discipline until 1879 when German physiologist, Wilheim Wundt opened a laboratory that focused exclusively on psychology.

Psychology has grown since then, with many contributors and schools of thought. One of those schools, of particular relevance to us for the purposes of this chapter, is that of transpersonal psychology. According to Mariana Caplan:

> Although transpersonal psychology is relatively new as a formal discipline,...it draws upon ancient mystical knowledge that comes from multiple traditions. Transpersonal psychologists attempt to integrate timeless wisdom with modern Western psychology and translate spiritual principles into scientifically grounded, contemporary language. Transpersonal psychology addresses the full spectrum of human psychospiritual development – from our deepest wounds and needs, to the existential crisis of the human being, to the most transcendent capacities of our consciousness.[96]

Some of the notable scholars in this field include Carl Jung, Abraham Maslow, Ken Wilber, and Timothy Leary.

THE MIND VS. THE BRAIN

TRUE WISE ALLAH, C'BS ALIFE ALLAH, & SUPREME UNDERSTANDING

"Time is to clock as mind is to brain. The clock or watch somehow contains the time. And yet time refuses to be bottled up like a genie stuffed in a lamp." – Dava Sobel, Longitude

The brain is almost useless to study when striving to understand the human mind. The reason for this is because the mind cannot be adequately explained by studying the brain organ.

When we look at the history of brain studies, we see that those studies were great and very informative, yet they usually fail to tell us specific information about the mind. This isn't because the scientists were stupid! It is because the mind is so dynamic and complex one must know its properties instead of its property (the brain).

Studying the brain to understand the mind has a long failed history. To understand why, consider the following analogy: Let's say you wanted to know how much food I had, so you look in my refrigerator. Every time you look into the refrigerator you see no food. Yet, whenever I as the owner of the refrigerator want something, all I need to say is "orange juice" and the carton appears in my hand. Confused, try to put a head of something into the refrigerator, and it vanishes before your eyes. It becomes a part of the refrigerator. Its existence is spread throughout the refrigerator rather than existing in one single place. To be even more accurate, the carton of orange juice was never place in the refrigerator. It was created within the refrigerator based on events going inside and outside the refrigerator. It was an "emergent property." To be specific it has been proven that thoughts, memories, and learning have no specific location in the brain. They simply exist along with the brain.

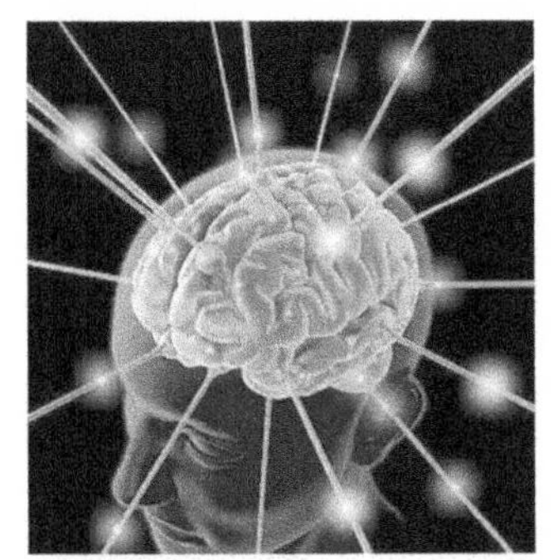

In the early 1900's, Karl Lashley conducted research to find the location of thoughts, memories and learning within the brain. He trained many rats (probably in the hundreds or thousands) to travel through a maze. Since they learned the maze – he thought – this learning must exist somewhere in their brains. This is a reasonable thought. What Lashley did was spend the next few decades of his research programs teaching rats and then cutting up their brains. After each rat successfully learned a maze, he would either sever a part of their brain (to destroy a neural connection) or actually remove a chunk of their brain. He did this systematically, so that he could determine which exact part of the brain was actually holding the memory of the maze. After years, months and days of training rats and carefully cutting pieces of their brains out, he was forced to acknowledge that memory had no specific place in their brains. No place in their brains!

John Lorber came to similar conclusions when he found that people barely needed a brain to be intelligent. Working with people suffering from hydrocephalus, many of his patients had less than 50 to 150 grams of brain mass left inside their skulls (that is less that 3-10% of a normal brain). Some of them were retarded; however some of them were highly intelligent.

Many other studies on both animals and humans have shown that the brain is extremely dynamic in terms of its compartmentalization. Compartmentalization of the brain means that specific areas or regions of the brain are responsible for different things. The fact that this is dynamic means that those regions can be reassigned if necessary. This goes against what was previously assumed about the brain: that all functions, thoughts, memories, learning, feelings, etc. were stored in specific areas of the brain. Since this is not true, it forced scientists to reconsider the relationship between the mind and the brain.

The relationship between the mind and the brain was eventually shown to be "distributed processing" meaning that any given 'process' is not existing in one location, but rather 'distributed' across a number of locations. This means that one 'thought' does not match up to one neuron in the brain or location of the brain. That thought could exist almost any place in the brain at any given time. This means that the mind does not depend on the brain in a way that studying the brain can tell us about the mind. A certain aspect of the mind (visualization, audio information, memories, body movement, etc) can exist anywhere in the brain where it needs to exist. The mind, in this sense, is beyond the brain.

The mind can exist beyond (yet within) the brain because it an emergent property of the brain. In other words, the brain does not create the mind; instead, the mind creates itself while living inside the brain. The brain is a network of firing neurons which causes an environment where self-organization can take place. Self-organization is the dynamic process through which the universe manifests itself. The Mind we know is a powerful property that emerges naturally using the brain as a platform.

THE ABILITIES OF THE BRAIN

With the technological advances of the 21st century, it appears that we are living within a science fiction novel. Things that we dreamed about when we were children have actually come to pass, with much more right around the corner. Living in such a world of wonder often blinds us to an essential fact: The world in front of us was constructed using the hardware and software of our brain. In focusing on the external, we tend to miss the internal. So we often do no exploration on the capacity of our own mental computers. Fortunately people appear now and then to remind us of the extent of our infinite potential.

Memorization is an ancient art. Before the advent of written language to record and transmit ideas and stories, the brain was utilized to preserve information in a dynamic way, so that it would be retained for future use. In fact, many myths and old stories preserved such information as animal migration routes, astronomical events, seasonal preparation, geological catastrophes, and scientific techniques. It is often forgotten that even religious texts such as the Torah or Qu'ran were retained orally before being written down. Many cultures have people whose station in society is that of the oral chronicler. In West African societies this person is the griot. Some griots are said to retain thousands of years of history.

There are many methods to expand, retain and exercise memory. They are grouped under what is called mnemonic devices. Two of the most noted traditional techniques involve drawing ideograms (or mental maps) and repeated dictation (or chanting). This is why indigenous cultures have so many graphic symbols (the art isn't "just" art for art's sake) and why, when most religious texts are recited in a tradition's setting, they are either sung or chanted. Yet is memory the only amazing ability of the brain?

Born in India in 1939, Shakuntala Devi was a "human calculator." In 1977, she calculated the 23rd root of a 201 digit number (without paper) faster than the Univac 1108 computer. Several others from India have performed such feats, such as Priyanshi Somani, who currently holds the World Record in "Mental Square Root" for completing 10 tasks of 6 digit numbers in less than three minutes. Why India? Because this is tradition that has long roots in the region, so people have built up their "mental calculation" abilities there. Yet, there are many others, like Thomas Fuller, the "African Calculator" who – despite being born into slavery – could give the diameter of the Earth in inches despite interruptions. Fuller's abilities drew considerable attention and he was challenged to compete with a white man of reputedly similar abilities. They were asked how many seconds a man would have lived at age 74, seventeend days and 12 hours. Fuller took only a minute and a half to answer 2,210,500,800 seconds and beat his opponent (who had a paper and pencil).[97] In research centers such as CERN (European Organization for Nuclear Research), such mental calculators were in huge demand before the advent of the modern calculators and computers.

Let's not forget visual-spatial reasoning (the kind of math that figures into geometry). Imagine being able to draw a detailed map of central London or all of New York City after just taking a helicopter ride above it. This (and more) has been done by Stephen Wiltshire, a Black

DID YOU KNOW?
If you saw *What the Bleep Do We Know?* you're familiar with Dr. Masaru Emoto's research on the effects of thoughts and words on the structure of water. In books like *The Hidden Messages in Water*, Emoto shows that words and sounds with negative intent produce blurry or distorted structures in ice crystals, while positive thoughts produced beautiful symmetrical crystals, like little snowflakes, In 1994, Rein and McCraty of the Institute of HeartMath also found that subjects in a meditative state were able to intentionally influence the molecular structure of water. It was a small, almost undetectable amount of change, but it showed that consciousness does have some interaction with real things at subatomic level.[98]

man of West Indian descent who was classified as autistic. There is also Alonzo Clemons, a Black man of Colorado who suffered a brain injury in his youth. He is developmentally challenged, yet he developed the ability to sculpt anatomically accurate versions of anything the sees a two dimensional picture of. It appears that sometimes brain injuries give some parts of our brains a rest so that we can truly see the vastness of its potential. In some instances, brain injuries can allow people to access functions of their brain that go beyond the brain itself. They are able to tap into the Mind beyond the brain.

THE POWERS OF THE MIND

If you've understood what we've said so far about the Mind vs. the Brain, you'll understand when we say that the Mind actually has no powers! The Mind doesn't "do" anything. It's not a personal God, a weapon, or a computer, at least not at the level at which it dictates the structure of the reality. Man is conscious, but yet one Man alone does not have "the" Mind. At the same time, Man IS Mind, so he has it all. But what exactly does he have? And is the Original Man the origin of the Mind, rather than the other way around? Some scientists think we have constructed reality in reverse through our consciousness, meaning everything that led up to this point is from our present-day consciousness. The ancients sometimes described the Mind of God as the All-Seeing-Eye? Is the All-Seeing-Eye the product of all eyes seeing?

Understanding the stage of the Mind at which reality is determined is very difficult, because we want to conceive it in familiar terms of active consciousness and it may not work that way. This is why the Hindus say Brahman is what cannot be comprehended, yet is the means by which we comprehend. The Qur'an says that Allah cannot be comprehended by man, yet elsewhere notes that he is nearer than our jugular vein.

What we can be sure of is that the Mind is not an active agent in the world. Man is. Man activates consciousness and thus accesses the Mind at increasingly higher frequencies, or stages, as we'll explain later. The different stages of consciousness allow for us to "tap into" the bird's eye view of the Mind. At our base level, all we can see is what's in front of us, but there is scientific evidence that some of us can see what others can't, predict what is coming within the near future, sense when we are around other conscious agents, and either know or influence the thoughts of others. Some of these phenomena may be tied to cues on the physical level, or information our brains can process (like predicting the future by calculating possibilities). Others may be tied to quantum-level (subatomic) phenomena, where the waters get murky. But, as we'll explore in this book, much of it may be sub-quantum, meaning there is a layer of "meaning" beneath all of what we can see and measure. After all, we can barely measure the quantum world! What's the nature of reality when we get to the root of it all? Simple: The Science of Self.

What is the Self?

C'BS ALIFE ALLAH & DIVINE RULER EQUALITY ALLAH

THE SELF HAS NO BEGINNING, AND NO ENDING

So what is meant by the idea of the Original Man having "no beginning, no ending"? Can we be like the Highlander and never die? Are we immortal? Will we reincarnate into a new form? If I say "til death do us part" to my mate do I need to make sure that I really mean it, as it'll never actually end? The simplest answer to all of the above is that the self is continuous over time and space. To understand this you have to be able to digest a few concepts:

- ❒ The Self is not equivalent to your ego
- ❒ The Self is the meso- aspects of your existence that are parallel to the macro- and micro- aspects
- ❒ The Self, in and of itself, is a singular organism

"AS ABOVE SO BELOW"

It is no mystery that the pattern of the Universe (macro) is repeated on the Earth (meso) and then on the sub-atomic level (micro). The pattern is the mind. Therefore we aren't a reflection of the Universe, it is a reflection of us. One version of the Anthropic Principle in physics explains that the universe is fine-tuned to promote life because it is, itself, alive. On a parallel level in ecology, the Gaia hypothesis describes the Earth as a macro-organism. If the Earth is a gigantic planetary su-

DID YOU KNOW?
The adult body is made up of 100 trillion cells, 206 bones, 600 muscles, and 22 internal organs. Because of the interconnections between the body's organs, the body can be considered as having between 9 and 11 systems. The easiest set to remember is: Circulatory System; Respiratory System; Skeletal System; Excretory/Urinary System; Muscular System; Immune System; Endocrine System; Digestive System; Nervous System; Reproductive System; and Integumentary System (the skin, hair, and nails).

per-organism, the Earth's organic "body" weighs six sextillion tons and has a surface area of 196,940,000 square miles. We can recognize something that large as a living body.* Therefore, we can use the word "body" in various ways. We can use it to refer to the whole Universe, we can use it to refer to the Original Man's primate body (as an individual or as a collective), or we can use it to refer to the Original Family.

The Self is the meso- aspects of your existence that parallel the micro- and macro- aspect of existence. So your collective unconsciousness and collective consciousness (or superconsciousness) refer to a continuum that don't begin or end with your person(ality), i.e. your ego.

Ramana Maharshi, an Indian yogi, suggests that we are ignorant about the true knowledge of our self, and confuse the real self with the ego:

> ...the final goal (of yoga, or life) may be described as the resolution of the mind in its source which is God, the Self...The mind and the breath spring from the same source. They arise in the heart, which is the centre of the self-luminous Self...Where the 'I' thought has vanished, there the true Self shines as 'I.'...The 'I,' the Self, alone is real. As there is no other consciousness to know it, it is consciousness.[99]

The body assembles or disassembles itself according to mathematics (physics). You can see this in the presence or absence of certain nutritional elements as a person is developing from a child into an adult. They must have certain vitamins and mineral to develop at different stages. You can predict the outcomes of development that occurs when particular elements are missing. It is the same with any formative process.

The body breaks down during decomposition according to mathematics. This is why CSI can determine time of death by the time signatures of various insects and chemical changes. We are composed of a multiplicity of interacting organisms and elements. In fact, within the Black man's body, you will find not just every one of the 94+ natural elements, but all of the most abundant natural compounds required

* In "The Black Woman is the Earth," True Wise Allah describes the abundant parallels between the Original Woman and the Earth. Again, the same patterns are reproduced on every scale.

DID YOU KNOW?
In Jungian theory, the Cosmic Man is an archetypical figure that appears in many of the world's ancient creation myths. He also represents the oneness of human existence, the universe, or both. In Chinese legend he is Pangu, in Jewish myth Adam Kadmon, in Persia Gayomart, and in India there is Purusha. In most traditions, this Cosmic Man becomes the Universe, with different parts of his body becoming different parts of the comos, some parts of him producing the features of the Earth specifically, and his own spirit giving life to the first man and woman. In many myths, the Cosmic Man is not just the beginning but also the final goal of life or creation. Jung believed this is not necessarily a physical event, but referred to the identification of the conscious ego with the self. We think those two ideas sound like the same thing.

for any form of life on Earth.*

The *Chandogya Upanishad* of ancient India notes as much:

> As large as the universe outside, even so large is the universe within the lotus of the heart. Within it are heaven and earth, the sun, the moon, the lightning, and all the stars. What is in the macrocosm is in this microcosm…All things that exist...are in the city of Brahman.[100]

The Original Man is the highest form of living mathematics and the universe is his macroscopic body. The universe is an extension of our personal body. On a celestial scale, this is seen as the Sun, Moon, and Stars. They have analogues (parallel forms) found on the personal scale as Man, Woman, and Child. Your family is your universe. Our body grows and takes shape through the process of generating family, babies, seeds, and future generations. I'll explain. Your physical body was generated by your father's sperm meeting your mother's egg and gestating in her womb. You are composed of their physical material. Yet your physical material did not spontaneously generate. It came from their sperm and egg. So your physical body is an extension of their bodies across time and space.

Now add to this physical inheritance the notion of collective unconsciousness and consciousness. You can literally transfer some of the mental information in your brain by teaching others. This is how data is shifted from one to another, again through time and space. Thus a thought can outlast a person's personal body. And contrary to popular belief, thought isn't just energy; it is the pattern that energy takes. So the basic form and structure/pattern of this energy (information) is preserved.

---

* Common convention is that there are 92 naturally occurring elements, but we now know that at least 2 more (neptunium and plutonium) are also produced in nature. See the article on the African Nuclear Reactor (in the Appendix) for details. Recent studies have found up to 98 of 118 possible elements being produced naturally on Earth through nuclear reactions and radioactive decay. For details, see *Nature's Building Blocks: Everything You Need to Know about the Elements* by John Emsley.

This pattern, which has no beginning and no ending, is reproduced on the quantum and on the interstellar level. Planets, People, Particles, Positions and Perspectives (thoughts), all reducable to PATTERN. They all follow a pattern of original mental order which is a format, a structure, an intrinsic process of growth and manufacture, even though specific instances and examples seem not to be ongoing.

This is akin to what is known as "Object Oriented Programming" in the classification of objects. So for instance you'll have a superset of something, like Cats. Within that superset you'll have subsets such as domestic breeds and wild breeds. Within those classifications, you'll have specific examples such as tigers or siamese cats.[101] Basically, there are sets within the superset. Jung saw them as diverse manifestations of deep archetypes.* As Michael Schneider notes in *A Beginner's Guide to Constructing the Universe*, the ancients knew that such archetypes were tied to number, and linked them together in their cosmologies.

In this case, the mathematical archetype, or superset, is the principle of reproduction which is mirrored on the micro, meso, and macro levels, reflecting a commonality of structure throughout the universe.

Neither mental nor physical is discontinuous. Both are merely transformed in personality (living, dying, birth, etc). There is no process by which a personality can "reincarnate," but the Self, the "true" Self, cannot die. If you want to lay your stake on immortality, have children and/or teach. Become immortalized through memory. This is no cliché. The root of the word "memory" is related to the roots of men, mental and mind. Transferring information to be retained by others is akin to transferring your mind into other personalities.

I DON'T WANNA BE ABSORBED INTO THE SUPERBRAIN!

Fear not. We're about to put it all in context. Western societies are hyper-individualistic, often at great expense. They traditionally only emphasize the ego. Conversely, most indigenous cultures have sought a middle road of communal harmony. This is part of the reason that the west was so fearful of what it called the "Red Scare" (i.e. Communism). The problem wasn't the system of governance. It was that can't see being part of anything without having "control" over it. Indigenous cultures, even those that transferred to a city-based way of life, still saw themselves as the stewards of the planet. So while the West

* An archetype, as defined in this text, is the original pattern or model from which all things of the same kind are copied or on which they are based; a model or first form; the prototype. For a full discussion of archetypes and symbology, see Volume Three.

> DID YOU KNOW?
> The Upanishads are a collection of Indian scriptures that represent the nature of indigenous pre-Aryan thought in India. They touch on, with no exaggeration, all of the topics covered in this text. They describe absolute reality as "unknowable" (in the same senses described by quantum mechanics) and merely an emanation (of varying appearances) from the ultimate consciousness of Brahman, which can be translated as God, the Mind, or the Self. One of the oldest and most important of the Upanishads, the Brihadaranyaka Upanishad, describes humans as the synthesis of the organ of speech, mind, prana (vital energy) and the twin cosmic desires of differentiation and unison, concluding "From infinite or fullness, we can get only fullness or infinite." In essence, the Absolute or Brahman manifests (in totality, not reduced form) as the individual self of Man, who – realizing this – knows he is the Infinite as well.

appears fearful of "communism" what they are really fearful of is community. Community has built-in checking mechanisms to prevent exploitation and egocentrism. It is the community which is an aspect of the organism that we refer to as Self, when we say the Self is one.

"It is important for this country to make its people so obsessed with their own liberal individualism that they do not have time to think about a world larger than self." – bell hooks

While some may think that becoming part of what Na'im Akbar has called "A Community of Self" will turn us all into the Borg from Star Trek (everyone thinking the same thing), I think of it more like The Smurfs. Yes, the Smurfs. There was a unified community, yet all had different roles to play. We can see this with the 99+ attributes of Allah yet all are still Allah. As the Qur'an declares, "Say Allah is one." Looking back to the science of the superorganism, consider the bee, which the Qur'an also speaks highly of. Bees may operate in a colony, but the "hivemind" doesn't mean they're each mindless. Through no specific genetic mechanism present in the individual, different bees nonetheless perform different roles, each of them vital to the success of the hive. This is why 10,000 bees can move as one.

"As pure water poured into pure water becomes the very same, so does the Self of the illumined man or woman verily become one with the Godhead." – Katha Upanishad

When Original Men are united mentally, that means that all are dealing with a single unified mentality, which is simply the Supreme Mind. In this mode, we can use each other's skills for the benefit of the whole, just as each organ in the body employs itself for the benefit of the whole, and just as bees, ants, or plants tend to do. When my brother is speaking, he is not just a set of organs making noise. He is speaking from the Mind. I think there's a lot to gain from the following excerpt from the *Brihadaranyaka Upanishad*:

> This universe, before it was created, existed as Brahman. 'I am Brahman;' thus did Brahman know himself. Knowing himself, he became the Self of all beings. Among the gods, he who awakened to the

> knowledge of the Self became Brahman; and the same was true among the seers. The seer Vamadeva, realizing Brahman, knew that he was the Self of mankind as well as of the sun. Therefore, now also, whoever realizes Brahman knows that he himself is the Self in all creatures. Even the gods cannot harm such a man, since he becomes their innermost Self. Now if a man worship Brahman, thinking Brahman is one and he another, he has not the true knowledge.[102]

We are the Mind. Man means Mind. That means we are the Highest Form of Living Mathematics, (i.e. Growth, Order, and Direction). "God" is not just an idea. God is both the idea and the physical representation. When people say "God is man," they may not know that they are summing up the results of thousands of years of scientific data in a three-word sentence. The Original Man who uses his Mind and recognizes the Community of Self enjoys an interesting duality. He is one of many, merely one personification of a Universal Self that spans all space and time. Yet because he IS this Mind (or Self) that provides the pattern for the Universe and all that comes with it, he is also the Supreme Being, who by virtue of his intelligence can mold and shape the environment around him. Being the Highest Form of Living Mathematics, (i.e. Growth, Order, and Direction), the Original Man is both subject and object in the sentence of reality.

For another look at some of these ideas, see "Quantum Physics and the Black God" by Divine Ruler Equality Allah, in *Knowledge of Self: A Collection of Wisdom on the Science of Everything in Life.*

CONCEPTIONS OF SELF ACROSS CULTURES

SUPREME UNDERSTANDING

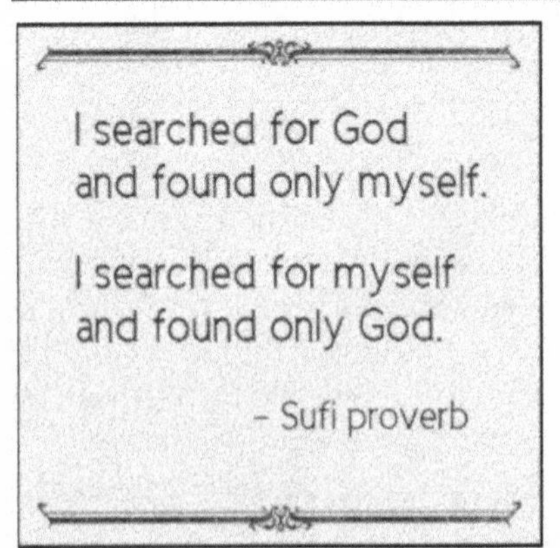

Although there is substantial agreement that all things derive from a single original source which was two before it became many, Original people have never conceived of the "Self" as *only* a union between Mind and Matter alone.

When Europeans encountered indigenous thought, they adopted their cosmologies but, through Western philosophy, reduced it to these elements, discarding everything else as "pagan witchcraft" and the like. As Placide Tempels notes in *Bantu Philosophy*,

> DID YOU KNOW?
> In the Yoruba cosmology, the Self is called *orí*. It represents free will and self-determination. The *orí* is one's "personal God," and has a greater role in their success than any external force. Even an *orisha* (a force of nature, or other cosmological force personified as an external divinity) cannot intervene in a person's life without permission from the *orí*. Although *orí* is a "free will" choice before birth, it is through *esè*, or life's struggle, that we see what one's *orí* will manifest. *Iwàpele* is the moral compass by which the *orí* must navigate *esè* to become a "good person," which is the ultimate form of success and self-actualization. In fact, the Yoruba word for human, *ènìyàn*, translates loosely as "entities in the world chosen to do good."

everything is an emanation of what the Bantu call the vital force, but Western philosophy either misinterprets African concepts by saying there are too many gods and spirits, or by trying to reduce the diversity of the spectrum to its polarities.[103]

The people of India, for example, conceive the chakras, kundalini (a "serpent" energy), and various conceptions of Self (Atman, Brahman, Purusha, etc.). The Bantu call the chakras *kijambas*. The ancient Egyptians and many other African cultures have between 3 and 7 different "components" of the Self. Some of these components of Self are more eternal and non-local than others, and some are more temporary and localized. Some are tied to one's personal self, while others are considered the essence of the ancestors. These cosmologies form the foundation for Western psychology and psychoanalysis, which would later reduce these components to things like Freud's ego, id, and superego, or Jung's conscious, subconscious, and unconscious mind. Examples from the African Diaspora include:

- ❒ Zulu: the body, aura, Law, and *uqobo* (essence).
- ❒ Nupe: *naka* (the body), *rayi* (the soul), *fifingi* (the shadow soul), and *kuci* (the personal soul).
- ❒ Shona: *muvuri/mutumbi* (the body), *mweyi* (soul or vital force), *bvuri* (shadow), and *njere/pfungwa* (mind).
- ❒ Yoruba: *ara* (the physical body), *emí* (the soul), *orí* (the inner head or divinity), and esè (individual effort). Some Yoruba also include a shadow, *ojiji*, as well as the *iye* (mental body) and *oka* (heart-soul).[104]
- ❒ Akan: *nipadua* (the body, literally "person tree"), *adwene* (conscious mind), and *okra*, an "active particle of the Supreme Being." There is also *mogya*, the mother's blood in the Akan's matrilineal culture, and the *sunsum*, the distinct personality that comes through the father's semen. Only the *okra* persists after death to become an ancestor.[105]

There are many other examples. For example, the "spirit" or life force has as many names as there are cultures in Africa: *ntu* (Bântu-Kôngo); *chi* (Igbo); *nyama* (Bamana, Dogon); vibration (Rastafari); *loa* (Haitian

Vodou); *vodun* (Fon); and simply "Spirit" among Black Christians.* Indians call it Prana. The Chinese call it Chi. The Egyptians called it Ka.

Among both scientists and Five Percenters, it's simply known as "energy." Scientists have attempted to narrow it down to a particular wavelength of the electromagnetic spectrum, but have yet to identify any specific "vital force" in man. Five Percenters, on the other hand, have sought to reunite the scientific concept with the indigenous concept, rendering "energy" as "Inner G" (as in God), or anchoring both concepts to yet another indigenous tradition, as "Inner Chi."

Indeed, Chi is not only a Chinese concept, but is also the vital force of the Igbo people of West Africa. As my Igbo brother Age once explained to me:

> Chi is a break-off of the word Chineke, which is the name of the creator God. And we are all described to be made up of Chi. The oneness of man and God has been understood for a long time in Africa.[106]

Among the Bantu people, *Muntu* is the essence of man:

> "Muntu" signifies then, vital force, endowed with intelligence and will. This interpretation gives a logical meaning to the statement which I one day received from a Bantu: "God is a great muntu." This meant "God is the great Person"; that is to say, *The* great, powerful and reasonable living force. The "bintu" are rather what we call things; but accorting to Bantu philosophy they are beings, that is to say *forces not endowed with reason, not living.*[107]

SO WHERE IS THIS VITAL FORCE?

Where is the Self? Where is the Mind? Where is consciousness?

We've already said quite a bit on the question of "where is the Mind?" Yet we've also said quite a bit about the brain. There's more to the story. Psychologist Christopher Holmes, author of *The Heart Doctrine,* contends that the consciousness that structures the Universe is not

---

* We avoid the word "Spirit" not because of what it "could" refer to (denotation), but because of how most people use it (connotation). That is, "Spirit" literally means the force that moves you. It derives from the same roots as both inspiration and respiration. In fact, the Hebrew word is Ruakh, meaning breath, which clearly moves us and gives us life. Nothing spooky there. When we think of being in good or bad spirits, those are emotions that move us. Nothing spooky there either. Even wine is called a "spirit" because of how it can "move" us. Still not spooky. What IS spooky is the idea that we are life-size puppets being moved by an external force that uses "Spirit" as the means by which some spooky external force compels us to act. That, unfortunately, is the way "Spirit" is conceived by most people. Even among those of us that do not see it this way, the word carries that meaning to those who hear us, in a way that "life force," "energy," "Chi," or even "electromagnetic force" do not convey.

> DID YOU KNOW?
> An August 1984 article from Science Digest discusses the unique character of the human heart: "About 70 times each minute, more than 2.5 billion times in a lifetime,* the heart beats on. What keeps it going? The heart, it seems, has a life of its own. The muscle fibres that make it up differ from those elsewhere in the body in that some of them generate their own electricity without receiving signals from the brain. In fact, the fetal heart begins beating before it has even formed nerve connections. The heart's pacemaker is a group of self-triggering cells called the sinoatrial (SA) node...What initiates the current in the SA node remains unknown, but its cells behave more like neurons than muscle fibres."[110]

centralized in man's brain, as many suppose, but is non-local, with a more energetic epicenter at the heart:

> The quantum Self at the heart of being is the origin of life and consciousness within the material body, with light emanating and radiating from a central Sun. The Self is the self-illuminating Sun of the body, while the mind, like the moon, reflects this light. Accordingly, mystics claim that the head brain does not have a light of its own, but simply reflects the light of the Self originating within the Heart.[108]

The heart does, indeed, have a stronger electromagnetic field than the brain, and emits much more energy.[109] And indigenous cultures throughout the world have identified the heart over the brain as the "seat of the Self." It has typically been the work of Western scientists to localize consciousness at the brain, but we take the position that the Mind, the Self, the Mathematical Law, the Infinite Consciousness (or whatever you want to call it), is not localized at either the brain or the heart, but is a non-local continuous phenomenon that we interact with and activate at varying degrees at different times. The Original man is structured (purposefully) to activate the highest living form of mathematics and consciousness, and has many anatomical mechanisms through which this occurs. The scientists of ancient India described these anatomical "seats" as the Chakra system.

Timothy Leary and Robert Anton Wilson described it as the Eight Circuit Model of Consciousness. Looking beyond the brain and heart alone, Leary and Wilson, both professional psychologists dedicated to mysticism, proposed that each of the circuits represents the activation of (or access to) a stage of increased cognitive function and consciousness. The circuits, like the chakras they were derived from, activate in ascending order, providing a link between the evolution of an individual human (from base-level needs fulfillment up to omniscience or God-realization) and the evolution of intellectual thought

* Interestingly enough, studies have found that nearly all living animals experience 2.5 billion heart beats in a lifetime. The beat's frequency may increase with the decreasing lifespans of smaller creatures, but the 2.5 billion beat feature is nearly universal.

through human history.

Many other models for consciousness and vital energy exist. Most of them are multifaceted. Most indigenous models synthesize vitality (or life force), and consciousness (or Mind), as well as various aspects of the electromagnetic universe and the forces of nature.*

In the Five Percent tradition, such distinctions have been conceptualized as the seven planes of energy,† and the five stages of consciousness. In my opinion, however, the Five Stages of Consciousness Model is the most useful one I've encountered. The Five Stages of Consciousness can be best summarized as follows:

- Conscious: The base stage, at which we know what's in front of our faces and make waking decisions about how to fulfill our wants and needs.
- Subconscious: The stage at which our thoughts are processing, but we may not be aware of them. The thoughts at this stage may be subtly "aware" of events beyond our personal awareness.
- Superconscious: This stage represents a greater sense of self than one's own personal being. It is the "Community of Self" in consciousness, the type of consciousness found in superorganisms, including human communities that think with a unified consciousness. In some senses, it is not so much "groupthink" as it is thinking in terms

A QUICK NOTE ON LEVITATION

Do you believe that people can levitate? If so, why? Have you seen it, and tested it, with your own eyes? We know that magicians like Kris Angel do levitation type stunts using wires and camera tricks, but no one seriously believes Kris Angel is "really" floating over Las Vegas. Except maybe the people who think that wrestling is real, or that the wild-eyed dude from Ancient Aliens is a sensible-sounding guy with a nice haircut. So what intrigues us about the idea of Indian yogis and Tibetan monks floating above the ground in lotus position? Maybe it was the kung fu movies. We know that seeing things in film tends to make things more believable when we are told about them in real life. But levitation isn't an indigenous practice. It's a magician's trick. And yes, it goes back to indigenous magicians who put on shows for their audiences, which included European visitors who brought these tales back home. Nowadays, there are even African holy men who put on levitation shows for attention. But consider this: We have been on this planet for millions of years. We know that creatures with legs live on the ground, and that gravity acts on any physical object with mass. There's no scientific mechanism by which a man's body will come 3 feet off the ground unless his butt is made of pure magnets and the ground is also totally magnetized. But, there's good news: You're technically levitating right now. Just not 3 feet off the ground. You see, when you sit or stand, the particles of your body (or clothes) never actually TOUCH the particles of the ground (or chair) beneath you. It is the electrical repulsion between the particles that keeps us from falling through the ground, and holds us a few nanometers apart from any surface we sit or stand on. By this rule, we don't technically every touch anyone or anything else, either. We just connect on an energetic level.

of what is best for the group's good. The Egyptians called this ideal the "summum bonum" or greatest good. Freud called this the Superego, which provided the moral conscience of man, and Jung called it the unconscious or collective unconscious, a thought pattern shared by a human community unified by archetypes.

- Magnetic conscious: At this stage of consciousness, we're not simply interacting with each other, but with the environment. From what we know about the Electromagnetic Universe, we know that electromagnetic fields are dynamically tied - somehow - to human consciousness. Magnetic consciousness is the mind we share as a people, with the common potential of transferring thoughts (or information). This would be the stage where we can sense each other's thoughts and intentions. It is by no coincidence that Shamanic States of Consciousness and ESP are both closely tied to significant changes in electromagnetic activity.
- Infinite conscious: This is the consciousness that extends across space and time. It is the ultimate sense of Self.

In addition to assigning these stages anatomical "seats" or "activation points" within the personal body, some of our ancestors used celestial bodies to represent such models on a macrocosmic scale (e.g. the Earth=Conscious, Sun=Superconscious, Moon=Subconscious). What we can take away from all this, and what it is essential we NOT misunderstand about indigenous thought systems, is that man exists firmly at the center of all of the above ontologies and cosmologies. The idea of an African world-view where man is a lesser being is the byproduct of Europeans romanticizing the "noble savage" who sought favor in the heavens to the point of not seeing himself as an active agent in the world. That is not us.

As Denise Martin – drawing on the research of Mbiti,[111] Fu-Kiau,[112] Thompson,[113] and several others – has explained:

> In the Bântu-Kôngo ontology and indeed most African ontologies, the person is at the center. The person is priest or priestess of the universe and the fullest expression of creation. The person is a part of creation like animals, trees, and nature but distinct through empowerment by choice and the ability to consciously direct the energies flowing through all creation.[114]

Because it is one of the elements consistent throughout the diversity, this "centeredness" must represent one of the essential elements of African thought, and indeed all indigenous thought. Another consistent element – despite the many different components of Self noted by cultures across the world – is that the most determinant two elements are Man's Mind and life force, both of which emanate from the Mind or Energy that created the Universe. Some cultures use these terms synonymously, while others draw distinctions. Some concepts clearly correlate to forces within human physiology, while others correspond to processes within nature. Throughout this text, we'll draw those connections wherever possible. However, it should be said that, with there being so little research into the intersections between indigenous Black thought and science, there remains much work to be done. Martin notes:

> To determine if African understandings of vibration and energies operate on an Einsteinian, quantum, or sub-quantum level would involve a synthesis of knowledge from esoteric African traditions and cutting edge physics and medicine, interesting work for the scholars up to the challenge. [Cheikh Anta] Diop proposed that this type of work requires a new philosopher who is able to go beyond the mechanistic view of the naturalist and simultaneously ground the physicist's calculations and equations in the material world. This philosopher "undoubtedly will integrate in his thought all of the above-signaled premises, which barely point to the scientific horizon in order to help man reconcile man with himself"

This is important work, and much remains to be done. This text is not only the first of a five-volume series; it is the beginning of a new era of scientific research into ourselves.

Other Sciences of the Self

SUPREME UNDERSTANDING

Throughout this book, we'll have quick discussions of various topics, encapsulated in grey boxes so as not to take away from the rest of the chapter's content. Some of these will be commentaries on various "esoteric" or "nontraditional" ideas that deserve more than a passing mention in the text. These "other sciences" range from indigenous renderings of verifiable scientific knowledge to rumors passed around on Facebook that don't have a leg to stand on. Some commentaries are simply quick discussions on topics that could take up a whole text on their own. Topics include Levitation, ESP, Metaphysics (in general), Telekinesis, Reincarnation, and so on. Wherever possible, our ultimate goal is to explore the African roots of valid concepts. One concept that we'd like to begin with, although our discussion of it is far too large for a grey box, is the popular idea of the "Third Eye." Is there such a thing as a third eye? Let's find out.

A Quick Note on the Third Eye

For thousands of years, Hindu tradition has spoken of a "third eye." This unseen eye is often depicted in the center of one's forehead in traditional Indian art, and is occasionally associated with the sixth chakra (also called Ajna or the third eye chakra in yoga). In *Egypt: Child of Africa*, Ivan Van Sertima details evidence that the ancient Egyptians also recognized a third eye, depicted as the Uraeus (a serpent shown rising from the forehead), or the Udjat, also known as the Eye of Horus. Van Sertima proposes that – over 1,000 years before the Greek Herophilus (who lived and studied in Egypt) would be credited with "discovering" the pineal gland, and over 3,300 years before Western medical science would "re-discover" it – the Egyptians had already identified its location, function, and deeper significance.

DID YOU KNOW?
During the European Renaissance, French philosopher Rene Descartes called it the "Seat of the Soul," saying it was unique in the anatomy of the human brain as a structure not duplicated on the right and left sides. Yet, under a microscope, we now can see that it, too, is divided into two fine hemispheres. Descartes was credited with developing the school of thought know as Cartesian dualism, which proposes a distinction between mind and matter, and between the self and reality. It was Descartes who said "I think, therefore I am." But Descartes was no visionary. The Indian Upanishads, which reveal what may have been India's most ancient theological tradition, discuss the same ideas. The Upanishads themselves were mostly composed between 1200 and 500 BC, but their ideas are believed to date back further than the oldest Hindu scriptures, over 5,000 years ago. Perhaps this connection also explains how Descartes arrived at his "unique" ideas about the pineal gland being the seat of the soul.

The pineal gland is located in the center of the brain's mass, much like the Great Pyramid is located in the center of the Earth's landmass. The gland itself is tiny, tucked away in a tiny crevice behind and above the pituitary gland, in direct line with the space between your eyebrows. As Robert Bailey explains in his article on the pineal gland in *The Hood Health Handbook, Volume One*:

> Located near the center of the brain, the pineal gland gets its name from the pinecone, which it resembles. It is reddish-gray and about the size of a pea or grain of rice, yet, together with the pituary gland, it has a profuse blood flow, second only to the kidneys. In fact, unlike much of the rest of our brain, the pineal gland is not isolated from the body by the blood-brain barrier system, allowing it secrete melatonin directly into our bloodstream. What's even more unique about the pineal glad is that the "third eye" label may be a literal reality.

Schwab and O'Connor, in a 2005 article in the *British Journal of Opthalmology*, write:

> If the development of the third eye seems mysterious, the function is even more obscure. Most observers believe the organ to be a solar dosimeter [a way to measure exposure to sunlight] useful for photoperiod recognition for circadian and seasonal rhythms, but there may be more to this murky organ than first meets the eye…

In other words, in our pre-human ancestors the pineal gland was once literally a "third eye."

In fact, a study by Lucas and colleagues, published in the journal *Science,* suggests that our pineal gland, even encased inside the brain, can still somehow "see" and respond to light and darkness, even in the absence of the other two eyes. Another study published in 2008 in *The Journal of Experimental Biology* discovered that some eyeless fish actually navigate using only their pineal gland.

Our pineal gland is activated by light, ideally sunlight, and it controls the various biorhythms of our body. It works in harmony with the hypothalamus gland which directs the body's thirst, hunger, sexual desire and the biological clock that determines our aging process. Van Sertima adds:

> The outstanding works by the modern African scholar Dr. Yosef ben-Jochannan, *The African Origins of the Major Western Religions, Black Man's Holy Black Bible*, and *We, the Black Jews*, have revealed how much of the world's religions, including Christianity have borrowed from the religions of Africans, particularly the ancient Egyptians. Thus it should come as no surprise to find in Genesis 32: 27-32 that it was in Peniel (pineal gland) that Jacob (mental slave, neophyte, dwelling in ignorance), symbolic of the undeveloped unconscious or entrapped within the physical body (the contained mind, soul, spirit) met the Angel of God (soul, spirit). During the wrestling bout which ensued, he ascended from his former state of lower consciousness and was transformed into Israel (liberated mind. soul, spirit). It was at Peniel that Jacob saw God face to face and his life was preserved or renewed: "And Jacob called the name of the place Peniel, for I have seen God face to face, and my life is preserved." Phylogenetically, the pineal gland is actually in humans a modified eye. In the lower life forms such as amphibians and reptiles...the pineal is present as a parietal or third eye on the top of the forehead. In the higher life forms, such as mammals, the pineal withdrew into the head converting from a physical eye into a light transducer which converts light into hormonal signals that can actually change the form and function of the physical body and levels of consciousness (spriti, soul, mind, body).

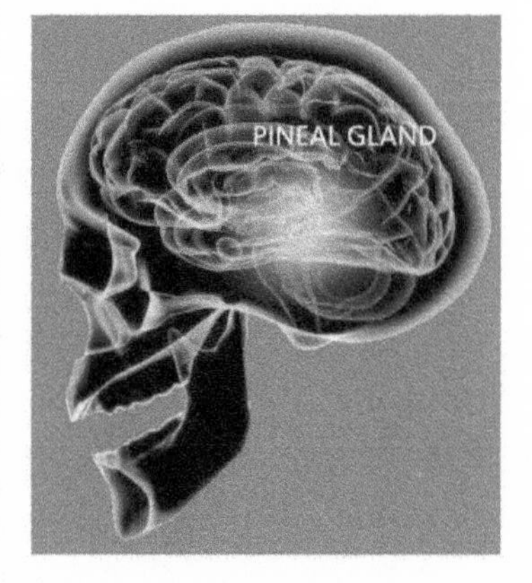

> The pineal hormones melatonin and serotonin operate as hormonal keys that unlock the door to the unconscious mind, dreams. Melatonin is known to initiate dreams by activating the locus coeruleus. It is known that ancient Egyptians did practice dream analysis and hypnosis. The recent re-discovery of the concept of the unconscious,

> hypnosis, and dream analysis, partially attributed to Sigmund Freud, should also be credited to the ancient Egyptians. A photograph of the desk of Sigmund does reveal a large number of Egyptian statues including one of the God Osiris, God of the Underworld or unconscious, confirming an African knowledge of the unconscious thousands of years before the time of Freud.[115]

Some people believe this "dormant" organ can be awakened to enable "telepathic" communication, the higher realms of thought, and many of the experiences associated with transcendental psychology and altered (or shamanic) states of consciousness. Some accounts suggest that a traumatic head injury can "activate" the "Third Eye" function of the pineal gland. Others suggest that there are much safer routes, such as regular meditation or a natural lifestyle. At any rate, what we know is that, YES, it is relatively dormant, and yes, some people will have an easier time "activating" or using theirs than others. After all, some pineal glands are more calcified than others, ranging from Original people having little to no calcification to white people having 60-80% calcification or greater. Adeloye and Felson's 1974 study found that a calcified pineal was twice as common in white Americans as in Blacks in the same city, but that Black Americans had more calcification than Blacks in Africa.*

THE SCIENCE OF DEATH

SUPREME UNDERSTANDING

IS THERE LIFE AFTER DEATH?

One interesting thing about death is that electromagnetic system of the body, the bio-energy, doesn't "die" immediately with the physical body. At the same time, all of our senses are found in our physical body. Without them, there is no way to perceive anything in the sense that we know it. In fact, our physical body and nervous system is what gives us our perception of time. What that means is that – outside of a physical existence – there's no "time" to run forward. No taste, no sight, no sound, no feeling, and no sense of time or place. That's not living, is it? So forget about life after death. Whatever comes after death is probably a lot like whatever came before life. And unless you

---

* For specifics, see Chapter Three of *Geographical Neurosergery*, available online at book.neurosurgeon.org For more on the pineal gland, its connection to melanin, serotonin, and melatonin, and details on calcification, check out *The Hood Health Handbook, Vol. One*. And keep reading to see the connection between the pineal gland and the electromagnetic field of the Earth, man's brain, and the Universe at large.

remember being part of the electromagnetic flux that exists throughout the universe, and whatever wavelength it exists on beyond that, I don't think there's any Pearly Gates for us to look forward to. In fact, most of the phenomena associated with near death experiences are directly related to known brain phenomena. That white tunnel of light? Yup, that's a neurochemical thing. For an in-depth roundtable discussion of what comes after death, *see Knowledge of Self: A Collection of Wisdom on the Science of Everything in Life.*

THE SCIENCE OF REINCARNATION

In *Knowledge of Self*, we described the Mind that designed the universe as "an infinite, timeless consciousness" something like a sea of water, of which we are each individual cups, containing all the elements of that sea. Well, let's say you pour that cup back into the sea. Do you think you'll get the same exact contents out the next time you draw a cup? Not likely. This concept is familiar to most of the oldest populations and traditions on the planet, such as that of the Efé, a "pygmy" people of the Congo. As reported by Jean-Pierre Hallet:

> The *balimo* [essence of the dead] does not retain any memories of its earthly existence; it is like a computer whose banks have been cleared. When a child is begotten, a balimo leaves the land of the ghosts and joins the other components that the lunar angel traditionally assembles to produce a new human being.[116]

Hallet notes that the Efé understand very well how father and mother contribute to the baby, and the "lunar angel" is simply the personification of the process by which the child's development is regulated. He then asks:

> "What if all the elements of a person are rejoined? Does that person live again?" I asked the Efé elder Mwenua. He replied, "It is possible that the same combination may be repeated once in a while. Such a person would still be new. He would not know what the former person had seen and felt." Here we see the germ of the reincarnation belief. The total person of Efé philosophy cannot, however: be reincarnated: he is the irreproducible result of all his elements combined with his experiences.

In fact, the idea of "reincarnation" only became a premise where human "souls" could be reborn in whole, one life after another, because this justified the caste systems they accompanied. In Hindu India, the Aryans perverted indigenous ideas of the universal self (the Atman) to that of an individual self that would be born into the same wretched caste repeatedly, unless one was humble and pious, and accepted their conditions without resistance to the status quo. This, as in Christianity during slavery, served to maintain the social order.

In other words, go hard with the life you have, because when you go, you're not coming back. Or as the adherents of the ancient Indian philosophy of Carvaka would say:

> While life is yours, live joyously;
> None can escape Death's searching eye:
> When once this frame of ours they burn,
> How shall it e'er again return?[117]

THE SELF THAT DOESN'T DIE – IT'S NOT YOUR EGO

Also, the idea of you (personally) living again after dying takes away from the sense of Self that is not ALL ABOUT YOU. In order words, what you think is "you" is mostly your own ego, not necessarily your true, eternal, non-local Self. That Self is not compartmentalized or individualized. It just is. As Isaac Asimov once noted:

> The molecules of my body, after my conception, added other molecules and arranged the whole into more and more complex forms, and in a unique fashion, not quite like the arrangement in any other living thing that ever lived. In the process, I developed, little by little, into a conscious something I call "I" that exists only as the arrangement. When the arrangement is lost forever, as it will be when I die, the 'I' will be lost forever, too. [118]

The Original Indian conception of death, before the Aryan invasion, is best detailed in the Upanishads. One of the most famous of these scriptures, the *Brihadaranyaka Upanishad*, notes:

> When a man dies, what does not leave him? The voice of a dead man goes into fire, his breath into wind, his eyes into the sun, his mind into the moon, his hearing into the quarters of heaven, his body into the earth cheerfully, his spirit into space.

This sounds very much like the African conception of death. The African worldview states that although the ancestors are dead, they continue to exist in different forms, which play various roles in affecting life in the present day. Birago Diop captures this in his poem in Kofi Asare Opoku's book, *West African Traditional Religion*, which begins:

> Those who are dead are never gone:
> They are there in the thickening shadow,
> The dead are not under the earth;
> They are in the tree that rustles,
> They are in the wood that groans,
> They are in the water that runs,
> They are in the hut, they are in the crowd,
> The dead are not the dead.[119]

In fact, these sentiments are echoed by modern science. The carbon cycle, nitrogen cycle, phosphorus cycle, and many other aspects of the Earth's ecology (and possibly the cosmos beyond as well) are means by which man "lives on" after Death. Indeed, the dead ARE in the trees and water.* Our ancestors also live on through us. Not only do we carry the history of everyone who preceded us in our DNA (and I mean everyone, going all the way back to our universal common ancestor), but we are physically composed of particles that come from Imhotep, George Washington Carver, and billions of others, right down to the earliest life forms.[120] In effect, even those who didn't make it (no surviving descendants) still became part of us.

"If you look deeply into the palm of your hand, you will see your parents and all generations of your ancestors. All of them are alive in this moment. Each is present in your body. You are the continuation of each of these people." – Thich Nhat Hanh, Vietnamese Monk and Activist

But, as we've already seen, the integral component of Man is Mind. There is no better mechanism through which we live on (mentally) than by the transferrance of information. Thus, teaching others is a form of reproduction and – after our deaths – reincarnation.

Finally, it's no coincidence that these indigenous conceptions of death are so reminiscent of indigenous creation myths, where the first man, or God (such as the Chinese *Pangu*), becomes the cosmos via the decay of different parts of his original body. That this build/destroy cycle is present in both instances tells us – as we'll see again and again – that there's no beginning or ending to any of this.

REVIEW

As we'll see in Volume Three, there is intelligence, or consciousness, to be found at all scales of life, from single-celled micro-organisms to whales and sharks. You'll find some sense of growth, order, and direction at inorganic levels as well, from photons to galaxies. The Mind being the "substance" of all things is the root of indigenous thought systems like animism. But if everything from atom to Adam has some degree of consciousness, what makes man so special? Not only is man the highest form of living intelligence, man doesn't simply have intelligence. We don't just have consciousness. We have SELF-consciousness. Not self-consiousness in the sense of being concerned about how our hair looks, but in the sense of "knowledge of self."

* And, through the consciousness that cannot die or decay or in the electromagnetic field that envelops us all, it can be said that the spirit lives on, but (as we noted earlier) there's no way for personalities to survive in either the Mind or the electromagnetic field.

According to Sir Karl Popper and Sir John Eccles's comprehensive work *The Self and Its Brain*, published in 1977, the highest mental experience is "knowing that one knows," also known as metacognition, self-awareness, or self-consciousness.[121] It is the most fundamental characteristic of the human species[122] and emerges from levels of linguistic communication not shared by non-human animals.[123] Even those super-intelligent chimpanzees don't have access to THIS level of consciousness.[124] Jared Diamond agrees, noting that man is unique among even his closest cousins on the evolutionary tree of life.

Why is this? To understand why the world is full of consciousness and intelligence, but the Original man presents the apex and epicenter of this consciousness, we'll have to explore how all this came to be. To understand man's place (and role) in the Universe, we have to study the Universe itself.

Even the simple is complex...yet the complex is simple.

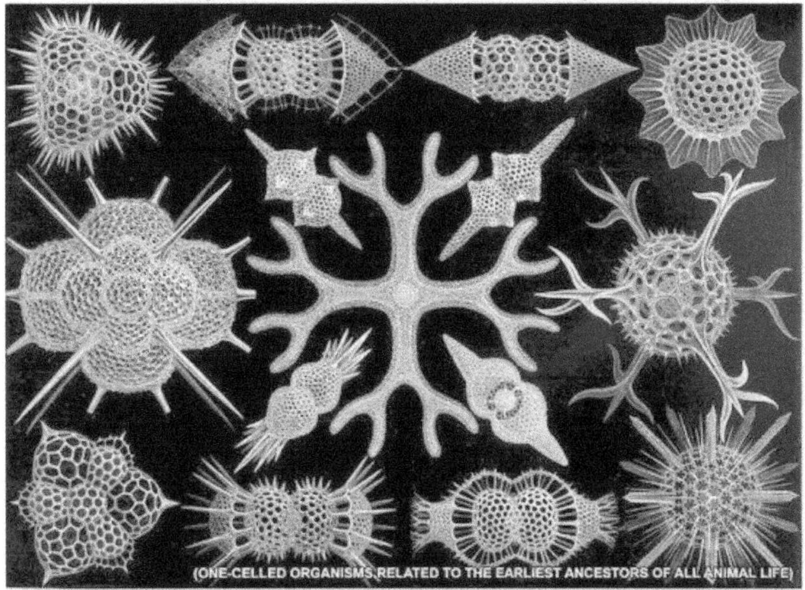

(ONE-CELLED ORGANISMS RELATED TO THE EARLIEST ANCESTORS OF ALL ANIMAL LIFE)

WWW.THESCIENCEOFSELF.COM

Take a break.
Put this book down for a minute.

Do NOT go back to reading this book until you do one (or more) of the following things:

- ❑ Call somebody who is going through some rough sh*t and make sure they are okay.
- ❑ Eat something that your body is telling you it needs, or drink some water.
- ❑ Wrestle, spar, or slapbox someone to make sure you "still got it."
- ❑ Take a walk through your neighborhood and see if somebody needs help with something.
- ❑ Clean up a part of your house, or organize some f*cked up part of your life.
- ❑ Tell somebody about this book and what you're learning. Invite them to come read it.
- ❑ Give this book away to somebody who needs it and get another copy for yourself.
- ❑ Cook something good, and make enough to share. Invite people.
- ❑ Check yourself out in the mirror and pick something to improve.
- ❑ Identify ten positive things about your life and stop forgetting them when you're stressed.
- ❑ Tell somebody you love them, cause it might be your last chance.

This has been a PSA from 360 and SDP.
Once you're done, carry on.

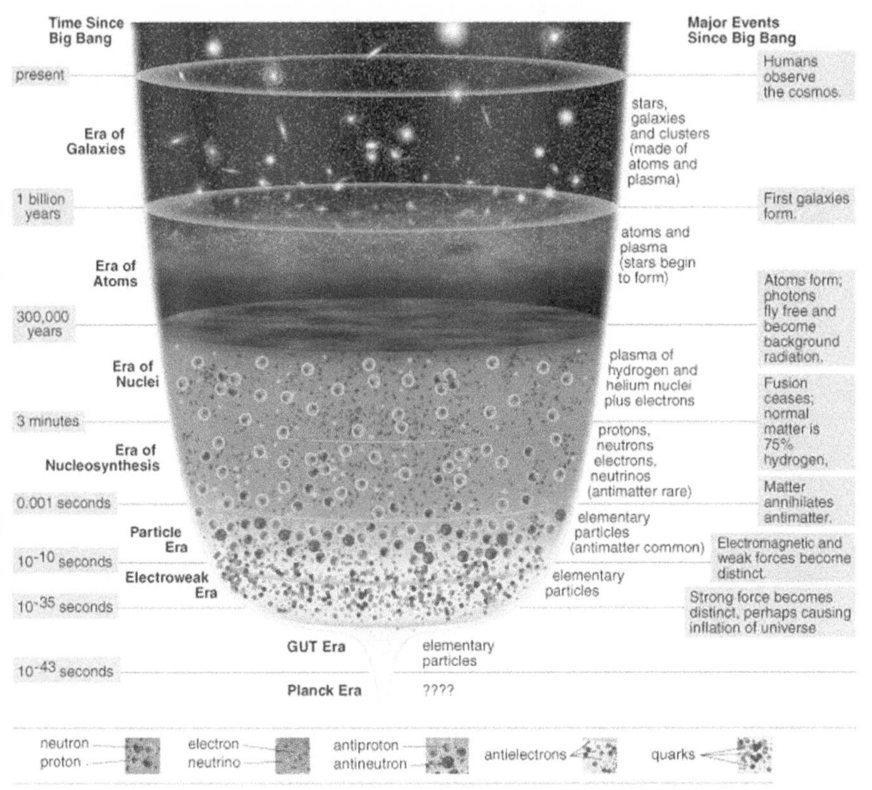

Formation of the Universe according to the Standard "Big Bang" Model

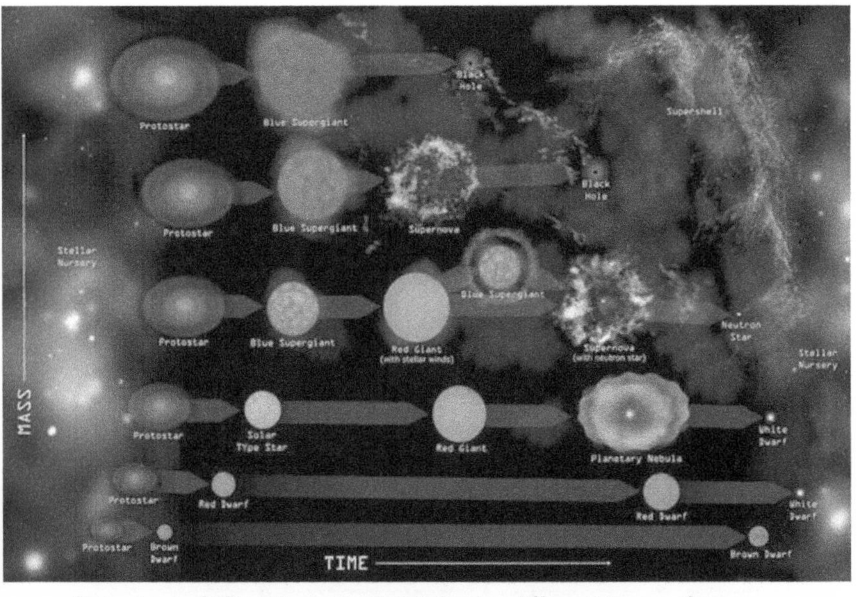

This chart of "Star Evolution" details the different types of stars in our Universe and what becomes of them over time.

SELF-CREATION OF THE UNIVERSE

HOW IT'S MADE…BY US

> *"We are part of this universe. We are in this universe. But perhaps more important than both of those facts is that the universe is in us." – Dr. Neil DeGrasse Tyson*

You've seen how science is the lens by which we can understand the universe with some degree of certainty. You've seen how the universe is structured (or ordered) according to natural laws that are inherent at every scale of existence, from subatomic particles to galaxy clusters. Everything operates according to these laws of growth, order, and direction. In the last chapter, you've seen how the Mind is the source of these laws, and how this Infinite consciousness is, as the Qur'an says of Allah, "nearer to man than his jugular vein."

In this chapter, you'll see how and why man is seated at the center of the Universe, and how this Universe came to be. We're going to start at the beginning that was really no beginning and follow the growth, order, and direction of the Physical Universe from infancy to maturity. We'll cover everything from the Big Bang to the formation of our planet. And everywhere we look, we will find ourselves.

WHAT YOU'LL LEARN

- ❐ Why and how the Universe is essential Black.
- ❐ How the Sun, Moon, and Stars came to be.
- ❐ Why the Earth was once red, black, and green.
- ❐ Why we are literally composed of stardust.
- ❐ How the Sun and Earth relate to Man and Woman.
- ❐ The story of the formation of the Universe.
- ❐ How gender is pre-programmed into all life and matter.
- ❐ Why man is at the center of the Universe.
- ❐ The relationship between the electromagnetic brain and the electromagnetic fields of the Earth, Sun, and Cosmos.
- ❐ How our ancestors understood the Cosmos better than some scientists do now.

THE UNIVERSE IS BLACK

SUPREME UNDERSTANDING

TRIPLE DARKNESS

The Buddha taught that all our ideas about the universe and reality were maya, or illusion (no relation to the Mayan people of the ancient Americas, who were quite real). This world we see, he said, was not real. It was the product of our consciousness. This sentiment has been echoed to varying extents by the teachings of the Jewish Kabbalah, Christian Gnosticism, the Hermetic teachings of Egypt, the Indian Upanishads, and even some elements of Islamic Sufism. Many of the world's people have proposed that there is "nothing but Allah." Or, there is no objective reality without the subjective viewer.

So let's consult science. According to science, exactly how "real" is the universe we inhabit? Well, based on measurements of the mass and movement of everything we can measure, we know that much of this universe is not made of matter. 25% of the universe is described as "dark matter," a poorly understood phenomenon detectable only by its gravitational pull on other material. 70% is "dark energy," another enigmatic entity that seems to play a role in the accelerating expansion of the universe. Dark matter is the theoretical force that structures the universe in its current form together while dark energy is the theoretical force that is pushing everything apart.

DID YOU KNOW?
In some cultures, the color Black may represent death because our bodies turn black when they are cremated, and often when buried. But no matter how our bodies are disposed of, all of our flesh and bones will eventually become black material known as fossil fuel. This is all because of carbon. There's even a pigment known as "bone black" made by burning carbon based life-forms (from carbon to carbon). So you were born black, and you'll die black – because carbon is what you started with, and another form of carbon is what you'll become. (See "The Carboniferous Period")

That means 95% of the universe's content is unknown. Ordinary matter – that atomic stuff that you, the Sun, and this book are made of – accounts for just 5% of the universe. Yet it would STILL be inaccurate to think of this 5% matter like specks of dirt in a glass of water, because matter itself is mostly "nonexistent." That is, within any individual atom, there is more "empty space" than there are actual particles. 99.99% of an atom is empty space![125] In fact, the reason why nobody uses that "solar system" model of the atom (seen below) is because the space between the innermost "cloud" or "shell" of electrons (really the field of probability where the electron "could be") is the size of a massive football stadium compared to the nucleus being a tiny football in the center. Or maybe even a fly in the stadium.

DARK MATTER ISN'T ACTUALLY DARK

To be clear, neither dark matter nor dark energy are physically "dark." The word "dark" simply means unknown, because they're very theoretical. As Divine Ruler Equality Allah argues later in this chapter, dark matter and dark energy are only theoretical values used to explain the unexpected level of structure in our universe.

But assuming dark matter and dark energy do exist, we should know these things aren't "black" in the visual sense of the word. But in another, perhaps deeper sense, as in the fact that they represent the "unseen" fabric of the physical universe, they are essentially "blacker than black." Like the "triple darkness" we discussed earlier, and the carbon that forms the basis for life, these things are not "physically Black," but "essentially Black."

Why not "essentially white" or any other color? Because "Black" is not a color so much as it is a state of existence. It represents the "formless void" of potentiality, containing everything else in the spectrum. If something is Black, it contains the whole spectrum. If something is white, it either rejects (reflects) or radiates (emits) all light. We'll get back to that. For now, let's focus on Black. Black, like white, enjoys a duality of nature. It represents life, and in some sense it also represents a "return to the essence." This is why among some cultures – depending on their orientation with death – Black can also represent death.

Black is the life-giver, and life-taker, it is the absence and the presence, it is the source and the destination, it is everything and it is nothing. Because of this, Blackness is – on every level, including the human presence on Earth – both the origin and final destination of all things. So when we talk about conceptual forces like dark energy, we're not ascribing it a visual color if we call it "Black." We're simply saying the "essence" of it parallels the essence of Blackness in its many other occurrences in the universe. This "Blackness" – at its deepest level – may be consciousness that informs all of the above, from dark energy to electromagnetic energy, from dark matter to black matter, from the womb of the Earth to the mind of modern man.

Ultimately, the universe became tertiary (three parts) once it added a component outside of the binary relationships we find in matter vs. antimatter or positive vs. negative charge. This third element was life. Thus, as Mary Baker Eddy once wrote:

> Truth has no consciousness of error. Love has no sense of hatred. Life has no partnership with death. Truth, Life, and Love are a law of annihilation to everything unlike themselves, because they declare nothing except God.[126]

Different renderings of the "map" of the universe, based on Cosmic Microwave Background radiation as determined by the COBE satellite. Note "design" in the top rendering.

This is because there is no complement to life. There is no anti-life. Death is simply the end of life's cycle, not an opposite reality. Life is quite special. Life is the product of information and self-organization, but unlike plain matter (which is also organized by those properties), life takes ownership of those properties and makes it own decisions. At no point in the spectrum of life is this more true than among man. (Later in the book, we'll have a full discussion on the meaning of life.)

"Intelligent life is neither incidental nor insignificant but has a place in the universe so special it could not even have been imagined before the invention of modern cosmological concepts. By understanding the universe, we begin to understand ourselves." – Joel Primack, View from the Center of the Universe

So, yes, there isn't much to the universe. It's only 0.01% real! But with

DID YOU KNOW?
You can actually SEE some of the cosmic microwave background radition for yourself, just by turning off your cable box and turning your tv to an unused station. Some of the white "snow" you see on your screen comes from photons, still hot from the Big Bang, which could never fully interact with atomic particles and which are still cooling even now. This residual heat is still present in the world today and because television can pick up stray electromagnetic waves, a TV set can show the residual radiation of the universe. TV static is anywhere from 25-33% cosmic background radiation.

the presence of life, the universe is made real by our consciousness. This is the most basic understanding of the implications of quantum mechanics.[127]

Let's talk about this universe we live in, how it started out, the processes it went through, the blueprint by which these process occurred, and how the question of whether or not the universe is "really real" relates to us today.

DEEP BEGINNINGS

Scientists are able to trace the universe back to its origins by studying something known as cosmic microwave background radiation. This is radiation left over from an early stage in the development of the universe, before the formation of stars and planets, when it was much smaller and much hotter. By studying this radiation, scientists can work backwards to calculate the age of our visible universe and many other details about its early conditions.*

Cosmic microwave background radiation reveals that our observable universe is 13.7 billion years old. The further you go back – from now to then – the universe gets smaller and hotter. In, fact there is a point where things are so small that everything is tinier than an atom itself. This is known as Planck time, the smallest time measurement that will ever be possible. Its 10^{-43} seconds. In other words, it's not 0.001 or 0.00001 of a second. You need more zeros before that one, 43 actually. It represent the length of time it takes one photon to cross the distance of a Planck length, which I'm sure you can imagine is just as ridiculously small. Basically, there is no amount of space or length of time smaller than the Planck scale, because that's the only point where anything can exist. Anything less and there's no laws that can apply. Before this point, it's just a static state of essentially nothing beyond that. No time, no space, no light (or energy/matter of any kind). Triple darkness.

* If you're interested in all the details of how this works, I recommend reading books like *The Left Hand of Creation* by John Barrow or *Cosmos* by Carl Sagan. I'm going to focus on a history that is mostly "general knowledge" on the history of the universe. When I get to parts that are hotly debated, I'll let you know that I'm either presenting the theories that make the most sense to me, or that we just don't know.

DID YOU KNOW?
Many indigenous myths record the Creator God battling serpents or monsters in the heavens before emerging victorious and creating the planets and life. Could these traditions describe the annihilation of antimatter? And if so, how would our ancestors know? We know that oral traditions were passed down since time immemorial, but how did the originators of these traditions have any idea about how the Universe came to be? When we look at all the mythical parallels (literally, hundreds of indigenous cultures telling similar stories that describe the early history of the cosmos better than Western scientists could in the 1800s), it seems possible that all of this knowledge is already "pre-programmed" into our DNA or inherent within our consciousness, and all it takes is for one to be wise enough to tap into it.

So this point, 13.7 billion years ago, is quite literally a point of singularity when time and space began. After this time, gravity emerges as the classical background in which particles and fields evolve following quantum mechanics. This "seed" of the universe is only 10-33 cm across. Scientists call this the Quantum Gravity Era, the Planck Epoch, or "Time Zero."

After this point, the universe launches into a phases known as cosmic inflation, where expanded acceleration takes place for a very brief moment before the normal "Big Bang" expansion commences. Next is Baryogenesis, where a small difference betwee n the reaction rates for matter and antimatter leads to a mix with about 100,000,001 protons for every 100,000,000 antiprotons (and 100,000,000 photons). This is major. It's a "tilt" in the universe that shouldn't be there if everything was even.

The universe seems to have gone from nothing (triple darkness) to singularity, to a binary universe of black and white polarities. Before the duality of positive and negative charges, or that of particles and photons, the first binary relationship was that of matter and antimatter. But in a process known as Supersymmetry breaking, the universe was defined by this split NOT being perfectly symmetrical. When the universe first phased into existence, if matter and antimatter had come out even in those first moments, they would have instantly destroyed each other, leaving an empty cosmos. But they didn't. Matter clearly won. After all, here we are. But why? Scientists using the Tevatron particle smasher in Illinois have found that the physical universe – for reasons they still cannot explain – shows "favoritism" towards the existence of matter over antimatter.[128]

In layman's terms, the universal petri dish is tilted to favor the emergence of – ultimately – you and me. This asymmetry is close to 1 percent, but that 1% was all that was needed to eventually produce all that we see today. In fact, much of the fine-tuning that favors the emergence of life is of similarly small and seemingly negligible magnitude. For example, the rotation of the Earth's inner core is only 1.1

degree faster per year than the Earth's crust. But without this slight difference, we wouldn't be here.

Within the first few seconds of the Big Bang, the universe grows and cools very quickly. Matter is produced by the annihilation between matter and antimatter, but with a very large number of photons per surviving proton and neutron.

100 seconds after the Big Bang, the temperature has cooled down one billion degrees Kelvin. Electrons and positrons annihilate to make more photons, while protons and neutrons combine to make hydrogen and helium. The final result is about 3/4 hydrogen, 1/4 helium, and very small traces of other elements. There are about 2 billion photons per proton or neutron. So there's a lot of light in this early universe. This matter, again, isn't evenly distributed throughout the space of the universe. The small fluctuations in its distribution allow for the law of gravity to take over and bring particles together. The more particles clump together, the more mass this clump or cloud has to attract other surrounding particles. It is entirely due to the law of gravity and the uneven distribution of matter that these early particle clouds formed, which eventually congealed even further to become the first stars, 100-

A Quick Note on "It's Only Real If I Think It Is"

No. No. No. This is what happens to people who get all of their scientific understanding from The Matrix, What the Bleep Do We Know?, and folks like David Icke. We've got to do better. Common sense tells us that – even if we think REALLY HARD about our bus coming in 5 minutes – unless that bus left the station 20 minutes ago, it ain't happening. You can't MAKE a physical event occur because you thought about it (or prayed, focused, meditated, visualized, or whatever other word you want to use that doesn't mean "action"). Because that physical event has very firm causative laws that make it happen. Those laws don't change because you personally want them to. Even on the quantum level, there are very clearly defined deterministic laws that tell you where something can be and where it can't be. It's not just chaos at that level, you know. Either way, the only place you need to connect with the quantum world is at the level of consciousness. And you're already there. And those laws we discussed. You already personify them. You're supposed to USE those laws to get what you want out of life. And the REAL "law of attraction" doesn't bring together unrelated phenomena at very different scales of existence. In other words, "visualization" doesn't attract "money." No, "working for money" attracts "money from work." Again, that's why we have arms and legs. No amount of meditation will fix widespread social problems with multiple causative factors. If it could, you'd totally unravel the way the world works in doing so. Instead, you seek out a causative factor in the problem you want to address, and then you REALLY (as in physically) work on that problem! And I promise you'll get faster results if you actually go out and get involved with a community service organization than if you visualize it happening every day while sitting on the computer looking at inspirational images.

200 million years after the Big Bang. These early stars reionized the Universe, and eventually created all of the other elements when such stars died. The first supernova explosions of dying stars spread carbon, nitrogen, oxygen, silicon, magnesium, iron, and so on (up through uranium) throughout the Universe. Meanwhile, the law of gravity (and that uneven distribution) allowed galaxies to form from stars attracting other stars. Then clusters of galaxies formed. Within these galaxies, stars were producing planets and moons, again through gravity causing pieces of matter to aggregate (or congeal) into solid masses that went into rotation and became round. Our Sun and solar system formed about 4.6 billion years ago, as a part of the Milky Way Galaxy.

A MATHEMATICAL NARRATIVE OF OUR UNIVERSE

1. First there was triple darkness, time zero. Then the singularity emerged. Within this singularity, a structure for growth, order, and direction already existed. As Primack and Abrams have noted, "the blueprint for the universe existed before the Big Bang."[129]
2. The consequent expansion produced particles and anti-particles which should have eliminated each other, leaving nothing. But the assymetrical nature of the universe's blueprint is what it takes to produce life.
3. In mathematical terms, 0 became 1, 1 split into 2, but then 2 added 1 to become 3. Three represents the emergence of matter due to this assymetry.
4. This matter became organized into systems and clusters due to the laws of gravity and the forces of dark matter.
5. Some of these aggregations of "spacedust" became more powerful through this process, growing into stars that emitted light and energy, with strong gravitation (and electromagnetic) fields of their own.
6. These stars reionized the universe, effectively distributing the balance of energy in the universe, along with transferring much of their energy to the planets they formed.
7. These planets embodied many different characteristics. Only one, it appears was "fine-tuned" for life.
8. Other planets and celestial bodies, including the moon and distant nebulae, played mitigating roles in the circumstances that favored life. On this planet, the law of entropy allows for the emergence of life at the expense of disorder exported to the external envi-

ronment.

9. And through these processes, this planet – in a universe that is 93 billion light years in diameter – would birth intelligent life. That planet, 93 million miles from the Sun that produced it, is our planet, the Earth.* This is where biological reproduction occurs, allowing all other cycles to follow, including the patterns that produced man.

And to be clear about all this, "1" didn't emerge at the beginning of the universe's formation as singularity and then disappear. Just as when a quantity goes from 1 to 2, and you still have 1, the universe never "left" 1. Just as Eastern cultures teach that the "dot" from which the Universe emanated is not just there at the beginning and end of time, as the alpha and omega points, but instead "exists throughout," all of the mathematical principles that predate our Universe's formation continue to exist at all times and across all dimensions. These laws are the means by which fields of possibility become matter, by which matter becomes organized into molecules, by which molecules form into solar systems, by which our planet produces conditions favorable to life, and by which live evolves into us. These laws never go anywhere. They are the field in which we exist.

As Divine Ruler Equality Allah has observed, "The creation of the universe is as easy [or difficult!] to determine as a rug's pattern when you have only a handful of the threads it was made with. The principles have been present from the beginning; only the expressions of them change."

And 1, or any other aspect of the blueprint (from 2-9, plus the 0 that represents a complete cipher), is found at every scale. 1 is the Mind that made the universe 13.7 billion years ago, but 1 is just as much the Mind of the Original Man today. It's all the same laws, no matter what scale it is expressed on. On its most basic level, this can be seen in pure number theory. Everything is based on a progression from 1 to 9. You can't imagine imagine a universe governed by a mathematical sequence like 2154609837 or any other syntax.

AN INTRODUCTION TO COSMOLOGY

ROBERT BAILEY

Cosmology (from *kosmos* meaning universe and *logia* meaning study) is

* For another, more technical look at the mathematical structure of the universe based on Information Theory, see the Appendix.

the study of the universe. The science focuses on its origin, development, nature and our place in the universe. Physical cosmology is the branch which focuses on the nature, physical origins and evolution of the Universe. Metaphysical cosmology is the branch concerned more with the totality of space, time and all phenomena; it answers question with philosophical methods. Cosmologists use many different sciences such as astronomy and physics to study the universe. They test experimental models and hypotheses about the universe and make observations of the parts of the universe which are visible from Earth.

Throughout history, we have developed numerous cosmologies and theories of the universe. As in many sciences, observation plays a key role here. Early people observed the heavens and knew them well. Nomads in the Sahara and tribal peoples in the American Southwest used stones to mark important celestial events like the summer solstice. They had to have precise knowledge of the seasons to know when to plant and harvest. Plenty of ancient civilizations built colossal stone structures aligned with the risings and settings of the Sun, Moon, planets, and some stars. The *Dresden Codex* is just one book that exemplifies the cosmological understanding of the ancients. Modern cosmology is said to have begun in 1917 with the final modification of Einstein's General Relativity. The Big Bang theory, proposed by Georges Lemaître, is the current prevailing model of the early development of the universe, being the most accurate. The basis of the Big Bang theory is that the universe was once in an extremely hot and dense state that expanded (a "Big Bang"). This expansion caused the infant universe to cool and resulted in its present continuously expanding state. In 1964, background radiation was discovered by satellites, which moved cosmology from a speculative science into a predictive science, as the discoveries matched predictions made by a theory called cosmic inflation, which is a variation of the big bang theory.

Mos Def once said, "Man thinks he got some place in space. He do! It's called Earth." This brings the point back home, literally. In other words, study the universe, but for YOUR benefit, by looking for how everything relates to you and where YOU reside. We made this planet our home for a reason.

THE SUN, MOON AND STARS

C'BS ALIFE ALLAH & DIVINE RULER EQUALITY ALLAH

Our solar system's Sun is a third generation star, only about a third the

DID YOU KNOW?
The Jesus of the Gnostic Gospel (considered by many to be the most authentic representation of the historical Christ's teachings) may have considered himself both the "Son of Man" and the "Sun of Man." And rather than attribute this quality to himself alone, he encouraged his disciples to kindle this divine spark within themselves. The Gnostic Gospel of Thomas relates a story of Christ telling his disciples to "shine":
There is light, within a man of light, and he (or it) lights up the whole world. If he (or it) does not shine, he (or it) is darkness...Jesus said, "If they say to you, 'Where did you come from?' say to them, 'We came from the light, the place where the light came into being on its own accord and established itself and became manifest through their image.'"[130]

age of our Universe. Yet, this Sun was, in fact, the grandson of the first physical Sun. And all of these Suns are derived from the Sun of Man. I'll explain. (Warning: Heavy reading ahead!)

In the beginning, the "point" it all came from was uniminably hot and energetic. This was a manifestation of a primordial sun. It is from this point (the singularity) that one became many. This early state of the universe has been referred to as the Dirac Sea. It was a point where the vacum was flooded with only energy.* From that point, the grand expansion (or Big Bang) took place.

The primordial medium of the Dirac Sea represents the original structure and physical composition of Man and his physical Universe incarnate as the electromagnetic field and mitigating particles. You can find a parallel of this in many indigenous cosmologies which speak of a primordial sea from which the Sun, Moon, and stars emerged.

At this point, although the universe was literally ablaze with light energy, it was still dark. This is because there was no matter to absorb, reflect or re-emit the flood of photons, the fundamental particle or substrate of the universe out of which all other particles of energy and matter arose. Energy had not yet been transformed into matter (as Einstein's famous $e=mc^2$ equation illustrates). So again, we had another stage of the primordial sun.

Once the universe had cooled enough for energy to be transformed into the subatomic particles which form the foundation of matter, the "making" of matter began. This process is referred to as photon pair particle production, the name given to the cycles of particle annihilation which give rise to matter. The residue of the production is not directly absorbed unless through determined detection such as Casimir Effect (manufacture of capacitors with plate distances of only a few

* Specifically, it was the ground state energy and fluctuation of harmonic oscillators that are the true building blocks of matter.

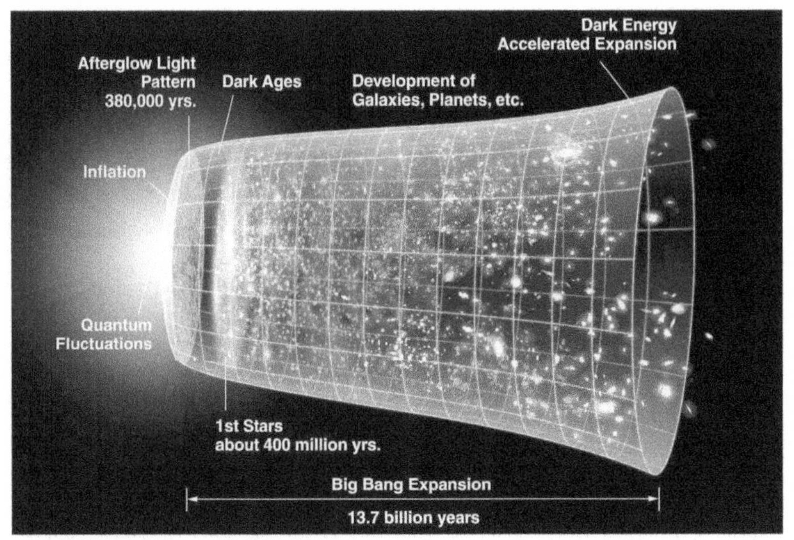

nanometers).

The first element produced was hydrogen. It is one proton with an accompanying electron. So hydrogen is the elemental base of matter. Some of the lighter elements such as helium and lithium were built up during this period also. Some of the heavier elements are synthesized in the later manifestation of stars, supernovae, or planets.

The formation of sun, stars, moons and planets starts with these elements condensed in the forms of gaseous clouds. Through gravity these elements would coalesce to form collections of these celelstial objects which are known nowadays as galaxies. The command/processing centers are the sun and stars. They also work as cosmic furnaces that transmute elements. The field of operations and assembling centers are the planets. The moons can work in various capacities yet commonly work as stabilization centers.

Through gravity and nuclear fusion the primordial hydrogen becomes concentrated into mass that grow into stars such as our sun. The process of nuclear fusion (which works on the principle of unification by the way) fuses two atoms of hydrogen (one proton and one electron in most cases) together to become one atom of helium (2 protons, 2 neutrons and two electrons). The repulsion that Hydrogen atoms have because of the electrons in their periphery is overcome by the high pressure at the center of the sun, so that the hydrogen can unify. This nuclear fusion gives off energy of all sorts. There's alpha radiation (high speed Helium nuclei), Beta radiation (high speed electrons), and Gamma rays and all the other forms of radiation (photons).

The sun is grown through the unification processes described above (condensation and fusion). The rest of the solar system is made as by-products of the sun and other stars. Just as with human evolution or agriculture, this development can occur through a slow, natural process (e.g., the accretion of the Earth) or a more forceful and immediate process (e.g., the expulsion of the Moon).

Thanks to the the elements produced by our sun, as well as the heavier elements expelled by dying (older) suns in other systems, our planets had the ingredients they needed to form. The planets also have the

A Quick Note on Astrology

There's nothing wrong with studying astronomy. And if you'd like to draw correlations between the properties and movements of celestial phenomena and the properties and movement of people on Earth, that's certainly one way to make sense of the universe as a web of connections. But here's the problem. Most astrological references are wrong, based off calendars that were composed long ago and – with the precession of the Earth's axis – have become terribly outdated. In other words, someone who thinks they're a Virgo may actually be a Leo. This is a point noted by Syed Amir Hassan Shah, author of the *The Book of Ilm-ul-Jáffár*, a handbook of Shiite Muslim esoteric knowledge:

"Let it be clear that, out of those *Alloom* (plural of *ilm*, meaning knowledge) which provide the information regarding the future and the past, the *ilm* of astrology is the oldest. First, for the reason of sacredness, it is not acceptable; secondly, it is a branch of mathematics and for this reason one needs to work very hard in this field. For most, it is not only difficult, but also almost impossible. In India, generally, the Brahmins have learned a few *Asloaks* (rules) and have tried to become *Jotish* (fortune-tellers/astrologists) and have brought a bad name to Jotish. Even if they master the Jotish, more often than not what they say turns out to be wrong, because the foundation upon which they laid their Jotish has long since rotted. Between now and then there have been many changes in the Solar system. For example, in the olden days, the point of balance of *Rabbei* (a rabbah means square) was in first degree of Aries, and the entrance of the Sun into this point, called *Nau-Rose* (the new day), was on the 21st of March. Now the Sun enters Aries on the 13th or 14th of April, but comes to the point of normality or balance on the 21st of March."[131]

In other words, that astrology app on your phone is *way off*. Not to mention that the modern-day characteristics assigned to the different zodiac signs came long after the zodiacs of the ancient world were composed. Before that, these qualities weren't meant to be so rigid and clearly defined. In recent years, people trying to work their way around such rigid characterizations of people (which often don't apply) will use every other celestial body in the heavens to explain their unique characteristics. Going back to the original point, there's nothing wrong with astrology if you can meet people and figure out when they were born by looking at their characteristics. It's not so cool when you do it in reverse order. "Oh you're Virgo, and that's why you're so critical." "No, dummy, I'm critical because my mother was critical as hell, and no, she wasn't a Virgo." Don't put people in boxes! We'd all like to understand each other, but there's an easier way that connecting people to their celestial twins. It's called *communication*.

elements (whether they are gaseous or terrestrial planets) to experiment with chemical combining and recombining them into new forms and structures. In a form of what we call the "strings on our fingers," these celestial developments embodied various symbolic traits which our future scientists could identify and interpret to activate different elements of later civilization.

In essence, without the living sun (Man) and dying suns of eons past (Man's ancestors), there would be no planets, no life on earth, and no civilization. The sun is the primary organ and avatar of the original man, wherever it exists, but particularly in our present solar system. In the macrocosmic body of the Universe, the Sun, Moon, and Stars are all organs. It is only through a psychological separation, removing ourselves from the bigger picture, that we fail to see the qualities we share with the functions of the Universe and its organs, or what this bigger really means for us here on Earth.

Another Look at the "Big Bang"

C'BS ALIFE ALLAH & DIVINE RULER EQUALITY ALLAH

The most popular theory among mainstream physicists and cosmologists concerning the nature of the universe and it's "origin" is the so-called "Big Bang" theory. The traditional Big Bang theory says that there was once nothing: no space, no time, and no energy. All of a sudden, out of nowhere, space, time, and energy emerged like a giant explosion. All energy was concentrated in a infinitesimally small point, which then spontaneously expanded in the form of nearly pure radiation, or light. Light or radiation consists of particles called photons, which always travel at the speed of light, and which have a zero "rest" mass. That basically means they are totally weightless.

In the standard Big Bang theory, this radiation eventually under goes a change into particles with a non-zero rest mass, in a process called pair production. Somehow, because of imbalances in symmetry in the interaction of these pair-produced particles of regular matter and anti matter, which naturally annihilate or "destroy" each other when they collide, there was a net abundance of leftover regular matter and hardly any antimatter.

Big Bang advocates further explain that although there is an abundance of energy in the universe from matter and radiation, the total energy of the universe is zero, because of "negative energy" contributed by the increasing rate of expansion of the observable universe. And the universe is not expanding from one singular point in space, but expanding in all directions, as if the entire universe was on the surface of an expanding balloon, with the skin of the balloon representing time and space.

In the earliest periods of time of the Big Bang model, the areas of the observable universe were more directly connected on a quantum scale of interaction. Quantum means a vary small scale, such that there are discrete levels of energy, space or distance, and time, with no "in between" region between these discrete levels of energy, distance, and time. The differences in these discrete levels of energy, distance and time are VERY SMALL. This means that everything that existed in our OBSERVABLE universe was all compacted into a size much less than the size of an elementary particle. Then all of a sudden, there was a great inflation which made the universe balloon in size at a rate of

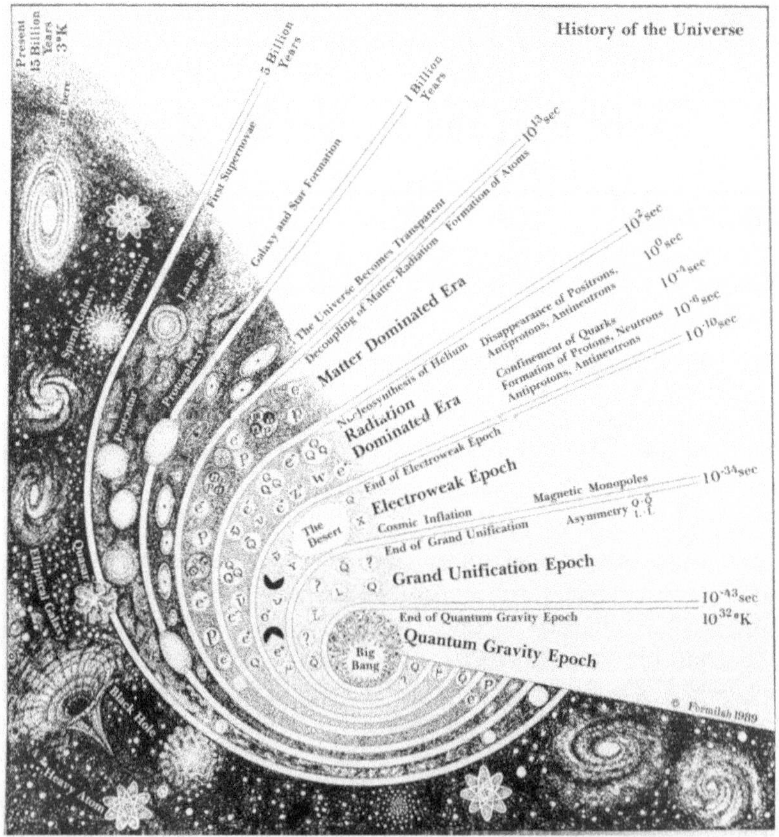

doubling its size every 10^{-37} seconds. This inflation lasted for 10^{32} seconds, and the observable universe ballooned to 10^{78} of its original size.

In the late 20th and early 21st century, evidence found by radio tele-

DID YOU KNOW?
In the cosmology of the indigenous Huichol people of Mexico (from where the slang word "cholo" derives), Grandfather Fire is the original light, the original knowledge, and the universe's own memory. In the beginning, he took the raw energy of creation and transformed it into vision by creating colors and images and into sound by singing. In this way he gave us human knowledge, and the Huichol are forever grateful to him, even though he is not seen as a "personal God." Grandfather Fire is considered to live on in every flame and spark, and fire is to be treated as an honored being. As you may have noted, there are clear connections here (fire=light=photons=mind=knowl edge). There also appears to be a commentary on how the rest of the electromagnetic spectrum derives from different vibrations (singing) of this original energy. In the Huichol cosmology, after Grandfather Fire, there are Grandmother Growth, Grandmother Ocean, Mother Earth, and Father Sun.[132]

scopes and other astronomy instruments point to a not so simple origin of the universe, but instead the reality of an eternal universe, of which the observable universe is only a small part. This evidence is the measurement of the cosmic microwave background, a seemingly constant low energy "echo" of microwaves from every direction you look in the observable universe, with a temperature of under 10 degrees Kelvin. The measurement of this background radiation seems to be "on the surface" very smooth, evenly distributed, or homogenous (the "same") in all directions. However there are small fluctuations over the whole measurement that point to this background radiation looking like a magnified small "particle system" with quantum like fluctuations, which get smoothed out over massive distances. Again, some patterns you won't be able to see until you zoom all the way, or zoom further in. These variations tend to point to a "preferred" direction in space which seems to be "cooler" than other directions, suggesting a slight "tilt" to the Universe.

These slight variations are echoes of the quantum fluctuations experienced when the materials in our observable universe were in a volume of space very much smaller than a proton (or, to be more accurate, smaller than a proton's scattering surface). This is supposed to explain why the cosmic back ground radiation seems to be nearly the same in every direction but there are still small fluctuations that point to it's origin from a more highly structured, compacted universe that was in touch with itself on a very small quantum level.

There are several flaws to the standard Big Bang cosmology. For starters, no one can use the Big Bang Model to explain why our observable universe is still expanding at a constant rate directly proportional to the distance of objects, although theoretical variables have been put out there to explain it, such as "dark energy" and "dark matter." There is also no mechanism in the standard Big Bang theory that explains how the universe underwent such rapid inflation in its early phase, to the point where areas of space-time were moving away from each other faster than the speed of light.

However, a newer theory of Chaotic Eternal Inflation explains how this is possible. Eternal Inflation isn't exactly the easiest concept to explain, but it's basically a very scientific way of saying that everything always existed in potentiality, and a decay, or phase transition, of this energetic field of possibility eventually begat the universe through its interaction with the vacuum in which the universe was born. There were many possibilities, but only one went through the necessary "layers" of transition to produce the embryo of our Universe. If this theory is correct, inflation is the reaction to a tightly curved space-time, something of a "cosmic egg" that pre-existed the so-called Big Bang (or expansion). This means that there is no such thing as a true "beginning" or origin, but that our current observable world or "Universe" is the unfolding of an earlier, pre-existing, more STRUCTURED, more ENERGETIC, semi-stable (more a realm of possibilities than "real") form of space and time.

Throughout the world, indigenous people – from the oldest cultures of ancient China to the Andaman Islanders – have myths wherein a celestial Black bird gave birth to a cosmic white egg that hatched (split) to spill out the cosmos, often through mathematical processes that replicate the processes involved in the formation of matter, the planets, life on Earth, and the Original Man. Perhaps they were onto something?

THE CONSCIOUS UNIVERSE

C'BS ALIFE ALLAH

Whether we're talking about life or matter, there is no beginning nor ending. This is the nature of reproduction. It's cyclical. Some will say there is a "line" that can be drawn between inorganic and organic "life," yet there are still processes of reproduction on both sides of that line. Reproduction is all about a neverending cycle. And whether we're talking about planets, particles, or thoughts, everything progresses in tune with the original mathematical order, which we've been calling the Mind.*

This is the closest we get to identifying an "origin" for anything, and this origin itself is timeless. The form and structure of the instrinsic

* Refer to our sections on the Mind to gain more insight.

process of growth, development, manufacture and reproduction has no birth record, even if the specific instances and examples seem not to be ongoing. There is an underlying format, structure and order that is timeless, even when we can date the earliest appearance of a particular example.

The blueprint of organization is merely a structure for observed reality. Observation is the method by which things, from atoms to planets composed of atoms, are brought into existence. Conversely, ignorance is the method by which things are annihilated. Imagination is the method by which things are made or grafted. Thus, the Original man is the supreme being, being that his mental faculties are what shapes all that we see.* The Original man is the personification of the structure that creates,† makes, destroys, and maintains the balance of reality.

These creation and destruction processes are continuous and result in equilbrium. Where does the Original man come from? He comes from the foundation for his existence: his own self, which has no beginning or end. He is the structure and order personified. That mind or structure is all there was, when the universe stood still at time zero.‡ On earth as in the heavens above, the Original man directs energy from a state of potential to kinetic in a predetermined fashion, which is mathematics. This is because determination is measurement and vice versa. That is, once you have measured a thing, you have specified its destiny.

The Original man is the measure of all things. And the original man has taught this power of measurement to others who sought to imitate him. This is why quantum mechanics and cognitive science are now being synthesized into one science by Western scientists, with all of the above conclusions. Yet none of this is a new discovery to the Original man, who codified the same science into the cultural traditions of Africa, Asia, Australia, and the Americas.

The mind is the mathematical structure. The body is the universe. This is why there is no way to destroy the original man. We have a personal body that we control, but our entire body is the universe, which repli-

---

* More in depth discussions of observer centered reality are within our section dealing with quantum mechanics.

† Some will argue that due to the etymology of creation (the making of something from nothing) that words such as create, creation, etc have no place in scientific discourse and actually enforce a religious view of reality. Though matter and energy have "no beginning nor ending" thus cannot be "created" forms that this energy and matter take which may not have had any predecessors can be created. So certain structures are the result of creation.

‡ The theoretical point of the first singularity, where notions of time and space had no relevance.

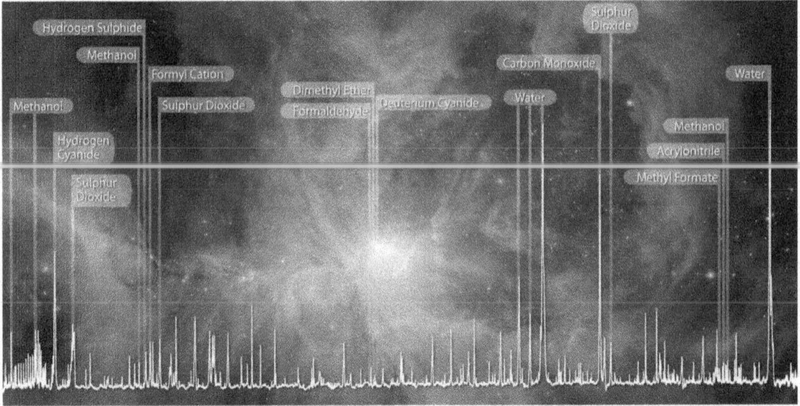

Most of the ingredients for life have been found in the Orion Nebula. By finely separating the spectrum of incoming light, astronomers are able to detect the chemical fingerprints of molecules like methanol.

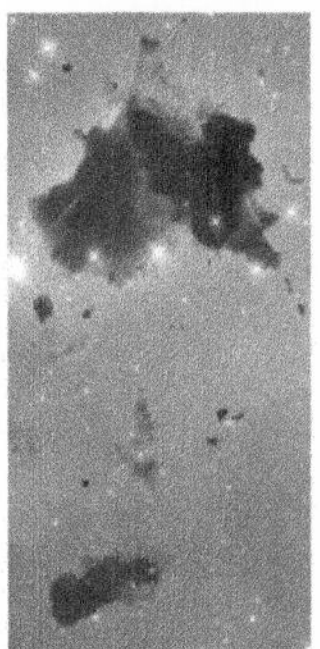

Bok Globules

Murchison Meteorite

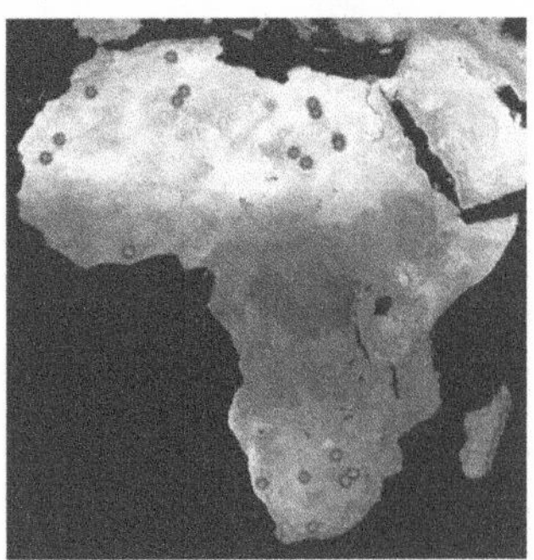

Impact Craters in Africa

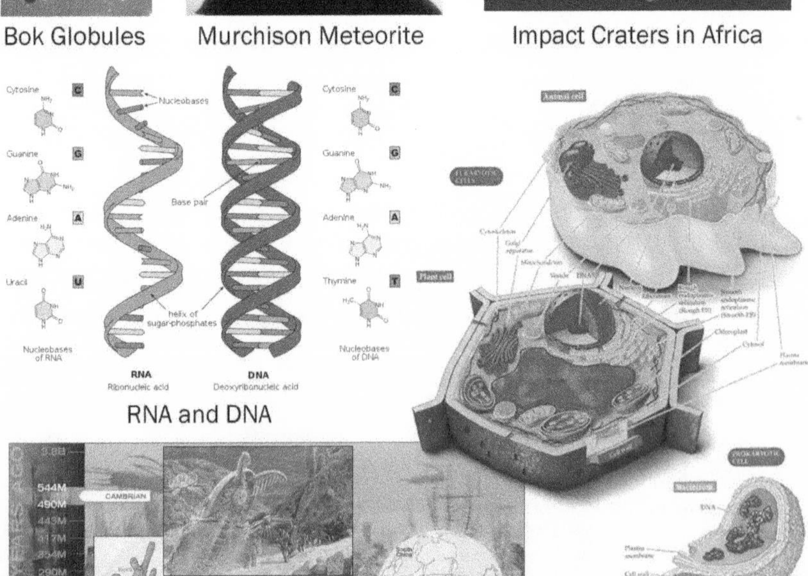

RNA and DNA

Eukaryote and Prokaryote Cells (above)

The Cambrian Period (left)

> DID YOU KNOW?
> Although the Earth is not the center of the solar system, we can rightfully say that we are at the center of the universe. Because the universe is a constant state of expansion, where everything is moving away from everything else, from our vantage point, everything is moving away from us. This has a lot to do with the unique structure of the observable universe, which is not a two-dimensional surface to say the least. Unless there are other intelligent observers elsewhere in our universe (and the odds are startlingly low) we are at the center of the observable universe, which extends 46-47 billion light years in every direction away from us.

cates the form and function of our personal body on a much larger scale. The same structure is replicated at various other scales. Thus the original man is the sun. The original woman is the earth. We are together within the atom. At all of these scales of existence, you will find reflections of the original man's family. The universe contains representations of family, and we reproduce the process of "creating a universe" when we create and maintain our families. As noted before, creation is an ongoing process.

"As far as we can discern, the sole purpose of human existence is to kindle a light in the darkness of mere being."
– Carl Jung

In other words, the universe is the original man's macroscopic body. The universe is an extension of our personal body. Everything is in the original man's universe or family (moon and stars). These are part of our universal body, which grows and takes shape through the process of generating families, children, and future generations. The cycle of life is the same in the heavens above and the earth below.

Man is the Center of the Universe

SUPREME UNDERSTANDING

Earlier in this book, we talked about the scales of existence. We know that the fractal patterns of mathematics in nature are replicated across every scale. Yet some of us can get lost in the grandeur of this massive Universe, or the unfathomable depth of the subatomic world. We look above and below and it seems like we're an insignificant speck. Yet it appears that it's just the opposite.

At the center of it all – as Joel Primack and Nancy Ellen Abrams point out in *The View from the Center of the Universe* – is us. Humans sit at the center of the scales of existence (which they depict in a cyclical "Cosmic Uroboros" swallowing its own tail) halfway between the scale of the subatomic world and the size of the visible universe. And the reason for this center-stage are tied directly to man's consciousness:

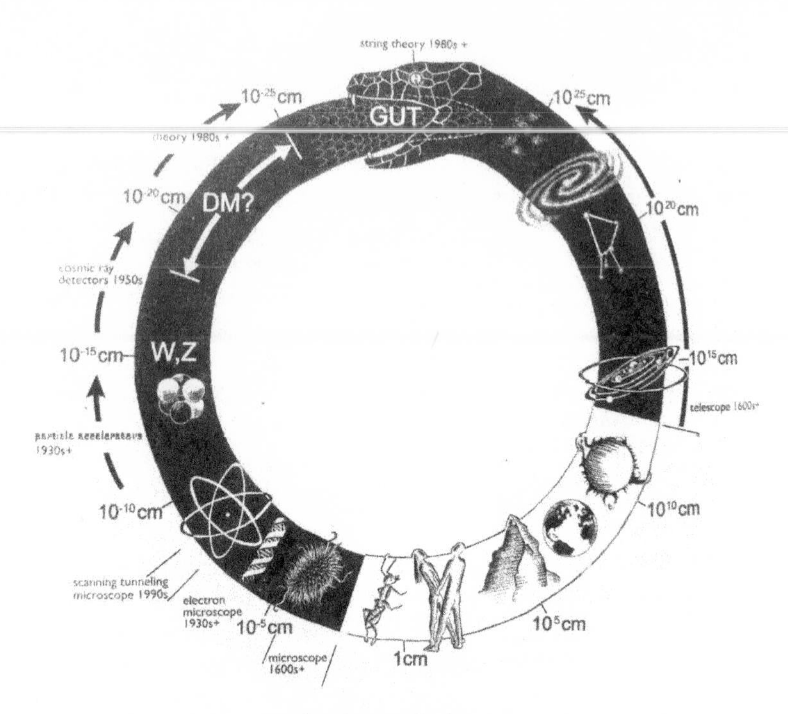

The Cosmic Uroborus
(from Primack and Abrams, The View from the Center of the Universe)

> Intelligent creatures in the universe have to be midsized. There is a kind of Goldilocks Principle: Creatures much smaller than we are could not have sufficient complexity for our kind of intelligence, because they would not be made of a large enough number of atoms. But intelligent creatures could not be much larger than we are, either, because the speed of nerve impulses – and ultimately the speed of light – becomes a serious internal limitation. We are just the right size. You might expect that a galaxy-scale intelligence would think at a fabulously deep level. But in fact the number of thoughts that could have traveled back and forth across the vast reaches of our Galaxy in its roughly 10-billion-year lifetime is perhaps the number an average person has every few minutes. The speed of light seems dizzingly fast to us, but on the scale of the visible universe it is excruciatingly slow and would prevent the parts of any large intelligence from communicating with each other in a reasonable amount of time…Thus the cosmos can't have a central brain or government. Thinking must be decentralized to make any progress, given the limit of the speed of light.[133]

In other words, Man is Mind. They continue, demonstrating that man's place and the emergence of his consciousness are no coincidence:

> Real thinking is the job of our size-scale...We humans exist on the only size scale where great complexity on the one hand and immunity from relativistic effects (like the speed of light) on the other are both possible. Our consciousness is as natural a blossoming on this special scale as a star is on its size scale or an electron on its own.
>
> Not only do intelligence creatures have to be approximately the size we are, but the universe had to be more or less the size and age it is to have produced us.[134]

Indeed, the Anthropic Principle argues that one reason why the universe must be so massive is that life requires it. But there are about a dozen variations of this theory. Some versions of the Anthropic Principle say that if the Universe weren't the right size and parameters for life (and man), then, well we wouldn't be here to talk about it. But the Final Anthropic Principle argues that the purpose of this Universe is to produce conscious man.[135] The Original Man, in effect, is the Alpha and the Omega, although he situates himself at the center.

GENDERATION (PART ONE)

C'BS ALIFE ALLAH

"We are connected to each other biologically, to the Earth chemically, and to the Universe atomically."
– Neil Degrasse Tyson

There are aspects of the universe (on the quantum, astronomical, chemical, and biological levels) that we refer to by gender. Some people are quick to object that gender should only describe strictly biological states (in reference to male and female animals or plants). Yet gender can be found at every scale of existence, and doesn't simply refer to "biological sex." It is in this manner that male and female gender can thus be applied outside of the realm of biology.

THE GENDER OF YIN AND YANG

In ancient ways of thought, like Taoism, there already existed a system of non-biological gender typology (the study or classification of types) in the Tao. The Tao itself is absolute, but – in its manifestations – there is a feminine aspect (*Yin*) and a masculine aspect (*Yang*). Looking at it from this point, explanations of biological sex and even social constructs of gender can be seen as outgrowths of systems like the Taoist *I Ching* ("Book of Changes"), which codified the manifestations of (and relationships between) Yin and Yang. The binary nature of the *I Ching* (based on Yin/Yang) has been noted in the West as early as 1703 in Gottfried Leibniz's *Explication de l'Arithmétique Binaire* (Explanation of Binary Arithmetic).*

Yin	Yang
Feminine	Masculine
Moon	Sun
Earth	Heaven
Cold	Hot
Wet	Dry
Soft	Hard
Empty	Full
Inside	Outside
Back	Front
Solid	Open
Near	Far
Slow	Fast
Small	Large
Rest	Act
Conserve	Spend
Dark	Light
Valley	Mountain
Water	Fire
Relaxed	Excited
Passive	Active
Receptive	Creative
Sad	Happy
Dense	Diffuse
Substance	Energy
Gravity	Levity
Mystery	Revelation
Absorptive	Radiant
Concentrative	Expansive
Local	Universal
Stillness	Movement
Curved	Straight

The *I Ching* has antecedents in the *Tablet of Destiny* of the Black Akkado-Sumerians of Elam-Babylonia (circa 2800 BC), and is also similar to the binary "divination" system of West Africa known as Ifa.[136] In fact, as we explored earlier, binary thinking systems can be found among indigenous cultures across the globe. Wherever we find these ways of thinking, we find conceptions of masculine and feminine? Why? Because we have always known about the binary nature of the universe, and how this ultimately manifests as human gender.

To be clear, when we refer to certain quantum or astronomical states as being masculine or feminine, it does not mean we think that certain atoms have a penis or certain planets have a vagina. It means that qualities that resonate with men and women appear in these objects and/or processes.

When we find the practice of science within ancient cultures, we don't see it isolated. We find it intertwined with philosophy, ethics, aesthetics, social etiquette, art, oral tradition, and other cultural norms. For example, one of the well known ways gender is applied outside of biological sex is in language. Grammatical gender is a system where classes of nouns trigger specific types of inflections in associated words, such as adjectives, verbs, and others. Most indigenous languages are gendered. While Old English (Anglo-Saxon) had grammatical gender, modern English lacks it, outside of its gendered pronouns. Grammatical gender is not the same as the biological and social notion

* Although many reductionists see the *I Ching* as simply a "divination text," its primary function is as a compendium of ancient cosmic principles.

of natural gender, although they interact closely in many languages.[137]

The English language does, however, retain some roots, that can tell us about gender on other scales. For example, there is a dichotomy of dyads to be found in any terms related to opposing polarities (repulsion) and their quest for union (attraction). So in the English language you have the prefix *di-* or *du-* (utilized in such word as divide, doubt, diverge, duel, distinct, etc) which is used to emphasize separation. You also have the prefix *tw-* (utilized in such words as twist, twin, twilight, etc) which implies a blending or union. Both prefixes are rooted in dyad ("twoness") language.

Yet perhaps, despite all our references to gender, our language lacks the means to accurately represent the masculine and feminine principles that are all around us. Much of this is due to social baggage. If we reduce our conceptions of gender to their most essential aspects, we can find gender at every scale. In doing so, we can find both the Original Man and Original Woman in their essences.

MACULINE ENERGY AND FEMININE PARTICLES

"This whole wide world is only He and She" – Sri Aurobino Ghose

In thousands of origin myths across the world, the first light (or energy) is masculine, and the dark matter it shines upon (to spark creation) is described as feminine. The dominant forms of energy in

A QUICK NOTE ON GENDER CONFUSION

Now, some of us are reading this section and trying to figure out how it's wrong. Despite the fact that nothing in this essay is meant to promote or put down one gender over the other, we are so screwed up over male vs. female that some of us will LOOK for that and find it where it doesn't exist. I know there are many women who are sick of the "dude laying on the couch," just as there are many men who can't stand the women who "don't need a man for anything." Let's accept this much: We all have issues, and some of us are further from our natural roles than others. But, clearly, our ancestors made it work for millions of years without debating gender roles. Most indigenous societies envisioned man as the active principle and woman as the receptive principle, merely because – despite whatever other social conventions they had – it was obvious that this was how human biology worked. You just gotta look down your pants (or skirt) to see Yin or Yang. It's not that complicated. We spend so much time trying to fight things on a philosophical level that we don't get anywhere. Everything masculine has some element of feminine, and vice versa. Every male has a little bit of estrogen, and every female has a little testosterone. If not, we'd be very screwed up. The bottom line is, there's no good to one without the other, and there's nothing "better" about being the Sun or the Earth. They're not fighting about who's who. The Earth isn't trying to set itself on fire so it can emit photons. Why? Because she loves who she is, and the role she plays in the bigger picture. What about you?

our past are the photons, gluons and gravitons. These are particles of radiation, meaning energy. Radiation is an aggressive agent that travels at terrific speeds (186,000 miles per second), no matter or when.

The universe was at a high temperature in the past because it consisted of a great density of these high energy particles of energy, in a very small pocket of space. As our universe expanded, other particles started to emerge when these radiation particles decayed and turned into particles with mass. Later particles resulted when energy (as photons) were reflected by matter (particles with a mass), or re-emitted by matter. Stop for a minute and consider where you see the correspondences between cosmic gender and human gender. There is a reason why the ancients described the early cosmos in terms of gender. Because gender was there.

It is for the same reason that the word matter comes from *mater* meaning "womb," and is related to such feminine words such as matrix, mater, matriarch, etc. while the ancient Nile Valley personification of solar energy has the name Ra which makes me think of "ray" and "radiation." But Ra was not alone. Most ancient personifications of solar energy (sun deities) were male, while ecological forces or processes were female.

Particles are the essential complement to the photons that would otherwise pass through each other indefinitely in a vacuum of nothingness. Without these particles, it wouldn't matter how bright the photons were, because they'd have nothing to reflect their light, so it would still be total darkness. They can also "hold on" to the energy given to them when they are created from radiation particles. They can "absorb" radiation particles and re-emit or "give birth" to other radiation particles after taking in enough energy. See the connection?

The relationship between radiation particles and particles with a rest mass (matter) corresponds to the gender relationship between men and women. Historically, men have been more engaged in energetic activities and travel more often than women.* This is whether you look at ancient migrations or indigenous hunting expeditions (not so

---

* Though there are many early migrations where male and female both traveled, it should be noted that it was generally males who expanded the horizon. This is noted in hunter-gather societies where males tracked game and in sedentary societies who had bursts of ages of exploration. This can also be seen in the Egyptian god Osiris (or Ausar), who Cheikh Anta Diop said was the Nubian who led the expedition north to settle Egypt. Meanwhile, the name Isis comes from Auset, meaning "throne" (as in Earth). In myth, Isis represented fertility and nature, while Osiris represented virility/agency and the cycles of life, death, and rebirth.

much if you're looking at some dude laying on the couch right now). Traditionally, women interact with other women, forming (and breaking) bonds. The male principle represents radiation, or emission, whether it is light, sperm, or knowledge. The female principle represents reception, nurturing the ideas that affect the universe mentally and physically. Just as the Sun emits light and the Earth bears life, the male sperm travels to provide the proteins in the genesis of a biological organism while the sedentary egg provides the vitamin and mineral landscape for its growth and development.

This is just the surface. There are many other analogies between the quantum universe and male and female. The Yin (male) and Yang (female) principles can be found in the interplay between energy and matter, waves and particles, electrons and protons, man and woman, the Sun and Earth, and even at the level of DNA.

There is a negative coil of DNA that corresponds to the wave aspect of matter (Yin), and a positive coil of DNA that corresponds to the particle aspect of matter (Yang). DNA is bonded by four amino acids (A, C, G, T), yet the T is replaced by the U when the genetic code is manifested. There are four forces that bond the universe, which correspond to the variations of Yin and Yang, and to our genetic code. The A acid corresponds to the negative polarity in quantum mechanics (passive yang). The G acid corresponds to the positive polarities (active yang). The U acid and C acid correspond to the two minimum values of the state vector in wave mechanics (passive and active yin). Gene mutations and energy transformations correspond to moving yin and moving yang. As the *I Ching* has 64 hexagrams that come from the four basic units, the genetic code has 64 "words" that come from the permutation of the four acids.

Some will say all of the above are arbitrary analogies. Others will say that they are evidence of mathematical structures enveloped on every level of existence. And finally, some will ponder Dr David Bohm's statement from *Wholeness and the Implicate Order:*

> Thus in a certain sense we "make" the fact. That is, to say, beginning with immediate perception of an actual situation, we develop the fact by giving it further order, form and structure.

We'll revisit gender (in a more biological sense) in a future chapter. For another look at the pre-biological science of gender, see "Quantum Physics and the Black God" by Divine Ruler Equality Allah, and "Optics" by the same author, in *Knowledge of Self: A Collection of Wisdom on the Science of Everything in Life.*

An Introduction to Geology

ROBERT BAILEY

Geology (from *geo* meaning "earth" and *logos* meaning "study") is the science which studies the history, processes, and physical composition of the Earth. It allows us to understand questions like how the Earth came to be, how it evolves and how to manage and locate our natural resources. Understanding earth processes like earthquakes, floods and landslides allows us to better predict future dangers and prepare for them. Geologists use a combination of field and lab work, as well as numerical modeling, to arrive at their findings.

Ever since we started shaping the earth with our tools, we've had some type of geological knowledge. The constant presence of rock tools and arrowheads suggests that they were deliberately made with a consistent technique. It also means that these early people had knowledge of where to obtain the rocks as well as which ones were suitable for specific uses. The earliest known mining activity in the Americas (12,000-10,500 years ago in Chile), shows early hunter-gatherer communities had well-developed geological knowledge.[138] Even today, indigenous people throughout the world have strong, systematic knowledge of soils and complex classification systems. For example, the Baruya people of Wonenara, New Guinea, have names for more than 20 different types of soil and know the comparative qualities of each.[139] Throughout the ages, there have been additional contributions to the science from Ancient Egyptian, Greek, and Islamic scholars. Modern geology was birthed near the end of the 18th century, with James Hutton hailed as one of the first modern geologists.

How the Earth was Made

SUPREME UNDERSTANDING

So how did the Earth come to be? As you read earlier, the expanding universe contained a distribution of particles that gradually clumped together into larger and larger masses with more and more gravitational force, eventually becoming stars and clusters of stars. Many of these stars, with their strong gravitational pull and steady emission of the basic elements, gave birth to planets.

Recommended Viewing: How the Earth Was Made

This 13-part series from the History Channel combines spectacular on-location footage, evidence from geologists in the field, and clear, dramatic graphics to show you how the forces of geology formed our planet. If you want to see what the Earth looked like when it was Red, Black, and Green, find Part One.

How? These stars were often surrounded by discs of particles sent out into space by the Big Bang and the byproducts of older stars that went supernova (exploding) and sent out a shower of particles created by such nuclear reactions. The disc of spacedust circling our Sun was composed of elements like iron, silicon, magnesium, aluminum, carbon, and oxygen. Based on recent studies, the complex organic molecules necessary for life may have formed in the protoplanetary disk of dust grains surrounding the Sun before the formation of the Earth.[140] Slowly, these grains came together and collected into clumps, then chunks, then boulders, then planetesimals, and then planets. The early Earth accumulated size, becoming increasingly hotter at its center, until the inner part of the protoplanet was hot enough to melt the heavy metal elements in this primordial cement mixer. These heavy, liquid metals began to sink to the Earth's center. This created the Earth's iron core, which separated from the semi-liquid mantle of hot molten rock around it. This all happened within the first 10 million years of the Earth's formation, setting the stage for the layered structure of Earth as well as its magnetic field.[141]

Our Earth experienced another dramatic event in its early years. While our planet was still hot and not yet fully formed, it lost a large chunk of its mass when it birthed the moon. Scientists have established that the moon was once a part of the Earth, but was blown from the Earth during its infancy. The question of HOW this happened is still a subject of debate. Many scientists believe that the pieces that became the Moon were knocked off the Earth when it was hit by another celestial body the size of Mars. In a recent alternative, Dutch geophysicists argue that it was created not by a collision, but a runaway nuclear reaction deep inside the Earth. Nuclear geophysicist Rob de Meijer and petrologist Wim van Westrenen propose that a georeactor – like the 2-billion-year-old nuclear reactor found in Gabon, West Africa[142] – went supercritical 4.5 billion years ago and blew 1/4th of the Earth's contents out from within. This georeactor, deep within the Earth (at the boundary between the liquid outer core and the inner mantle), would have launched out all the elements present the Earth's core, mantle, and early atmosphere (which the other theories can't explain). Then, theorizes van Westrenen, "a ring of debris formed around the Earth, out of which the Moon then gradually coagulated."[143] Our present day Moon continues to drift away from the Earth, ever since its formation 4.5 billion years ago, at a rate of 4 cm per year. It is one-fourth of the Earth's size, suggesting that the Earth – currently 24,896 miles in circumference at the equator – was once closer to 35,000 miles in circumference.

DID YOU KNOW?
Not only do women "sync" with the Moon, when women are closely associated – living together under one roof, for example – their menstrual cycles also synchronize with each other.[145] Association with males also affects the menstrual cycle, generally shortening and regularizing them (to increase fertility).[146]

What the Earth lost in mass, however, it gained in other areas. As we'll see when we explore the later history of the Earth, the Moon's role is a regulatory one. Without the moon, there would be no gravity to pull up the tides, no steady axis tilt to maintain our seasons, and possibly no life on Earth, as the Moon protects us from many of the objects that come barreling our way from the depths of space. That's why it wears so many scars! The Moon lacks life so that the Earth can have it!

The Moon is so intrinsically tied to the cycles of the Earth that the woman's menstrual cycle coordinates with the lunar cycle. Those who cycle as often as the moon (28-29 day cycles) tend to be the most fertile. The cycle even coordinates with the phases of the moon, with full moon being the time of greatest fertility. In a 1987 study of the subject, Winnifred Cutler and colleagues noted that this finding vindicates the many ancient cultures who appeared to know this fact before Western science: "Historical indications that fertility rites were scheduled with consideration for the phase of the moon may have been reflecting accurate perceptions which we have yet to discover."[144] In fact, the Moon's cycle doesn't just correspond to the menstrual cycles of women – it is intrinsically tied to the cycles of nature itself.

Within 150 million years, a solid crust with a basaltic composition had formed. About 3 billion years ago the planet acquired water, possibly through a chemical process on the planet, or through bombardment by a shower of meteorites that already contained water. After several million years of rain and condensation, the earth was covered with water. The volcanic Earth continued spewing lava on the surface. This lava cooled upon hitting the water, producing Earth's earliest landmasses.

The water was so full of iron that it was green, and the atmosphere was so dense with carbon dioxide that it was red. The igneous landmasses, like most volcanic rock, were black. Thus, 4 billion years ago, the Earth was red (the sky) black (the land) and green (the water).

About this time, superconducted water falling between the cracks of the crust helped produce granite from the basalt surface. Granite is lighter than basalt. Thus the continents, which were formed from granite, literally float over the crust. And almost as soon as the Earth's circumstances permitted, we have evidence of life in stromatolites,

DID YOU KNOW?
Most animal cells (including ours) contain a pair of centrioles, tiny turbine-like organelles oriented at right angles to each other that replicate at every cell division. Yet the function and behavior of centrioles remain poorly understood.[147] Could these spinning "turbines" in our cells parallel the spinning electromagnetic "motor" in the Earth's core? And could either of these phenomena be tied to the "bipolar" spin of elementary particles themselves?

formed by cyanobacteria that flourished on both land and sea. But here's what's deep. Life might have been here before the continents. In fact, as you'll see in the next chapter, life might have MADE the continents.

Soon after, micro-organisms living in the water produced so much oxygen they created the ozone layer, which would protect future life as it moved onto land. Other forms of early life set the stage for future developments. For example, the spread of a fungus across the earth helped both the spread of plantlife and the development of soil from its decay. And so it went on, with both population explosions and mass extinctions ultimately leading to the emergence of modern man. Without everything that came before us, there could be no us. But this shows us the cycle of life, not a pyramid. Plants die so we can eat them, but we die so plants can eat us. That's what we learn when we study the carbon cycle, the nitrogen cycle, and the phosphorus cycle. Plants don't only rely on the free energy provided by the Sun and water, they also thrive off nutrients provided by the death of the same animals that eat them (herbivores), and the animals that eat them (carnivores and omnivores)!

THE ELECTRO-MAGNETIC UNIVERSE

SUPREME UNDERSTANDING

> "From the smallest particle to the largest galactic formation, a web of electrical circuitry connects and unifies all of nature, organizing galaxies, energizing stars, giving birth to planets and, on our own world, controlling weather and animating biological organisms. There are no isolated islands in an electric universe." – David Talbott & Wallace Thornhill, *Thunderbolts of the Gods*

If you've ever done that science experiment where you wrap a wire that's hooked up to a battery around a nail, creating a magnetic nail, you know that when an electric current passes through a metal wire, a magnetic field forms around that wire. Likewise, a wire passing through a magnetic field creates an electric current within the wire. This is the basic principle that allows electric motors and generators to operate. And as you know if you've ever used a compass (a real one, not the one on your GPS), the Earth is magnetic. The Earth has an electromagnetic field that plays a major role in nature, evolution, and human consciousness. As we'll see, this electromagnetic field can be traced back to the Original Man and Woman.

THE MOTOR THAT DRIVES EARTH'S MAGNETIC FIELD

The Earth's diameter is 7,926 miles (at the equator; it's slightly less from north to south pole). Many of us think that the Earth's center is a hot gooey mess of molten rock, like one of those nasty chocolates with the cherry filling. But about 3,200 miles into our planet's center, there's actually an inner core made of solid iron, about 1500 miles in diameter. Surrounding this is a fluid outer core, which in turn is wrapped onion-like by the mantle and outer crust. But this inner core is rotating faster that the mantle and crust. It's what some scientists call "the motor of the world." And it's literally like one of those little motors you might have yanked out of one of your toys when you were a kind. You know, the kind where the inside spinning part is wrapped in wire and outside is a set of curved magnets? Oh you didn't take your toys apart like that? Maybe it was just me. Either way, it's the same science, apparently: As Song and Richards explain:

> Electric currents of about a billion amps flow across the boundary between the solid inner core and the fluid outer core that lies around it. In the presence of the Earth's magnetic field, these currents generate massive forces that tug on the inner core. And because the outer core has a relatively low viscosity, the inner core can spin freely.[148]

Translation: All this electromagnetic energy keeps the motor spinning, with the liquid outer core acting as lubrication. And this motor recharges itself. As we learn in the NOVA documentary *Magnetic Storm*:

> The liquid metal that makes up the outer core passes through a magnetic field, which causes an electric current to flow within the liquid metal. The electric current, in turn, creates its own magnetic field – one that is stronger than the field that created it in the first place. As liquid metal passes through the stronger field, more current flows, which increases the field still further. This self-sustaining loop is known as the geomagnetic dynamo.[149]

This "geomagnetic dynamo" also depends on the rotation of the planet to stay charged:

> The so-called Coriolis force also plays a role in sustaining the geomagnetic dynamo. Our planet's spinning motion causes the moving liquid metal to spiral, in a way similar to how it affects weather systems on the Earth's surface. These spiraling eddies allow separate magnetic fields to align (more or less) and combine forces. Without the effects caused by the spinning Earth, the magnetic fields generated within the liquid core would cancel one another out and result in no

DID YOU KNOW?
Despite the "idea" of North being "up," and South pole being "down," there's no "real" up or down when it comes to the Earth. It's all a matter of perception, because our only sense of down is being held down to the Earth's surface by gravity. Otherwise, the convention of making the Northern hemisphere the top of the world came from European mapmakers. These mapmakers have notably also distorted the map so that, not only are European nations on top, they are depicted as much, much, larger than they really are. These flaws in traditional maps have been corrected by more accurate maps like the South-Up Map and the Peter's Projection Map, but the bogus Mercator Map remains the one that everyone uses.

distinct north or south magnetic poles.[150]

And every so often, the Earth's magnetic polarity reverses. In other words, every compass pointing north would point south, as the south pole had become the new magnetic north pole. Geologists and paleomagnetists (yes, that's a real type of scientist) can look at the iron bands in layers of rock to tell when such magnetic reversals occurred. So how often does it happen? Reversals happen, on average, once every 250,000 years, and they take hundreds, or thousands, of years to complete.

But, as science writer Peter Tyson notes, there's still much to explain:

> It might sound as if scientists have all the answers regarding magnetic reversals. But actually they know very little about them. Basic questions haunt researchers: What physical processes within the Earth trigger reversals? Why do the durations and frequencies of both normal and reversed states seem random? Why is there such a disproportionately long normal period between about 121 and 83 million years ago? Why does the reversal rate, at

A QUICK NOTE ON THE AURA

Keeping things as simple as possible (for now), everything in this universe is a form of energy. Even matter. It's all just moving at different frequencies. All of these different frequencies occur along what's known as the electromagnetic spectrum. This spectrum covers all the wavelengths from radio waves to gamma rays, with the rainbow of the visible light spectrum taking up only a small space in the center. We'll get to all those different forms of energy at one point or another in this book, but let's just talk about the electromagnetic spectrum in general. You see, we live in an electromagnetic universe. Some things have a stronger electromagnetic field than others. What people typically call the "aura" is the electromagnetic field that surrounds the human body. What people typically call the chakras are energy centers within the human body. Are these things visible to the human eye? Not likely. Some people may claim to be able to "pick up" on the fields of energy emitted by the human body, but our eyes just aren't designed to pick up that end of the spectrum. Maybe if you were a bird or something, but not humans. Does that mean its not real? There is certainly a electromagnetic field to the human body, and to many of the anatomical locations identified with the chakras (particularly the brain), but anyone telling you they can see it, is either saying they can "feel it" with some extra-sensory perception, or they're just trying to sell you their vision.

DID YOU KNOW?
Saturn's rings have dark radial spokes that wax and wane about every 621 minutes. This is close to the rotation period of Saturn's magnetic field, which suggests that the rotating magnetic field of the planet interacts with the particles in the rings, forcing the density or reflectivity changes that we see as spokes.[153]

least during the past 160 million years, appear to peak around 12 million years ago?[151]

One interesting theory about the cause of geomagnetic reversals was proposed by geophysicists Richard A. Muller and Donald E. Morris. They suggest that when the Earth is struck by an extraterrestrial object (meaning asteroid, not spaceship), the results of the impact can be great enough to affect the rotation rates of the crust and inner core, throwing off the geomagnetic dynamo. It's just like hitting a swinging pendulum with a penny, I imagine. This leads to a slowdown or weakening, sometimes followed by a reversal, once the dynamo gets back to spinning at its regular rate. What's especially deep about this theory is that, basically, the Earth's motor has a battery that can recharge itself.[152]

But the Earth is not alone in this tendency to flip-flop. The sun's magnetic shield appears to reverse its polarity approximately every 11 years. Even our Milky Way galaxy is magnetized, and experts say it probably reverses its polarity as well.

And it appears that the sun plays a major role in the earth's global electrical circuit. Not only do solar flares affect the Earth's magnetic field, measurements of atmospheric electrical currents have revealed that

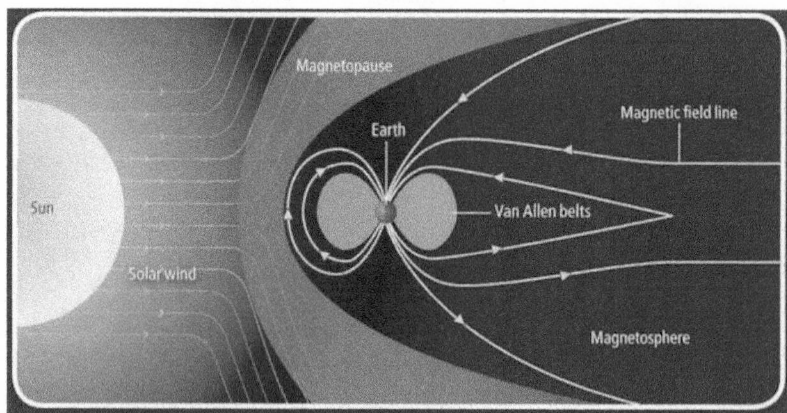

solar flares also stimulate large surges in the flow of electricity from the atmosphere to the earth's surface. What's deep is, this one-way flow of unseen electricity has to be balanced by thunderstorms somewhere else on the planet, suggesting that thunderstorms are often related to solar activity.[154] Seismologist Surendra Singh has also found evidence that geomagnetic activity triggered by solar flares hitting the

DID YOU KNOW?
One example of our planet's electromagnetic activity is the Schumann Resonance Phenomena, which describes the electromagnetic field in the Earth's atmosphere. Nikola Tesla proved Schumann's theory correct and led to him developing the idea of wireless energy transmission...in 1905. But don't be jealous of Tesla's genius. As you'll see in Volume Two, we had wi-fi figured out 5,000 years ago.[156]

Earth's magnetic field plays a role in microearthquakes.[155] Yet while a severe weakening or disappearance of the magnetic field would leave us open to harmful radiation from the sun, we're not sure if these "flips" inflict any direct, lasting damage.

THE MAGNETIC MOUNTAIN

The magnetic features of the Earth may seem invisible to us, but they're perceptible to sharks. A 1995 study focused on hammerhead sharks in the Gulf of California, where you'll find an underwater mountain named Espiritu Santo, also known as the "magnetic mountain." The study found that the sharks could find the magnetic mountain (which they circled for hours during their resting period), with incredible accuracy, often precisely following the same paths in what seemed a featureless ocean. Researcher Peter Klimley discovered that the sharks were actually using the magnetic field, following paths coincident with lines of high magnetic gradient. This "geomagnetic" sense is also found in some birds and mammals that have small particles of magnetite in their bodies, but scientists still don't know how they might be incorporated into a sensory organ. In fact, hammerheads and many other sharks are extraordinarily sensitive to electrical fields, responding to fields as low as 10-8 volt/cm. William Corliss has speculated, "Perhaps there's some interaction occurring between the shark's fractal "navigational programming" and the Earth's grid of electrical and magnetic fields."[157]

Birds navigate using some of the same senses. So it seems birds, sharks, and all kinds of other creatures can navigate without their physical eyes, using the Earth's geomagnetic field. Wouldn't it be amazing if people had these senses? Well, who said we didn't?

THE ELECTROMAGNETIC BRAIN

Our brains are electromagnetic. You probably have heard that your entire nervous system (of which your brain is a part) runs off of electrical signals being transmitted between neurons, synapses, and receptors. Your brain is conducting electricity right now, enough to power a lightbulb. In fact, your whole body conducts electricity. But don't worry, your body doesn't produce enough electricity to justify that "human battery" scenario from *The Matrix*.

Artist renderings of the early Earth

Magnetic fields of the Sun and Earth with related phenomena as seen on Earth

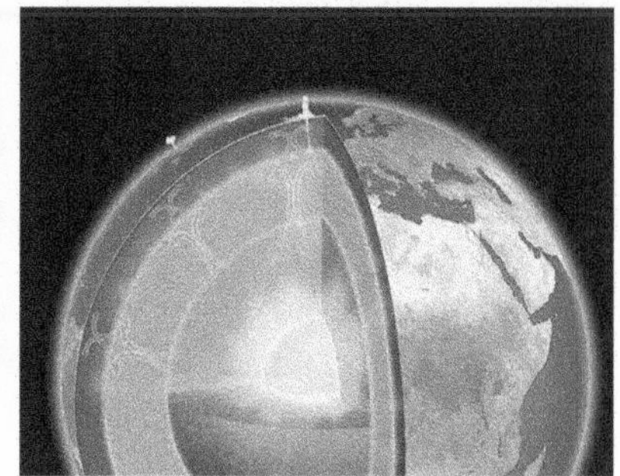

Cutaway view of the Earth

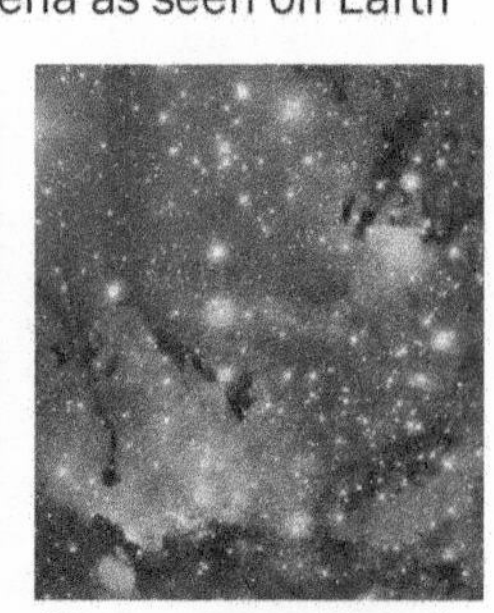

Hourglass Nebula Cygnus Loop Nebula

Two fractals in nature

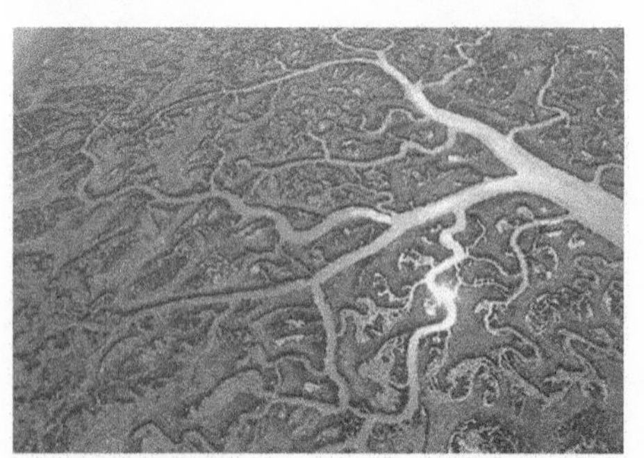
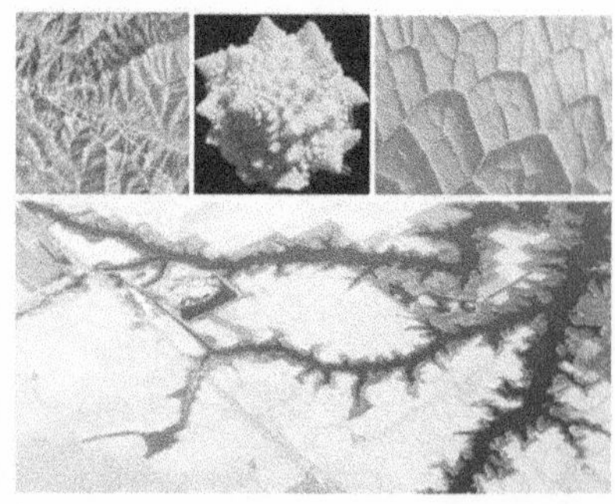

Other fractals in nature

DID YOU KNOW?
Homing pigeons use two different sensors: the sun's compass on sunny days and magnetic field GPS on cloudy days. A 1979 study discovered that pigeons, like sharks, have magnetite crystals in the back of their heads. The same tissues contained yellow crystals likely made by the iron-storage protein ferritin, which was probably used in the biological synthesis of the magnetite. These ferritin may be relics of magnetic bacteria, making them yet another ancient bacteria to have conjoined with another species.159

What's most important here is the relationship between your brain and the electromagnetic universe we live in. As you may have read in our Appendix, our brain waves move at different frequencies. Some experiments show connections between these brain states and resonant electromagnetic waves, raising the possibility that the human brain has evolved to be "in tune" with Planet Earth. A study published in the journal *Perceptual and Motor Skills* suggests that geomagnetic activity is related to Extra-Sensory Perception (ESP, also known as "psychic powers") and psychokinetic mental activity.[158]

In 1987, Canadian neurophysiologist Michael Persinger published research implicating the influence of a magnetic field on ESP. In his study, Persinger exposed a subject's temporal lobes (at the sides of the head), to a magnetic field of one, four, or seven pulses per second (measured in Hz). Then he showed the subject Zener cards, a common tool in tests of ESP. The 1 Hz and 7 Hz magnetic fields did nothing to help exceed the "random chance" rate of 20% correct guesses. But 4 Hz pushed the subject's correct guess rate up to an extraordinary, and highly unlikely, 50%. You'll see this 4 Hz rate again in the piece on Shamanism in Volume Three.

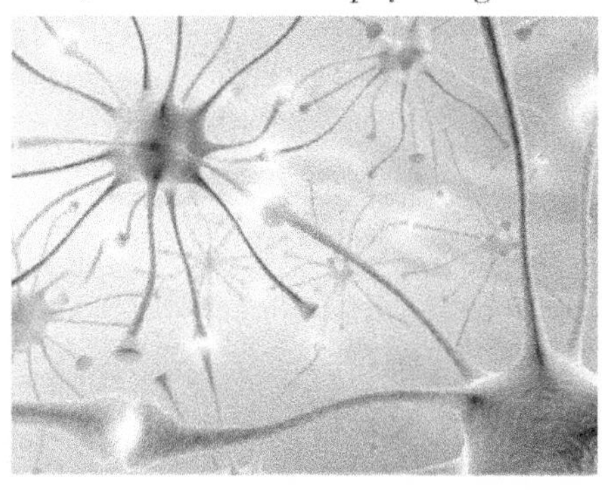

So, clearly, the brain's electromagnetic field is definitely susceptible to outside influences. In April 2010, neuroscientist Liane Young and her colleagues at MIT and Harvard University reported that they had altered people's moral judgments using transcranial magnetic stimulation, a procedure that briefly disrupts neural processing with a magnetic field induced by electric current. Young asked each of 20 volunteers to judge 24 scenarios that involved morally questionable behavior. Young saw that, when subjects were zapped, they were more likely to focus on the outcome (nobody died) than on the intent

(Grace tried to poison her friend).[160] Hmm...just consider the implications of such a finding – good and bad.

Michael Persinger also studied the effects of minute electromagnetic signals, such as those observed in the geomagnetic field, upon human consciousness and perception:

> Contemporary neuroscience suggests the existence of fundamental algorithms by which all sensory transduction is translated into an intrinsic, brain-specific code. Direct stimulation of these codes within the human temporal or limbic cortices by applied electromagnetic patterns may require energy levels which are within the range of both geomagnetic activity and contemporary communication networks. A process which is coupled to the narrow band of brain temperature could allow all normal human brains to be affected by a subharmonic whose frequency range at about 10 Hz would only vary by 0.1 Hz.[161]

This would help explain a phenomenon known as "Bushman Direction Finding." BDF is better than GPS. As noted in Encounter Magazine, a South African publication:

> It is claimed that the Bushmen has a sixth sense. They have a very highly developed and uncanny sense of direction, far superior to an European or African. A Bushman may turn, circle and zigzag for hours when hunting, but when returning to camp he will head exactly in the right direction.
>
> A tribesman was tested by blindfolding and leading him through various paths for several hours. When the cloth was removed, he pointed to the exact direction of his camp. Children too, never lose their way. Together with this "guiding instinct," they apparently see a vision of the trail ahead.[162]

Activity: Connect the Dots

The electromagnetic channels of our bodies are known to the Chinese as the Chi meridians used in acupuncture, Traditional Chinese Medicine, and the pressure points of many Eastern martial arts. The Chi meridians correspond very closely to the human nervous system, but Traditional Chinese Medicine recognizes pathways that Western medicine does not. Considering that the Chinese have been a relatively healthy people for the past 5,000 years (until Westerners introduced the American diet), they may know something Western science doesn't. It is the work of future researchers to connect the dots between the human body's electromagnetic field and the binary energy of Yin and Yang, and the pathways by which Chi travels through, and beyond, the body. In my experience with traditional acupuncture, we'd hear words like "low red blood cell count" from Western doctors, and then "your Chi is low" from the acupuncturist. So, dig into the scientific journals and textbooks that you can browse online (at scolar.google.com and at books.google.com) to see what the research has to say about the two systems. Perhaps you'll be able to connect some dots that have yet to be connected!

DID YOU KNOW?
Another experiment found that 5 Hz electromagnetic fields could double the inductive abilities of study participants. That is, these theta wave frequencies increased the amount of inferences they could make listening to a narrative, suggesting they were better at seeing the unseen or reading between the lines.[163]

This might sound like hype to some, but this ability was confirmed by an Australian study. In 1987, 35 subjects assembled at a university, where they were blindfolded and driven along a winding route about 12 miles long to a spot over three miles from the university. The sun had set and audible cues were suppressed. Very few of the subjects could guess the direction of the university, the spot from which the journey began. However the subjects were extremely accurate in "guessing" the directions of their homes. What sense of ours is at work here?[164]

This may have something to do with the human brain containing biomagnetite, which could give us (and a few other animals) an "electromagnetic sense" providing a link between our "electromagnetic" brains and all the universe's other forms of electromagnetic phenomena.

It appears that this magnetite may have played a significant role among Earth's earliest life. Inorganic magnetite is octahedral, while magnetite manufactured by bacteria take the form of cubes, hexagonal prisms, or noncrystalline teardrops. When we look at the magnetite in marine sediments, they seem to have been organically manufactured. This suggests that industrious bacteria have been busy producing magnetite ever since "lowly" life forms appeared over 2 billion years ago.[165] Now, if our single-celled ancestors were making magnetite that long ago, it makes me wonder how much they made, how it affected the magnetic field of the Earth, and how those facts are connected to all the facts we've learned above. William F. Corliss once suggested:

> [What if] magnetic bacteria, as agents of Gaia, actually constructed the earth's magnetic field for the specific purpose of erecting a shield against space radiation, and thereby allowing the development of more complex life forms on the planet's surface?

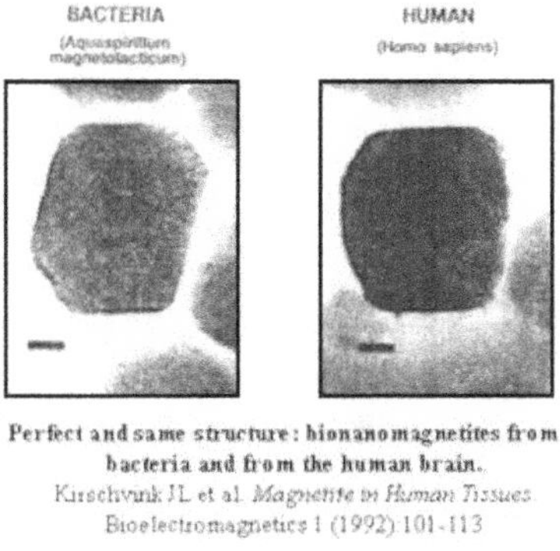

Perfect and same structure: bionanomagnetites from bacteria and from the human brain.
Kirschvink JL et al. *Magnetite in Human Tissues* Bioelectromagnetics 1 (1992) 101-113

And where is this biomagnetite found in the human brain? Few studies have been completed, but it appears the root of our connection to the electromagnetic universe is found in one of our oldest, yet least understood, organs: the pineal gland. The pineal gland, buried deep within our brains, is sensitive to external magnetic fields, including the Earth's geomagnetic field.[166] The uncalcified pineal gland may be the "port" by which we and the Bushmen "sync" our brains with the electromagnetic field of the universe.

BLACK AND WHITE

SUPREME UNDERSTANDING

There's a reason why the symbol for Yin and Yang is black and white. This duality represents the binary relationship we find in all perceived reality. As the Upanisads and many other indigenous traditions notes, there is no such duality at the source of it all. Absolute reality is absolutely unified. There are no parts or partners at that level. But as soon as anything can be considered a part of this physical universe, we've got duality. For example, you have the nature of life and matter, which – as we've seen throughout this book – is essentially Black. That is, all physical particles have a property of black body radiation,* and if not for their varying rates of reflecting light, everything would absorb all light and it would all be pure black. But some particles reflect light, so we get to see the other side of this duality: light and energy. Light and energy are essentially white.† The interplay between energy and matter forms the perceived reality of nearly everything we see in this physical universe. From the electromagnetic properties of particles, people, and planets, to the emission of photons (light) that occurs in the Sun, our

A QUICK NOTE ON GOING LEFT 4 TIMES INSTEAD OF GOING STRAIGHT

Recently, I came across a Facebook post discussing the electromagnetic field of the heart. The post itself had some interesting science to it, some of it based on the research of the Institute of HeartMath. But a commenter on the post went left with it (as we tend to do, when we're not grounded in the science of how things work) and asked "Is there a way to tune your magnetic field to protect against the world's poisons?" This was my response: "There are levels of effects based on scale and frequency. You can't use a magnetic field to reverse the effects of eating pork. You can do a full body detox and change your eating habits. You can't change your eating habits and expect your financial success to improve. And so on. If two things are on a similar frequency of the electromagnetic spectrum, I'm sure one can affect another, but too many of us want to use subatomic consciousness to move fifty pound objects. In plain English, there are very straightforward ways to deal with problems of any degree."

So, with that said, the bottom line, is don't go left four times instead of going in the most straightforward direction available. Because, if you think about it, you may find some interesting diversions along that route (or get lost), but you'll always end up further behind than if you'd gone straight. With such immediate problems before us, we can't afford to get too far behind.

skin, and our brains. This duality, as C'BS notes in "Genderation," forms the basis of gendered relationships as well. Mind and Matter, Light and Dark, Male and Female, and so on.

REVIEW

Ar this point, you've seen that there is a structured evolutionary process to the formation of our Universe, which we simply call mathematics, or the blueprint. Yet you may still have questions like "What IS this Universe, exactly?" and "What is it FOR?" We describe some aspects of it as conscious, others as quantum, and others as electro-magnetic. These are all different frequencies of the same eternal essence. We know, from the previous chapters, that the essence of it all is consciousness, which we call the Mind. But if energy became matter, and matter became the planets, what comes next? And why? In the next chapter, we'll show you how life came to be, and why the evolution of life was the most recent development in the process of, as the ancients would say, "the word becoming flesh."

skin, and our brains. This duality, as C'BS notes in "Genderation," forms the basis of gendered relationships as well. Mind and Matter, Light and Dark, Male and Female, and so on.

REVIEW

Ar this point, you've seen that there is a structured evolutionary process to the formation of our Universe, which we simply call mathematics, or the blueprint. Yet you may still have questions like "What IS this Universe, exactly?" and "What is it FOR?" We describe some aspects of it as conscious, others as quantum, and others as electro-magnetic. These are all different frequencies of the same eternal essence. We know, from the previous chapters, that the essence of it all is consciousness, which we call the Mind. But if energy became matter, and matter became the planets, what comes next? And why? In the next chapter, we'll show you how life came to be, and why the evolution of life was the most recent development in the process of, as the ancients would say, "the word becoming flesh."

Examples of Acanthophracta, a class of Radiolaria, a type of Eukaryote dating back to the Cambrian Period. They produce distinct mineral skeletons.

THE BOOK OF GENETICS

THE MEANING OF LIFE

> *"What Heaven has given is called the Law of Nature. To follow this natural way is to follow the Way. To nurther this Way is called learning. The Way must not be left, even for a moment. If it could be left, then it would not be the way." – Confucius, The Doctrine of the Mean*

Throughout this text, you've observed that everything proceeds according to a process. The blueprint of growth, order, and direction governs both organic life and inorganic matter, pulling both together through the property of "self-organization" and pushing both forward through an evolutionary process known. Evolution simply means growth and development, and it isn't simply a theory for how "man came from monkeys" (which no scientist has ever said, by the way). As astronomer Harlow Shapley has noted:

> Evolution is not chiefly limited to the relation of man to his anthropoid forebears. That phase is one of the minor steps in the development that pervades the whole universe...From our survey emerges an appreciation of the importance and magnitude of inorganic evolution.[167]

In this chapter, we'll explore the stories of how the Earth came to be a living planet, not just full of life, but alive itself. So we're going to tell the story of life on this planet. The narrative driving this story along is not evolution by random chance, however, but via a principled process that makes sense in light of the themes we've already addressed (Mathematics and the Mind). That is, this growth (evolution) came with order (properties like self-organization, aggregation, symbiosis, specialization, etc.) and direction (progress with the end result in mind). The Mind provided the mathematical structure, and the end result was the diversity of life we see around us today. But make no mistake, anything we see did not make it here by chance. The struggle for survival was not meant for everyone to win. In the end, only the smart survived. That's not a typo. In the story of life, as we'll see, survival and success depend heavily on conscious adaptation (sometimes pre-adaptation) to changing circumstances along with other strategies (like balancing cooperation and competition), not simply being strong enough to withstand's "nature's punishment." In fact, for those organisms who have "followed the law," Nature has not doled out punishments so much as it has provided opportunities for growth. For those who refuse to live in accordance with the way things work, Nature may not be so kind.

WHAT YOU'LL LEARN

- ❒ How and why the Earth became the home of life.
- ❒ Why evolutionary theory is not a threat to your self-esteem.
- ❒ How the planet's 5,000,000,000+ species all descended from a single common ancestor.
- ❒ Why Darwin might not have been racist, but he was still kind of a racist.
- ❒ What any living organism must to do to survive and become successful.
- ❒ What our earliest ancestors did to "prepare" the Earth.
- ❒ How evolution requires organisms to gamble.
- ❒ What happened to the dinosaurs, and why.

THE STORY OF LIFE

SUPREME UNDERSTANDING

WHAT WAS THE FIRST LIFE LIKE?

About 300 years ago, Swedish zoologist Carl Linnaeus put life in taxonomic order. He mapped out an elaborate classification system, assigning every species to a genus, every genus to a family, every family to an order, and so on, all the way up to a kingdom. For Linnaeus, there were only two kingdoms that a species could belong to: animal or plant. In the centuries that followed, scientists realized this model didn't work very well, and developed three new kingdoms to classify all the odd ones out. Then, genetic studies revealed these five kingdoms were also inadequate for explaining why plants and animals have more in common with each other than with most bacteria.

In 1977, microbiologist Carl Woese again reduced everything to two categories, Prokaryotes and Eukaryotes. Prokaryotes (with two domains, Archaea and Bacteria) are typically single-celled microorganisms whose cells lack a nucleus. Eukaryotes have cells with nuclei and organelles, making them more complex. Plant, fungus, protest, and animal cells are Eukaryotes.

Most of life's genetic diversity found in Woese's first two domains (Archaea and Bacteria).* A single quart of seawater can hold 60,000 different kinds of bacteria – more than 10 times all the species of mammals on Earth. And the differences among those bacteria are not superficial. Basically, we have more in common with potatoes than two bacteria that look nearly identical. What does this mean?

Some scientists in the "life has no meaning" camp have used this fact to argue for the "insignificance" of humans. They cite the late arrival of modern humans on the Earth (emerging within the last 500,000 years of a 4.5 billion year history), along with the tiny niche occupied by humans in the scope of planetary biological diversity (mammals only make up about 5,490 of the planet's 5-billion-plus species).

But neither of these facts suggests "insignificance." That's like saying our brain is insignificant because it only makes up 2% of our body weight. We have plenty of evidence that humans are anything but insignificant, from the role we've played in altering our environment (which we discuss later) to the anthropic principle we discussed earlier, which argues that everything that preceded man was "fine-tuned" to promote the emergence of man.

So why are there at least 10 million species of bacteria on the planet, with most of these simple organisms more different from each other than, say potatoes are from humans? It's the same principle that produces more genetic diversity between populations in Africa (and between Twa people like the San and the Jarawa) than there is between, say Russians and Chinese. It's because ancestral populations start off with a unique genetic signature, but split off into branches with various genetic differences from their ancestors. Recent branches in the tree will have more in common than branches that split off long ago and hundreds of millions to change. So this leads us to wonder about what the original, ancestral population was like.[168] And after centuries of classification, genetic analysis, and reinterpretation of the data, it's clear that t here was a SINGLE, universal, common ancestor for all of life.[169] The question then, is HOW?

A Layman's Guide to Evolution

C'BS ALIFE ALLAH

"If man came from monkeys why are there still monkeys? Man didn't come from monkeys!"†

---

* Woese's system is known for having three domains. The main difference between the first two domains (Archaea and Bacteria) is the chemical process by which they live.

† "Darwin said we come from monkeys. Nope. He never said that. This common

"It's called the theory of evolution. It hasn't been proven yet. It's a belief just like anything else."

One of the main aversions people have against science comes whens they are approached with something that contradicts their belief system. There's a fight-or-flight response when they feel something threatens their place in the universe or religious/residual-religious thinking. Thus, two of main science-related issues that people have a problem with tend to be the Big Bang and the evolution.

AIN'T IT A THEORY?

Overall, the main opponents of evolution are active Creationists, religious people who have a poor understanding of science, and supposedly nonreligious people who are still dealing with residual religious thinking. The religious traditions of the book (Judaism, Christianity, and Islam, and by extension, Zoroastrianism and Bah'ai) dictate that man was created fulled formed and placed on the planet.* The legacy of these religions impacts even those of us who don't identify with those religious traditions anymore. This is because not identifying with a religious tradition doesn't equate with one gaining a scientific education. All the above causes people to try to equate the statements of evolution with the claims of religion. In order to clear all of this up, the first thing that we have to deal with is the language of science, to focus in on what the "theory" of evolution is truly stating.

The first major word that we have to take to task is "theory." To most of us, the word theory means something that requires proof. It's something without facts, a hunch, a guess. Well guess what? There is also a scientific meaning to the word theory (just like "bat" has two meanings). When scientists use the word, it means that it has been strongly supported, strongly substantiated and well documented through observations. The theory is the umbrella which brings together all of the facts so that an explanation can be presented that fits all of the observations and can be used to make predictions. So in science the goal is to get to a theory which is the explanation.

---

misconception belies a profound misunderstanding of evolution. Saying we come from monkeys is like saying you are the child of your cousin. Darwin said that monkeys, apes and humans must have a common ancestor because of our great similarities compared to other species." – John van Wyhe, historian of science at the University of Cambridge

* Ironically, many of the mystical ANE (Ancient Near East) traditions of those same religions (such as Sufism, Kabbalah, etc) speak to stages in the development of the universe and of man.

A theory is not a law. A scientific law describes things. A scientific theory explains them. For instance, there's the law of gravity. Everyone knows if you walk off of a cliff, you will fall. That is a scientific law. It is a description. Yet to explain that law we first had Newton's Theory of Gravitation. Then later on we had Einstein's Theory of Relativity.

Now evolution is a fact (law). As Stephen Jay Gould notes in "Evolution as Fact and Theory":

> Evolutionists have been clear about this distinction between fact and theory from the very beginning, if only because we have always acknowledged how far we are from completely understanding the mechanisms (theory) by which evolution (fact) occurred. Darwin continually emphasized the difference between his two great and separate accomplishments: establishing the fact of evolution, and proposing a theory – natural selection – to explain the mechanism of evolution.

The most substantial element of the law of evolution is a change in frequency of alleles* in a gene pool over generations, variation, replication and selection. There is no denying that. The most prevalent explanation (theory) for that law (fact) is the Theory of Evolution by Natural Selection. Yet, Darwin's theory of evolution is only one version of evolutionary theory. There are plenty of other versions.

SO WHAT IS EVOLUTION?

> "Nothing in biology makes sense except in the light of evolution," because it has brought to light the relations of what first seemed disjointed facts in natural history into a coherent explanatory body of knowledge that describes and predicts many observable facts about life on this planet." – Theodosius Dobzhansky, evolutionary biologist

What I find amazing is that many of us act as if there's no evidence of evolution all around us. Last time I checked, there weren't any wild packs of chihuahuas roaming the forest, not one of us popped out of their mother's womb fully formed, and corn was domesticated through agriculture from a unique little grass seed in South America. The only constant is change and everything changes and transforms over time. So while it's easy to see that one organism changes over its lifetime, why does it seems so impossible for whole classes of plants, animals, bacteria, and so on, to transform over a much longer period of time?

All life on this planet started with a common ancestor. From that

* A gene is a molecular unit of heredity of a living organism. An allele is one of two or more forms of a gene or a genetic locus (generally a group of genes) that is located at a specific position on a specific chromosome. Chromosomes are the organized structure of DNA and protein found in cells.These DNA codings determine distinct traits that can be passed on from parents to offspring.

common ancestor (we're talking on the micro-biological level here), diversification began. This diversitification kept happening at every biological level. So when we're talking about evolution we're not just talking about animals and plants constantly evolving, we're also talking about DNA, proteins, and viruses evolving. Evolution deals with characteristics that are inherited, which transform over successive generations of defined biological populations.

The way to infer evolution is actually quite simple. The traits that are similar within a species point to a recent common ancestor. So, for instance, there is a gene for topical melanin (skin shade). The gene for European pale skin only appeared on the scene approximately 6,000-12,000 years ago.[170] So this means it was at that point (6,000-12,000 years ago) that people with topical melanin and white Europeans shared a common ancestor. You're basically just grouping biological organisms in clusters (species) and creating a ladder backwards. You also take into account the sequencing of DNA that can reveal things not detectable to the naked eye. These ladders can be used to construct evolutionary histories and extend all the way into the fossil record.

So before we start dipping into the various theories (explanations of the data) of evolution lets be clear on the facts that cannot be disputed (well they technically can be disputed, but it's like disputing the law of gravity by trying to walk off a cliff):

- More offspring are produced than ever survive.
- Traits vary among individuals which lead to different rates of survival and reproduction.
- Trait differences are transmitted through heredity. When a segment of the population dies, it is replaced by the descendents that were better adapted to survive and reproduce in the environment.
- How and why this takes place delves deeper into specific theories, but the above facts are easily observed and accepted.

OTHER EVOLUTIONARY THEORIES

C'BS ALIFE ALLAH

The theory of evolution that most of us think we're aquainted with is Darwin's Theory of Evolution. But when someone says they don't "believe" in Darwin's theory of evolution that's cool. No one in the scientific communicate advocates his theory in full anymore and his isn't the only player in the game out there anyway.[*]

What we have now is the modern evolutionary synthesis which utilizes Darwin's original theory as a beginning point (primarily his ideas on natural selection and gradual evolution). Several other elements have been integrated with his inital explanations (and some parts of his original theory have been cast to the wind). In the Modern Evolutionary Synthesis theory, additional explanation such as the role of genetics (due to the observations of Gregor Johann Mendel[†]), the role of mutation (punctuated equilibrium[‡]) and genetic drift[§] have all been integrated.

Now, while Darwin's theory is often refered to as "survival of the fittest," there are other theories out there which don't necessarialy follow that line of reasoning. One, the Mutual Aid theory of evolution, doesn't dismiss the notion of competition that Darwin advocated, but does, however, state that the element of cooperation has been underemphasized, noting that Darwin was influenced by Capitalism. The author of this theory likewise was influenced by Socialism. After examining the evidence of cooperation in nonhuman animals, pre-feudal societies, in medieval cities, and in modern times, he concludes that cooperation and mutual aid are the most important factors in the evolution of a species and its ability to survive.

One its leading proponents, Petr Kropotkin, considered cooperation a feature of the most advanced organisms (e.g., ants among insects, mammals among vertebrates) leading to the development of the highest intelligence and bodily organization.[171] Beyond the various theories

---

[*] Many believe that theories are often proven wrong or right, but more often, a working theory may cover some elements and not others. As Primack and Abrams note in *View from the Center of the Universe*, "New theories, even revolutionary ones, don't have to totally replace old theories that worked. They may simply encompass an old theory in a new and larger theory that explains things beyond the scope of the older theory." This has occurred with many theories, notably evolutionary theory, Big Bang theory, and general relativity (the classical theory of physics).

[†] Gregor Johann Mendel was an Austrian scientist andAugustinian friar who gained posthumous fame as the founder of the new science of genetics. Mendel demonstrated that the inheritance of certain traits in pea plants follows particular patterns, now referred to as the laws of Mendelian inheritance.

[‡] Punctuated equilibrium proposes that most species will exhibit little net evolutionary change for most of their geological history, remaining in an extended state called stasis. When significant evolutionary change occurs, the theory proposes that it is generally restricted to rare and geologically rapid events of branching speciation.

[§] Genetic drift refers to the change in the frequency of a gene variant (allele) in a population due to the random sampling that occurs in nature.

of evolution, there are also many hypotheses (a proposed explanation of something that can be testable), such as the Red Queen hypothesis, the Black Queen hypothesis, and so on.

There are also "goal-driven" theories of evolution (known as Orthogenesis) which argue that evolution ocurs in a directional, purposeful way. Many aspects of it have been disputed due to evidence that everything doesn't proceed in a strict linear fashion. Yet with increased understanding of recent developments in mathematics (chaos theory and fractals, for example), more complex patterns and sequences are being unveiled. In other words, directional evolution can proceed towards a destination even with multiple bifurcations (splits) along the way, in a non-linear format, something like how this book proceeds along a predetermined path but takes on many tangents and divergenet discussions.

The most substantial elements of orthogenesis can now be found in the theory of convergent evolution, which was developed to explain the acquisition of the same biological trait in unrelated lineages, such as birds and bats both developing wings to fly, or dolphins and sharks developing the same structure to swim. In both cases, they don't share any ancestors who had those traits, yet they each developed those convergent traits. Other ideas that correspond with this view are those of environmentalist Dr. James Lovelock and biochemist Dr. Alfred Rupert Sheldrake, who both argue for a coherent system of organization for evolution. In *Gaia: A New Look at Life on Earth,* Lovelock suggests that we look at the Earth as one large organism.[172] In *The Presence of the Past: Morphic Resonance and the Habits of Nature*, Sheldrake offers that there is an underlying field that plays a role in organizing genetic data.[173] There are many other notables in the study of evolution such as Dr Gregory Bateson,[174] Henry Bergson,[175] Jean-Baptiste Lamarck, James Watson and Francis Crick (co-discoverers of DNA), and Theodosius Dobzhansky. Biologist John A. Davison contends that the idea of convergent evolution requires a pre-existing structure:

> The so-called phenomenon of convergent evolution may not be that at all, but simply the expression of the same preformed 'blueprints' by unrelated organisms.[176]

One of the earliest proponents of goal-driven, blueprint-oriented evolution was actually a Jesuit priest, Pierre Teilhard de Chardin, who was also trained as a paleontologist and geologist.[177] His views were so controversial to the church that many of his books weren't published in full until after his death in 1955. Teilhard's interest in evolution was sparked by his reading of *L'Évolution Créatrice* (*The Creative Evolution*) by

Henri Bergson, which offered an alternate explanation for Darwin's mechanism of evolution, suggesting that evolution is motivated by an élan vital, a "vital impetus" that can also be understood as humanity's natural creative impulse. Teilhard studied paleontology under luminaries like Marcellin Boule and Henri Breuil, whose focus on the history of Original people must have sparked something in de Chardin to look beyond the 6,000 year history taught by his Church. Teilhard traveled throughout Africa and Asia and wrote works based on his growing understanding. The Roman Catholic Church soon banned his works and forbade him from teaching.

His landmark book *Le Phénomène Humain* (*The Phenomenon of Man*), was completed in the 1930s, but wouldn't be published until after his death because of the Church's repression. In it, he writes of the unfolding of the material cosmos, from primordial particles to the development of life, human beings and the noosphere, and finally to his vision of the Omega Point in the future, which is "pulling" all creation towards it. He explains that human spiritual development is moved by the same universal laws as material development, writing:

> Nothing is comprehensible except through its history. 'Nature' is the equivalent of 'becoming', self-creation: this is the view to which experience irresistibly leads us...There is nothing, not even the human soul, the highest spiritual manifestation we know of, that does not come within this universal law.[178]

Teilhard also argued that, in recent times, evolution (particularly among humans) was being inhibited by individualism and social isolution. He states that evolution requires a unification of consciousness and that "no evolutionary future awaits anyone except in association with everyone else."[179]

Teilhard explains the entire universe through its structured and goal-oriented evolutionary process, delineating the evolution of matter into a geosphere, then a biosphere, then into consciousness (in man), and ultimately to a supreme consciousness, which he calls "the Omega Point." Effectively, in Teilhard's explanation, evolution is the means by which the universe becomes the home of God, and by which it will return to a point when there is, again, nothing but God.

Might be tough for some of us to process, especially coming from a Jesuit priest. Maybe he had ulterior motives? Well, we know he wasn't pushing his church's agenda because they hated him. We also know he wasn't the only one to propose this view. In their 1986 book, *The Anthropic Cosmological Principle*, British astronomer John Barrow and American mathematical physicist Frank Tipler propose the Final Anthropic Principle, which says the conscious life that is now in the universe (and which, according to the Participatory Anthropic Principle, created the universe) will continue to evolve until it reaches a state of totality that they call the Omega Point. At the Omega Point:

> Life will have gained control of all matter and forces not only in a single universe, but in all universes whose existence is logically possible; life will have spread into all spatial regions in all universes which could logically exist, and will have stored an infinite amount of information including all bits of knowledge which it is logically possible to know...[T]he totality of life at the Omega Point is omnipotent, omnipresent, and omniscient![180]

In other words, man created the universe to allow man to become the Creator of the universe.

An Introduction to Biology

ROBERT BAILEY

Ever wonder about life, how it came about, how you KNOW you're alive, what composes life and questions of that sort? If so, you may take a liking to Biology.

Biology is a natural science – the science concerned with the laws that govern the natural world. More specifically, biology is the study of life (from *bios,* which means "life" and *logia,* which means "study of") and living things, which covers everything from origins, to function, how they evolved, their structure, taxonomy and distribution. It is a major science in that it can be divided into smaller fields of study which are generally categorized based on the type of organism studied. Much of biology is based on five basic principles: cell theory, evolution, genetics, homeostasis and laws of thermodynamics. Though biologists do ask questions about life, they're not the philosophical questions like, "what is the meaning of life?" or "why do I exist?" No, biologists focus more on the question of "how?" How did it happen? How did it evolve? How or what makes something alive? What are the characteristics of life?

While the origins of modern biology and its study is often traced back to ancient Greece, sciences similar to and included within it, have been studied as early as the civilizations of ancient Mesopotamia, Egypt, South Asia and China. The study of biology grew with improvements to the microscope which made it possible to observe life on an atomic level. At all levels of study, biologists learn about life in the same ways: by observation, questioning, hypothesis, testing and explanation. This is known as the scientific method, which can be thought of as "hypothesis driven." Due to new advancements, which allowed the entire genetic code of an organism to be mapped out, a new "data-driven" methodology has unfolded many important discoveries for biology. You're engaged in some facet of biology every day.

THE MEANING OF LIFE

SUPREME UNDERSTANDING

WHAT IS LIFE?

In *The Fifth Miracle,* Paul Davies contends that life is quite unique in our universe. "The second law of thermodynamics, arguably the most fundamental of all the laws of nature, describes a trend of decay and degeneration that life clearly bucks," he writes. Davies explains that life is not the inevitable result of Earth's chemical soup, but a structured, organic vehicle for information. Davies notes, "The key to biogenesis

A QUICK NOTE ON BREATHARIANISM

Breatharianism is supposed to be a lifestyle where one does not eat or drink, and only needs air to live. A variation is that we only need to gaze into the Sun to live. This is another example of European tomfoolery seeping into the consciousness of Original people via the Internet. As you've seen, one of the fundamental properties of life is that it must consume and convert energy via metabolism and nutrition. If it was meant to be another way, our bodies (and the organic chemistry of all other life) would be organized another way. There's nothing wrong with eating or drinking. That's nature. In the 19th century, Indian yogis practicing a life of ascetism (where they rarely ate, drank, or slept) encountered European visitors who were enamored by their amazing abilities. These Europeans developed mystical conceptions of these holy men, and some of these holy mean weren't so holy. Seeing an opportunity for fame and profit, some of them put on shows that made them seem more mystical than they were. The reports back to Europe and America exaggerated things even further. Since that time, holy men of various nationalities have announced various superhuman abilities, most of which are not traditional facets of their cultures to begin with. If you check out the Wikipedia article on breatharians, the frequency by which they're caught eating cheeseburgers and whatnot is hilarious. One of them, a white woman who renamed herself Jasmuheen, could not last more than a day without food, even though 60 Minutes was filming her! The sad part is that at least 3 of her followers have died trying to follow her teachings. The most recent case has been that of 81-year-old Indian ascetic Prahlad Jani, who was monitored (but very poorly). Now, it's certainly possible that, with enough practice and discipline, he could go without food and water for an extended period of time. But to stop completely? How would his body receive the nutrients we need for our organs to function properly? How would he produce the energy needed for his muscles to work? That's simply not how human physiology works, and – if it were – malnutrition and famine wouldn't be such serious problems affecting billions of Black and brown people across the globe!

was not a chemical combination but the organization of information. If the key was the formation of a logical and informational architecture, the crucial step was an information-processing system." In addition to information content, Davies identifies several other key properties of life. Based on his work, along with the research of several others in the field, here are the nine properties of life:

1. Information Content: Specified, contextual information that is needed to thrive, replicate, and have the following properties.
2. Hardware/Software Entanglement: Information alone isn't enough. There must be a medium, consisting of nucleic acids and proteins acting as the hardware where this information acts as the software.
3. Complexity: Even the simplest organisms are highly complex.
4. Organization: Living organisms aren't only complex, but organized. Organized complexity requires coherence rather than chaos.
5. Metabolism and Nutrition: To be considered "alive" an organism must do something. That something involves the processing of chemicals to produce energy and carry out the tasks below.
6. Permanence and Change: Life flourishes on Earth because of the balance between the conflicting needs of conservation and change. Davies notes, "This ancient puzzle is sometimes referred to by philosophers as the problem of being verses becoming." Another unique feature to the balance struck by life is that it is never a "perfect" balance. Life is never "perfect," as can be seen in the replication errors and mutations that allow organisms to change across generations. It is in fact the imperfection, this lack of "perfect symmetry" that allows life to thrive.
7. Autonomy, or Self-Determination: Living things enjoy a sort of freedom that we don't associate with nonliving things. Within their given limits, they can make choices to "do as they please."
8. Growth and Development: Both living and nonliving systems can grow and spread, but living things are unique because they don't simply grow bigger, they grow better, so to speak. That is, they develop and evolve. The law of entropy (the second law of thermodynamics) appears to be an obstacle to this mechanism of life building its own internal order, but the dissipation of entropy into the organism's environment allows it to build order by "exporting disorder."
9. Reproduction: A living thing should be able to replicate itself and – in its progeny – reproduce the replication mechanism it con-

DID YOU KNOW?
Formation of a carbon atom requires a nearly simultaneous triple collision of helium nuclei (known as alpha particlesı) within the core of a giant or supergiant star. This is known as the triple-alpha process. Any further nuclear fusion reactions of helium with hydrogen or another helium nucleus produce lithium-5 and beryllium-8 respectively, both of which are highly unstable and decay almost instantly back into smaller nuclei. So carbon couldn't form at the Big Bang. It had to be birthed inside of stars (bigger than our Sun) that provide just the right temperature and pressure to produce carbon, but not so much as to turn it to something else. In smaller stars, carbon acts as a catalyst that keeps the CNO cycle moving forward. The CNO (for carbon–nitrogen–oxygen) cycle is one of two fusion reactions by which stars convert hydrogen to helium and emit energy, the other being the proton–proton chain process that occurs in our Sun.

tains. In other words, living things produce other living things. And thus, the cycle of life.

LIFE IS ESSENTIALLY BLACK

Everything living on Earth is a "carbon-based life form." This means that the element carbon forms the backbone of biology for all of life on Earth. Carbon is "special" because it has four valence bonds, where the energy required to make or break a bond is just at the right level for building molecules which are not only stable, but also reactive to the environment. The fact that carbon atoms bond readily to other carbon atoms allows for the building of long and complex molecules. This is what allows simple carbon to easily become complex melanin. So complex molecules made up of carbon bond with other elements, especially oxygen, hydrogen and nitrogen, and provide the building blocks of life. In Paul Davies' book *The Fifth Miracle*, as in many other works discussing the Anthropic Principle, he notes that much of the "fine-tuning" in our universe is directed towards favoring the emergence of carbon.

Where does this carbon come from? The stars. Not our Sun, but stars that predated our Sun. Our Sun is a third-generation star, and may produce a lot of things, but the carbon that makes of most of our body predates our Solar System. In fact, carbon, nitrogen, and many of the other essential elements that make up the human body had to come from distant stars in the form of spacedust. Echoing Carl Sagan and Harlow Shapley, Primack and Abrams have observed, "Human beings are made of the rarest material in the universe: stardust. Except for hydrogen, which makes up about a tenth of your weight, the rest of your body is stardust."

Of course, there are literally a billion other factors that need to be present to produce the kind of life we have on Earth, but it all begins with organic matter made from carbon, which is typically black. Do you want to know the story of how this "essentially black" organic matter came to be the Black man? We're talking about over five billion years of history! If you're ready to take that journey, keep reading.

We're going to tell the story of life on this planet. The narrative driving this story along is not evolution by random chance, however, but via a principled process that makes much more sense in light of the themes we've already addressed. That is, this growth (evolution) came with order (properties like self-organization, aggregation, symbiosis, etc.) and direction (the end result in mind). The Mind provided the mathematical structure, and the end result was us.

In the Heavens Above

SUPREME UNDERSTANDING

WHEN AND HOW DID LIFE BEGIN?

People think Earth started out as a fiery environment where nothing could live. Scientist even named Earth's formative years the Hadean eon, after Hades, the ancient Greek word for Hell. But new research suggests the planet may have been suitable for life pretty quickly. Earth formed about 4.5 billion years ago, just after the Sun collapsed into being. After formation, planets take some time to fully contract and cool. But apparently, it didn't take long before there were oceans and continental crust similar to what we have today. "Our data support recent theories that Earth began a pattern of crust formation, erosion, and sediment recycling as early in its evolution as 4.35 billion years ago," Bruce Watson of Rensselaer Polytechnic Institute reported.

> Of course, even with the existence of water and crust, the Earth was not the friendly place we now know. The planet would still have been quite hot, and the atmosphere would have consisted only of carbon dioxide, water, and volcanic gases. But life may still have been able to exist in these types of conditions.

But if life originated on Earth, it had less than a billion years (4.35 to 3.5 billion years ago) to do so. The oldest fossils found so far are known as stromatolites, and are dated at 3.5 billion years old. If we work backwards and assume that still simpler life and pre-life chemical evolution preceded these fossils, the advent of life on earth must be pushed back much farther to allow time for gestation.[181] Not to mention the possibility that our planet may have been sterilized by asteroid bombardments one or more times, with life springing forth twice or even several times BEFORE 3.5 billion years ago.[182] In fact, as we noted earlier, there's evidence that life was present at the time the continents were formed 4 billion years ago. That scenario cuts the window in half.

Many scientists don't find this short window of time realistic. This is partly why no one can demonstrate how life could develop on Earth, given the chemicals and conditions present (and no conscious agent involved). But if life originated elsewhere, the time span expands.

The Wilkinson Microwave Anisotropy Probe says the Universe is 13.7 billion years old. However, there had to be at least one cycle of star birth/death to provide the carbon, nitrogen and oxygen essential to life. This couldn't have happened right away, and may have taken up to several billion years to produce sufficient quantities. This puts the earliest possible emergence of life in the Universe at about 12.7 billion years. Therefore, that allows for about 9 billion years (12.7 to 3.5 billion years ago) to produce life, which is more plausible. It may have literally taken that long for the first seeds of life to take form.

So, just as the primordial ocean between Earth's early peaks of land (before the continents formed) were the "home" of its first cellular life, the "thin soup" of space between the stars must have been, according to Hoyle and Wickramasinghe, the real "swamp" where the precursors of this life originated. In 1981, two decades before much of this was verified, Hoyle and Wickramasinghe wrote:

> Precious little in the way of biochemical evolution could have happened on the earth. It is easy to show that the two thousand or so enzymes that span the whole of life could not have evolved on the Earth. If one counts the number of trial assemblies of amino acids that are needed to give rise to the enzymes, the probability of their discovery by random shufflings turns out to be less than 1 in $10^{40,000}$.

Hoyle and Wickramasinghe concluded that the genes that control the development of terrestrial life must have evolved on a cosmic scale, where there has been more time and much more room for shufflings.[183] Hoyle's 1982 book *Evolution from Space* argued that the evolution of terrestrial life can be affected by the extraterrestrial inoculation of genetic material.

THE BLACKNESS OF SPACE

So what's out there – in the blackness of space – that could become life? When we look out into the night sky and ponder our origins, the depths of space can appear vast and mysterious. But in the blackness of space lies a clue. Space certainly looks black, but if you look close enough, you'll see spots that are darker than the rest of space. Some of these regions can only be seen with a telescope, but others, like the

Coalsack Nebula famous amoung Australian Aborigines, are visible to the naked eye. These black insterstellar clouds are known as dark nebulae, because they are so dense they obscure the light coming from the stars behind them. The smallest of these dark nebulae are known as Bok Globules. These dark clouds, considered the most likely location for the actual birthplaces of stars, may be composed of black particles related to melanins. It is in these dark nebulae that we might find the origins of the organic matter that these comets carry.

To really get a grasp of where I'm going with this, I need to quote you a long block of text from a paper by Italian researchers Bruno J. R. Nicolaus, Rodolfo A. Nicolaus and Marco Olivieri, titled "Speculation on the Chemistry of Interstellar Black Matter":

> Black matter is found universally, especially in the amorphous state. All the black matter known to date, from the lithosphere and biosphere to the cosmos, generally has the same chemical and physical properties...
>
> Radioastronomy has shown there are organic molecules in the black dust clouds in the Milky Way...Giant red stars also emit enormous amounts of carbon dust into the surrounding space...This implies that interstellar space may look black not just because of the lack of light, or because strong gravitational fields prevent light escaping, but also because of the presence of black matter in the solid state. This matter would be in continual transformation under the action of radiation...
>
> It is generally acknowledged that biological evolution followed on the heels of molecular chemical evolution, and this has led to the proposal that there was probably some synthesis of porphyrin-like substances in prebiotic times. The presence of porphyrin-like sites in the melanins suggests that the black particles found on the earth had some sort of catalytic role, in symbiosis with metals. Compared to minerals they would be ideal candidates for a prototype structure in the pre-enzymatic era (stereospecific sites, clathrates, ability to bind metals, photoprotection, etc.). The black particles may also have played a part in forming the primordial atmosphere on earth on account of their ability to trap and release gases.
>
> Black particles in general are an interesting feature in the evolution of interstellar and biological matter...The black particles might also have played a part in prebiotic evolution as atom and molecule assembly structures, or as generators of other molecular structures that have been annihilated. The shock waves produced by the supernovae might have the effect – like interstellar particles when they move faster than 25 km/second – of making the black particles in space explode. High-speed ions can cause fragmentation, and mixtures of simple products form. Some of the simple organic molecules found in black clouds are also among the products of pyrolysis and fast atom bombardment of sepiomelanin, melanin from hair, tyrosine-melanin, serotonin-

> melanin, and tryptamine-melanin...The Stardust, Space Technology 4/Champollion, and Rosetta space missions should transmit analytical data on black dust in the years 2005, 2006 and 2013 so we should then have more details. The information presented here, together with whatever we find out from studying terrestrial black matter (tar, melanin, synthetic black) could be helpful in analysing the samples collected in the various missions...
>
> As an electrical conductor, black matter can regulate the chemistry and the balance of ions, radicals and molecules within interstellar clouds. Black matter helps shield organic matter from radiation. The basic energy for organic synthesis – heat, ionizing radiation, ultraviolet radiation – comes from the stars. The smallest fragments resulting from the explosion of black matter probably give rise to organic molecules similar or identical to some already known on earth.
>
> The cosmochemistry of black matter may stimulate fresh interest on earth in research on the melanins, which so far has strayed along the wrong paths. Once we have straightened out problems of purification and extraction, further interesting developments can be expected in nanochemistry, nanobiochemistry and nanophysics of these black particles.
>
> Melanin is a conductor whose configuration varies under the action of electric or electromagnetic fields. Melanin assembles the simplest elements and can control the form and function of cell adhesion. Stellar melanin is a producer of organic molecules, while terrestrial melanin assembles organic molecules and macromolecules.[185]

That's a lot to digest, I know. But I think know you get an idea of where we're coming from when we talk about the "Blackness" of

A QUICK NOTE ON MELANIN

Melanin is a perfect absorber of light and essential to life because of the natural protection it provides against radiation from the Sun. But melanin isn't limited to humans. Melanin, or a closely related organic compound, can be found in all forms of life, just about wherever you find pigment. And this melanin finds its highest concentration among Original people.[184] Its unique properties remain the subject of considerable study, but one thing's for sure: Life itself is essentially black. For more on melanin, check out *The Hood Health Handbook, Volume One.* There, Robert Bailey takes you a tour of melanin's ability to conduct electricity, its role in neurological function and in regulating other body functions, its ability to resist x-ray diffraction, strong acids and alkaline agents, its response to light, sound, and other electromagnetic frequencies, the way it bonds to chemicals (specifically drugs that have historically been targeted at Black and brown communities), and 99 other attributes. There is one thing I'd like to add: Melanin is not magical. That is, you can answer every single question in the universe with "melanin." That's not how science works. There are thousands of other chemicals within your body, and all of them play a role. Carbon itself takes on many forms, and melanin is just one of them. Please stop using melanin to explain everything from how high you can jump to the way you play Spades.

space. You've probably never heard about any of this before, but this is not "made up." Other studies have corroborated these ideas. For example, chemist Pascale Ehrenfreund and astrophysicist Jan Cami published a study in the June 2010 issue of the journal *Advance* promoting the same theories, only they – like nearly everyone else except for the Italian researchers – never use the word melanin. They wrote:

> Carbon is found in space in all its allotropic forms: diamond, graphite, and fullerene. Astronomical observations in the last decade have shown that carbonaceous compounds (gaseous molecules and solids) are ubiquitous in our own as well as in distant galaxies. The first chemical enrichment of the universe may likely be connected to the first generation of stars. Large carbon abundances are already extrapolated from observations of the strong C[II] and CO lines in the hosts of the most distant quasars.
>
> Carbon in space was first produced in stellar interiors in fusion reactions and was later ejected into interstellar and intergalactic space during stellar collapse and supernova explosions. In the denser regions of interstellar space, the so-called interstellar clouds, active chemical pathways form simple and complex carbon molecules from carbon atoms. Circumstellar envelopes are regarded as the largest factories of carbon chemistry in space.[186]

They're saying the same thing as Nicolaus and colleagues, but they're not going to mention any connections to melanin. Either way, it's clear: Even in the Blackness of space, we find parts of ourselves, perhaps the "source" of our own Blackness. And Melanin isn't just some pigment. There's a reason it's such a key ingredient in life. It appears that melanin is the organizing molecule behind advanced life itself. In introducing a 139-page *Medical Hypothesis* paper titled "Melanin: The Organizing Molecule," Barr, Saloma, and Buchele write:

> (Neuro)melanin (in conjunction with other pigment molecules such as the isopentenoids) functions as the major organizational molecule in living systems. Melanin is depicted as an organizational "trigger" capable of using established properties such as photon-(electron)-phonon conversions, free radical-redox mechanisms, ion exchange mechanisms, and semiconductive switching capabilities to direct energy to strategic molecular systems and sensitive hierarchies of protein enzyme cascades. Melanin is held capable of regulating a wide range of molecular interactions and metabolic processes primarily through its effective control of diverse covalent modifications...To support the hypothesis, established and proposed properties of melanin are reviewed (including the possibility that (neuro)melanin is capable of self-synthesis)...Melanin's role in embryological organization and tissue repair/regeneration via sustained direct current is considered in addition to its possible control of the major homeostatic regulatory systems – autonomic, neuroendocrine, and immunological.[187]

If you understood all that, congratulate yourself. Either you've been reading the preceding chapters very closely, or you're already pretty darn knowledgeable. If you didn't understand and of it, just know that it basically proposes that melanin turns energy into life.

DID LIFE COME FROM OUTER SPACE?

But if all the organic materials for life on Earth can be found within the depths of the black clouds of space, how did it end up on Earth? 2,500 years ago, the Greek philosopher Anaxagoras proposed a hypothesis called "panspermia." Panspermia (Greek for "all seeds") proposed that all life, and indeed all things, originated from the combination of tiny seeds pervading the cosmos. This idea, which Anaxagoras probably learned when he traveled to Egypt to study the sciences, apparently makes a lot of sense even today.

Researchers have found that simple life-forms such as bacteria can survive the treacherous journey through space and the Earth's atmosphere, and that Earth-bound meteorites may have actually carried with them the "seeds of life" itself. In 2004, the Stardust Mission discovered a range of complex hydrocarbon molecules, the building blocks for life, inside comet Wild 2. Several similar discoveries followed. But nothing compared to what scientists found on a meteorite from Murchison, Victoria in Australia. A 2008 analysis of the Murchison meteorite indicated over 100 amino acids and other organic compounds, including uracil, an RNA nucleobase, and xanthine, another critical chemical.[188] These results demonstrate that many organic compounds which are considered the building blocks of life on Earth were already present in the Universe and may have played a key role in life's origin.

In all, scientists are certain that there are over 14,000 molecular compounds in the Murchison meteorite, with some estimates at 50,000 or more. "Extraterrestrial chemodiversity is high compared to terrestrial relevant biogeological and biogeochemical-driven chemical space," wrote the authors of one study. In other words, there's far more out there than we've found down here.

Of this incredible array of compounds, scientists have only identified a few hundred thus far. But the black color of the Murchison meteorite – and other carbon-based meteorites like it – is not a coincidence.

In fact, more than 70% of the Murchison carbon content has been classified as "(macromolecular) insoluble organic matter of high aromaticity."[189] This suggests the presence of melanin. It also contains alanine, a biological predeterminant for melanin production in hu-

> DID YOU KNOW?
> Aboriginal people have mythical recollections and interpretations of meteorite impacts, which is to be expected, as at least 20 meteorites have impacted Australia within the past 50,000 years. But how deep does their knowledge go? The Aranda creation story of *Tnoralla* in the Northern Territory tells how a group of women dancing in the Milky Way caused a baby in its wooden carrier (known as a "coolamon") to fall and crash into Earth. The coolamon, with infant intact, tunneled underground before turning over, trapping the baby inside a burial mound. How long did the Aborigines preserve this knowledge of meteorite impact and crater formation? And could the Black baby coming to Earth in his organic carrier have represented the meteorite that brought the seeds of life to our planet? I don't know exactly what the Aranda really know, but it's clear that – for thousands of years – they were way ahead of Westerners like American President Thomas Jefferson who – as late as the 19th century – refused to believe "that stones should fall from heaven."[191]

mans. Scientists believe the Murchison meteorite could have originated before the Sun was formed. The researchers say it probably passed through primordial clouds on its way to Earth, picking up organic chemicals.[190]

A similar find, the Tagish Lake meteorite, also presents organic compounds that appear to be billions of years old. NASA scientist Scott Messenger has said that such organic compounds could have originally been found in the molecular cloud that gave birth to our Solar System, or at its outermost reaches. Other studies have suggested that the early precursors of Earth life could be found in the swirling disc of dust grains that became our solar system, meaning some of the organic precursors to cellular life actually became the planets, including our own planet, Earth. In some sense, part of the planets themselves was alive as well. Referring back to the last chapter, it seems that one living energy made the planets, while another living energy became the planets.

Yet whether they were part of the early cloud or landed on a solid Earth as a shooting star (meteorite), these "seeds of life" could have easily come from much, much further away, since most comets orbiting our Solar System actually came from distant stars before being pulled in by the Sun's gravitation.[192]

THE ORIGIN OF LIFE

SUPREME UNDERSTANDING

Even if the early Earth was bombarded by black meteorites from black clouds, covered in black organic matter, how did this produce LIFE? After all, there's plenty of evidence that other planets and moons have been sprayed with the same black dust. What happened

DID YOU KNOW?
According to Paul Davies, it is possible that there was no origin of life, because life is eternal, with its contributing elements "spread around the universe...without having originated anywhere in particular."[194] This parallels the evolution of complex life through "mergers" involving elements of life from all over the planet.

on Earth that was so special?

"You'd be more likely to assemble a jumbo jet by passing a hurricane through a junkyard than you would to assemble human DNA by chance." – Francis Crick, co-discoverer of DNA

Well, that's a question scientists are still trying to figure out. We know that the Earth provided a hospitable environment for early life to develop. We know that the number of factors that were required for life to thrive are so many, that the chances of life being reproduced anywhere else in the universe are 1 in $10^{40,000}$, or one part in a quadrillion quadrillion quadrillion quadrillion. These "astronomical odds" suggest that, not only were the conditions on Earth "fine-tuned" for the development of life, the gigantic expanse of the Universe actually needs to be exactly THIS big (about 50 billion trillion observable stars) in order for us to exist. There are, of course, some scientists who think of it all as a matter of chance, but they have yet to demonstrate that life can originate anywhere else by chance. All of the evidence suggests that it would take another universe to produce life again through natural means.

"An honest man, armed with all the knowledge available to us now, could only state that in some sense, the origin of life appears at the moment to be almost a miracle, so many are the conditions which would have had to have been satisfied to get it going." – Francis Crick, co-discoverer of DNA[193]

So we know how unlikely it is to happen again or anywhere elese, but this still leaves the question of "How do simple organic molecules form a protocell?" unanswered.* We know that, even before there was life on Earth, there were chemical reactions occurring. We know that, because of the self-organizing properties of matter, these chemicals were already forming systems.†

The question, then, is how the primitive Earth cooked up interstellar organic matter into the proteins needed to constitute life. Leslie Orgel

* There are many theories, including some that propose that nucleic acids developed first ("genes-first"), while others propose the evolution of biochemical reactions and pathways first ("metabolism-first"). But the most recent (and reasonable) theories are those that take both possibilities into account (a "hybrid" theory).

† In all living things, amino acids – the "building blocks of life" – are organized into proteins. The construction of these proteins is mediated by nucleic acids like DNA or RNA. But the nucleic acids are themselves synthesized through biochemical pathways catalysed by proteins. It's basically a "What came first – the chicken or the egg?" scenario. Which of these organic molecules came first, and how they formed the first life, are the questions scientists have yet to answer.

from the Salk Institute points out that few of the chemicals that can "make it happen" are found on our planet at the time...except for carbonyl sulfide. Carbonyl sulfide is known to fume out of volcanoes today and was likely present in our planet's volcanic past. Orgel's study, published in the October 2004 issue of the journal *Science*, suggests life arose near underwater volcanic vents, which to this day support thriving, self-contained ecosystems.[195] If we envision a life-bringing meteorite actually entering one of these vents, it paints very unique yet familiar picture: **The Black "seeds" of life from the heavens above, inseminating a fertile Earth**.

Yet this brings us to the next question – How did the earliest organic systems became living cells? As we noted earlier, there are two types of living cells: prokaryotic and eukaryotic. Eukaryotes include nearly everything large and complex, including all plants, fungi, and animals. Prokaryotes are much simpler, covering two domains of single-celled life, Bacteria and Archaea. They lacked a nucleus and membrane-bound organelles. Basically, they were the simplest kind of living cell possible. So it's pretty much common sense that the first living organisms were prokaryotes.*

But answering these questions would still leave another question: How did the first living cell emerge from the organic brew simmering in Earth's primordial oceans? Scientists have lots of theories about what preceded prokaryotic life. One theory is that prokaryotic cells evolved from protobionts, which are simple aggregate "systems" of organic molecules surrounded by a membrane-like structure. Protobionts have many life-like properties, including the ability to replicate, but do so inaccurately† Scientists haven't made up their minds on how it happened, but what we DO know is that, in order for ANY form of early life to move from inaccurate chemical replication to precise reproduction and metabolism (which would be needed for life to thrive), they'd

* Others have questioned this conclusion, arguing that the current set of prokaryotic species may have evolved from more complex eukaryotic ancestors through a process of simplification. Meaning, sometime around 4 billion years ago, the direct ancestors of animal life gave birth to bacteria, rather than the other way around. Other theories have argued that the three domains of life arose simultaneously, from a set of varied cells that formed a single a gene pool.

† Protobionts are metabolically active protein clusters that display a number of other properties of life, and can self-assemble into microspheres, store and release voltage like a neuron, and – when enclosed by a protein lipid membrane – regulate their accumulation of some molecules and the exclusion of others. This property improved the ability of protobionts to survive and compete with others. Basically, it's a protocell. And all these features sound like the predecessors of life.

need some sort of genetic code, such as DNA. And that leaves us with MORE questions.

BEFORE DNA THERE WAS RNA

DNA is genetic code. Everything living today has DNA. Yet before DNA, there was a code known as RNA. RNA (ribonucleic acid) is the predecessor to both DNA and protein.[196] Yet, like a living fossil, RNA still lives in us, as rRNA, allowing us to study its properties. Because RNA can both store information like DNA and act as an enzyme like proteins, it may have provided another one of the "seeds" of early life. This is because life requires enzymatic properties, such as the ability to reproduce (or self-duplicate) and to catalyze the chemical reactions needed to form proteins.

RNA did all those things, and did them well enough to produce the first living, reproducing organisms. RNA would also need to serve as an information storage system, able to reproduce "from memory," so to speak. But – until DNA took this job – the use of RNA as an information storage system consumed a great deal of energy and was highly inaccurate (or mutation prone). This wouldn't work for modern life, but it allowed for primitive life to have the massive variety needed to produce at least one surviving line of descendents in the midst of millions of failed dead ends. Eventually, DNA, through its greater chemical stability, took over the role of data storage, while protein took over the job of catalyzing the needed chemical reactions. In a sense, DNA became the brain and protein the body.

RNA WAS ACTIVATED BY THE SUN

Like DNA, there are 4 amino acids required for RNA. They are adenine, cytosine, guanine, and uracil. We know that Earth was peppered with all kinds of organic matter in its formative years, including at least one of those 4 amino acids, uracil. But while scientists look for the other three in space, studies have proven that heating a simple terrestrial chemical known as formamide (which may also be found in space) in a mineral stew creates most of the ingredients for RNA. But formamide is missing something: Guanine. No guanine, no RNA. According to a new study, however, it takes a UV ray to activate guanine in the formamide brew. "A lot of things can happen when you put a photon into the mix," Georgia Tech physicist Thomas Orlando told *Discovery News*. In other words, organic matter from the depths of space + a chemical brew that only Earth could provide + a blast of UV light = Life.[197] Yet, even with this view, there's still something miss-

DID YOU KNOW?
Professor Andrew Watson states that life (intelligent, complex life) on other Earth-like planets is "possible in an infinite universe, but improbable." Watson looked at the probability of each of these steps needed for life to exist in relation to the life span of the earth." His model, published in the journal *Astrobiology*, suggests the chances of intelligent life emerging less than 0.01% over four billion years. In his analysis, Watson also noticed what appears to be a purposeful direction and structure to evolution: "Each step is independent of the other and can only take place after the previous steps in the sequence have occurred. They tend to be evenly spaced through Earth's history and this is consistent with some of the major transitions identified in the evolution of life on Earth."[199]

ing. As Stuart Kauffman has noted, even if all the ingredients to a living cell are present, something still had to "give it life": "A free-living cell, prokaryote or eukaryote, is in fact a collectively autocatalytic system – virtually no molecule, including DNA, catalyses its own formation."[198]

So what catalyzed it? And how? Kauffman continues:

> The dominant view of life assumes that self-replication must be based on something akin to Watson-Crick base pairing. The 'RNA world' model of the origins of life conforms to this view. But years of careful effort to find an enzyme-free polynucleotide system able to undergo replication cycles by sequentially and correctly adding the proper nucleotide to the newly synthesized strand have not yet succeeded.[200]

What does this mean? That the "life assembling by random chance alone" theory presents the most unlikely scenario for the formation of life through its various stages. There is statistically almost NO chance this could have occurred without direction. We don't have enough evidence to assemble together and form an accurate picture of the FIRST stages of life, but if you look at the stages that followed, you should be able to (a) see the patterns and processes that reproduce across scales of space and time, and (b) determine direction by looking at the trajectory of more recent events to estimate the orientation of their origin point. In other words, if you look at the timeline, you'll be able to see the growth pattern's order and know its direction.

THE TIMELINE OF LIFE ON EARTH

SUPREME UNDERSTANDING

FIRST WE MADE THE CONTINENTS

By 4.35 billion years ago, the Earth's crust solidified, leading to the formation of the oldest rocks found on Earth, and atmospheric water condensed into oceans. At this time, the Earth spun more rapidly on its axis, so a day was only 4-5 hours long, and the moon was much closer. Meanwhile, giant meteorites continued to smash into Earth – so the only constant "stable" environment was the deep ocean floor. The ocean provided protection from excessive UV radiation (in the absence of an ozone layer), along with a relatively stable temperature and pH, which allowed life to emerge. We don't know how many times it emerged, but we know that we don't have any evidence of survivors until about four billion years ago, when the bombardment of the Earth by meteorites and asteroids stops.

It's reasonable to think this life emerged on Earth about 4 billion years ago, since we have widespread evidence of life by 3.5 billion years ago, based on fossils (rock formations known as stromatolites) discovered in Australia. These microscopic organisms "breathed" carbon dioxide and synthesized their own food. In other words, these primitive life forms were not so primitive after all.[201] Later, they evolved glycolysis, a set of chemical reactions that extract energy from organic molecules such as glucose and store it in the chemical bonds of ATP. That's the same way we extract energy from the food we eat. In fact, glycolysis (and ATP) continue to be used by almost all organisms, unchanged, to this day. Although this primitive form of photosynthesis may have produced the Earth's continents, it didn't put oxygen into the air. Thus the early atmosphere, lacking oxygen and loaded with volcanic methane, remained poisonous to present-day life. Everything had to come in its own due time. These early organisms weren't oxygenating the atmosphere or building the ozone layer. That would come later. First, they had to build the continents.

A team of geologists led by Minik Rosing of the Geological Museum and the Nordic Center for Earth Evolution at the University of Copenhagen, Denmark, says the appearance of photosynthetic life 4 billion years ago could have converted solar energy into chemical energy, "cranking up the Earth's energy cycle and altering its geochemistry." This process would eventually produce granite (which our continents are made of) from basalt (which the surface of the Earth is made of). "Life might, in the end, be responsible for the presence of continents on Earth," Rosing said.[202]

After building the continents, it seems these organisms spread and mutated rapidly, giving rise to the ancestors of today's three domains: Bacteria, Archaea, and Eukaryotes. One of these ancestral lines, the Eukaryotes, would later develop into us. In essence, our ancestors made the continents.

SNOWBALL EARTH - SOUNDS LIKE FUN BUT IT WASN'T

But 2.3 billion years ago, nearly everything was destroyed again. In yet another instance of the "reset" feature built into life's mathematical nature, the Earth plunged into a global ice age that scientists call "Snowball Earth." This is no exaggeration. Some computer models suggest the planet was encased in a shell of ice at least a half-mile thick. But inside rocks collected near Elliot Lake in Ontario, Canada – rocks older than Snowball Earth – scientists have found oil trapped in water droplets, in the crevices of rock crystals. As we know, oil is a "fossil fuel" because it is the residue of the organic remains of prehistoric life. Astrobiologist Roger Buick and petroleum geochemist Adriana Dutkiewicz found evidence in the oil that suggests eukaryotes and bacteria were alive and kicking 100 million years before the planet froze over. And some of them, the ones who would go on to populate the planet, survived.[203]

WHAT CAUSED SNOWBALL EARTH?

As if by design, Snowball Earth was cause by the same bacteria who would survive it. Before the first snowball event, the Sun was only 85% as bright as now. But the atmosphere was loaded with methane, which acted like a blanket keeping things warm, much like carbon dioxide does today. When the Earth's volcanic activity finally decreased around 3 billion years ago, it led to a global "nickel famine" (a drop in nickel in the world's oceans) and the downfall of methanogens. This led to the rise of prokaryotic organisms known as cyanobacteria. These photosynthesizing bacteria had the ability to break down water and release oxygen. The oxygen initially oxidized dissolved iron in the oceans, creating iron ore, but the more cyanobacteria, the more the oxygen concentration in the atmosphere rose.[204] By 2.3 billion years ago, in what is referred to as the "Great Oxygenation Event," the anerobic (non-oxygen) forms of life were nearly wiped out by the oxygen consumers, who were now taking over the planet.

These oxygen-consuming organisms were our ancestors, you know. But the unrestrained spread of life sometimes brings about the need for elimination. This global boom in oyxgen caused the Earth to start cooling down, and glaciers moved towards the Equator, as they would many times in geologic history. And when the glaciers retreated back toward the poles, they scoured the land and released abundant nutrients into the oceans. The cyanobacteria, with their newly developed ability to make oxygen, fed off the fresh flow of nutrients, and their numbers exploded. The explosive spread of cyanobacteria destroyed the atmospheric methane, and global temperatures plummeted to minus 58 Fahrenheit. Ice at the equator was a mile thick. And that's when things went bad. Again.

Most life on our planet died. What little survived clung to the underwater vents or survived underground. Even today, life can be incredibly resilient, eating rocks, swimming in boiling water and enduring thousands of years in the deep freeze. Some of the organisms that did survive adapted to breathe oxygen, now that there was a lot of it. It was this ability to use oxygen that allowed life to evolve to more complex forms, the scientists say.

Then what happened? We had to thaw out what was frozen, and the only way to do that (quickly) is with heat. The Earth's new setup caused carbon dioxide, a greenhouse gas with less insulating properties than methane, to build up enough to generate another greenhouse period. There's also a possibility that a nuclear reactor in Africa went into operation, producing plutonium and generating immense amounts of heat. Yes, you read that right. For details on that, see "A Nuclear Reactor in Africa?" in the Appendix of Volume Two.

What we know is that temperatures soon climbed to about 122 Fahrenheit around the globe. In modelling the scenario, scientists say we were within inches of a permanent deep-freeze. And, they caution, it could happen again. As Kopp's supervising professor Joe Kirschvink explains:

> It was a close call to a planetary destruction. If Earth had been a bit further from the Sun, the temperature at the poles could have dropped enough to freeze the carbon dioxide into dry ice, robbing us of this greenhouse escape from snowball Earth...We could still go into snowball if we goof up the environment badly enough. We haven't had a snowball in the past 630 million years, and because the Sun is warmer now it may be harder to get into the right condition. But if it ever happens, all life on Earth would likely be destroyed.[205]

WE CAME TOGETHER TO BECOME MULTICELLULAR

During this era, the moon was still very close to the earth and caused tides 1,000 feet high. The earth was continually wracked by hurricane force winds. These extreme "mixing" influences are thought to have

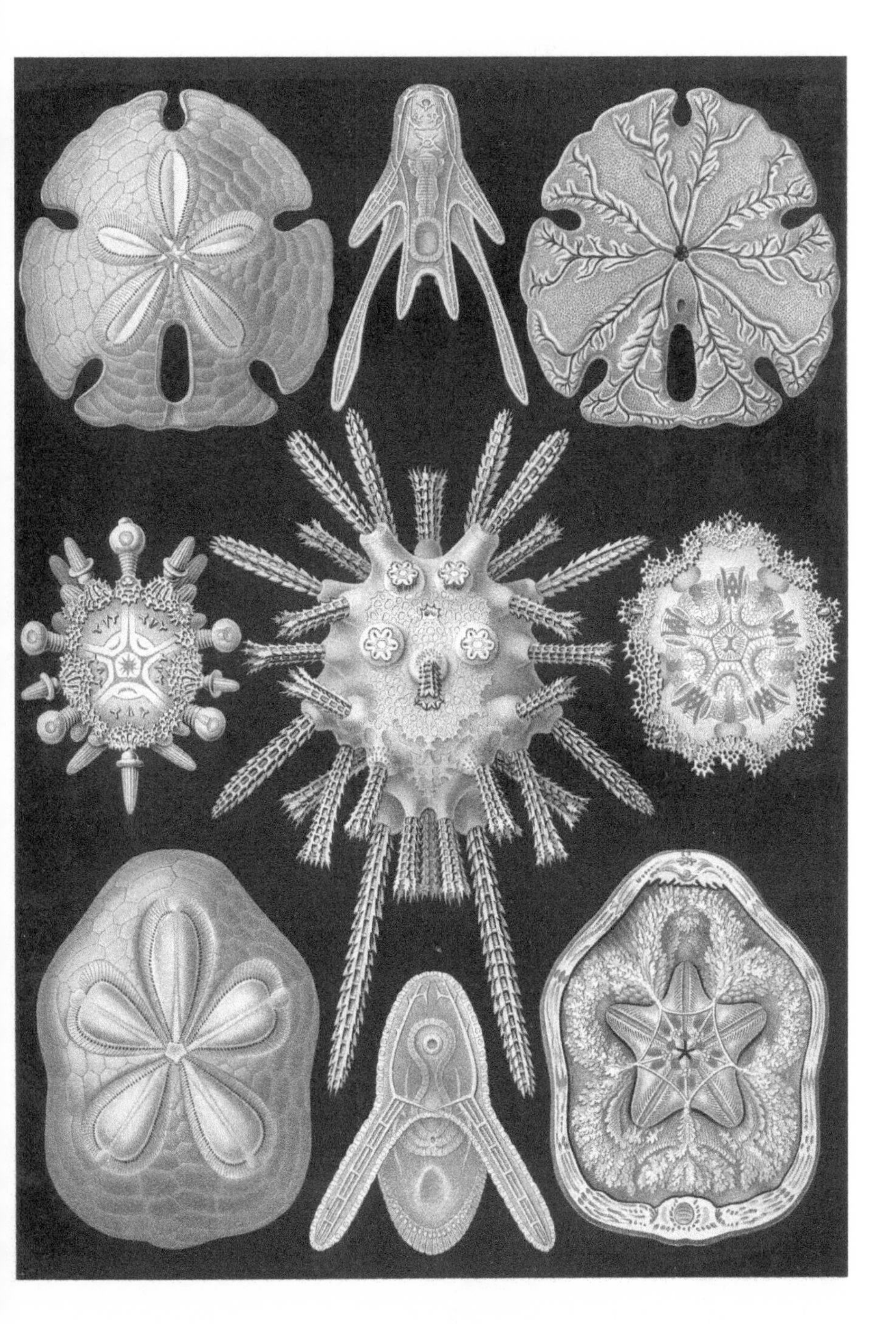

stimulated evolutionary processes. But, as always, it wasn't environmental influence alone. There was direction. Just as prokaryotes were formed from different things "coming together," it appears that the first multicellular organisms arose from cooperation and combination between prokaryotes. One theory suggests that two branches of the Prokaryote tree came together to become Eukaryotes, through a process called endosymbiosis, with some prokaryotes absorbed wholesale into these new and complicated organisms. The best-known examples of this absorption are mitochondria and chloroplasts, the structures that generate energy in animal and plant cells. According to Lake, the union likely took the form of endosymbiosis, in which one of the prokaryotes literally swallowed the other, and the two grew together.[206] Other theories argue that Eukaryotes resulted from the complete fusion of two or more cells, with the cytoplasm coming from a bacteria cell, and the nucleus from a member of the archaea, a virus, or some sort of prebiotic organism like the protobionts.

However it happened, we eventually ended up with more complex Eukaryotic cell organisms, posessing a nucleus and "organelles." Organelles are structures within cells that perform specific functions necessary for the evolution of fungi, protists, plants, and animals. These are the predescessors, on the microscopic level, to our present-day organs. But the story of how we got some complex single-celled Eukaryotes to modern man is just as deep as the story of life's origins. For beginners, it wasn't a straight shot, or line of ascent. As Carle Woese has explained:

> The eukaryotic stem on the phylogenetic tree of life spawns many branches before one gets to the split that separates the ancestors of plants from the ancestors of animals, which seems to have happened more than a billion years ago. There seem to have been many earlier branchings from the eukaryotic stem, all represented by unicellular eukaryotes (such as the slime molds, the flagellates, the trichomonads, the diplomonads, the microsporidia, among others.)[207]

Basically, we've got a huge family tree. Fortunately, some of our ancestral cousins have left direct descendants who can tell us a lot about our heritage. As biologist Robe DeSalle observed, the question of origins is the key understanding our present predicament:

> Evolution has done all these experiments, and when you reconstruct common ancestors, you're reconstructing the results of the experiment...If you want to look at the development of our brains, of our nervous system, of anything we have as a result of experiments that nature has done, the best way to do it is to reconstruct our ancestors.[208]

For example, studies have revealed that choanoflagellates (a single-celled organism with a flagellum, or "tail") like Proterospongia may be the best living examples of what the last single-called ancestor of all animals may have looked like. And they can tell us a lot about what life was like back then. They live in fractal colonies, and show a primitive level of cellular specialization for different tasks. This is important, but we'll get to that later.

After the Snowball Earth episode and the Great Oxygenation event, the rise in atmospheric oxygen allowed multicellular life to emerge from the consolidation of various Eukaryotes. Scientists once believed that such complex life-forms didn't develop until recently (within the last 600 million years). But in July of 2010, an international team found evidence of multicellular organisms dating back 2.1 billion years.[210] The research team analyzed 250 fossils found in clay deposits in Gabon in western Africa. The fossilized creatures, some of them as wide as the palm of your hand, appeared to have an organized internal structure composed of a network of cells capable of signalling and coordinated responses, suggesting complexity far beyond the simple bacterial structures they expected to find. Again, the "colonial" nature of these organisms is important, but we'll have to get back to it. There is more evidence of multicellular life from around 1.7 billion years ago, but multicellular life didn't experience a population explosion until the Cambrian Period over a billion years later. For this to happen, we needed the development of something new – sex.

DID YOU KNOW?
What some biologists consider the origins of sex, others describe in the language of conquest. Scientists think it's possible that the mitochondria in our cells (all cells with mitochondria, actually) were created by "invading" bacteria. Bacteria may have also become the chloroplasts in plant cells. Spirochete bacteria may have combined with cells to form flagella and cilia. Such "mergers" provided cells with metabolism, photosynthesis, and mobility, which they previously lacked (and would have been doomed without). Margulis and Sagan assert that plant and animal evolution would never have taken place unless one life form attacked another and the latter defended itself, all this followed by accommodation and the development of a symbiotic relationship. They write "It is in this light that we are beginning to see the biosphere not only as a continual struggle favoring the most vicious organism but also as an endless dance of diversifying life forms, where partners triumph." This remains as true today as it was in the distant past. For example, without the bacteria in your gut, you'd be perpetually sick. There's also evidence that a significant portion of our DNA is descended from a virus. We'll talk about THAT history in Volume Four.[209]

THEN WE STARTED HAVING SEX

Two billion years ago, even with the Earth steadily stabilizing towards its present-day conditions, the environment still presented a number of the challenges for life. To defend themselves against UV radiation, bacteria developed a new mechanism to ensure reproduction – sex. Until this time (and we're still not sure how early it was), single-celled organisms simple reproduced by binary fission (splitting into two) or budding. There was, as with everything, implicate gender before this point, but each organism contained both the Yin and Yang of gender.

> DID YOU KNOW?
> In order for two parents to produce a child, an amazing mathematical process must occur. The organisms must produce haploid gametes (sperm or egg cells that contain half the diploid, or full set, of chromosomes). To do so, a form of division occurs (meiosis), in order to remove half of the genes. Then, when the gametes fuse (i.e. when the sperm fertilizes the egg), they produce a zygote, which restores the diploid (full set) of chromosomes, with half coming from each parent. Mathematically, we make halves of ourselves to come together and become one again.

Now, organisms were having sex.[211] Sexual reproduction increased the rate of evolution, leading to the rise of multicellular organisms, and ultimately to biological gender. The evolution of sex did, however, come with its downside. When organisms require two partners to reproduce, each partner is only able to contribute only half of their genetic material to their offspring. Further, asexual reproduction allows populations to double every generation, equaling near exponential growth (which explains how micro-organisms spread so far and wide within the first 2 billion years of the planet). But sexual reproduction doesn't multiply at anywhere near that rate. So, in order to be successful, these organisms would have to selectively breed with suitable mates, or risk their offspring being weaker versions of themselves. Yet, sex was used to our advantage, as we conferred the best adaptive traits, leading to at least a billion years of cumulative improvement.[212]

THAT'S WHEN THINGS GOT COMPLICATED

As C'BS explores in "Generation, Part Two," this led to the emergence of more complex life-forms. Recent studies have found evidence of the world's earliest animals, sponges, as much as 850 million years ago, much earlier than previously thought.[213] They are among the simplest of animals, with partially differentiated tissues. Sponges are the oldest animal phylum still around today.

THINGS COOLED DOWN AGAIN

Around 850 million years ago, we also find evidence for another global ice age. It's possible that there were at least four extreme climate reversals during this period, taking the planet from global glaciation (ice) to global hot house (caused by greenhouse effects from volcanic activity under the ice) and back. Scientific opinion is still divided on how this affected biodiversity or the rate of evolution, but more recent examples show us that – even when a large percentage of life is killed off – this allows the remaining life to thrive and further develop.

In fact, some biochemists think this "Snowball Earth" episode is to thank for the evolution of modern animal life. Their research revealed that there was a spike in marine phosphorus (which is essential for life) levels about 750 to 635 million years ago, caused by erosion and chemical weathering from the global ice age. During this period, the same UV light that gave life its spark was preventing life from extending beyond the shrinking oceans. But higher marine phosphorus levels would have led to a more "oxygen-rich ocean-atmosphere system" through photosynthesis (by the ocean plant life).[214]

Thanks to the increased oxygen production of cyanobacteria (also known as blue-green algae), the accumulation of atmospheric oxygen allowed the formation of an ozone layer. Little specks of algae gave us that. The ozone layer blocked ultraviolet radiation, permitting the colonization of land. Basically, the cyanobacteria fought the battle to permit us an exodus from the ocean. Little is known about these early terrestrial microbes. Unlike animals, they don't leave behind much that scientists can find.[215] But these organisms prepared the way for more complex life by seeding the land with organic compounds that became soil.[216] As a result, more complex Ediacaran organisms appeared. These were the first large, complex multicellular organisms. The first fungi appeared around this time as well, helping create the soil that would be needed for future plantlife.

By 600 million years ago, animals still didn't move around much. Animal movement may have started around this time with cnidarians, possibly prompted by the environmental changes mentioned above. Cnidarians were the first animals with the ability to move freely due to the evolution of nerves and muscles. Simple eyes also evolved at this point. Cnidarians were also the first animals with an actual body of definite form and shape. They had radial symmetry. This is important, but we'll get back to it later.

By 550 million years ago, flatworms developed from cnidarians. Flatworms are the earliest animals to have a brain, and the simplest animals to have bilateral symmetry. They are also considered the simplest animals with organs. This development ushers in the larger creatures of the Phanerozoic Eon ("obvious life"), beginning with the

DID YOU KNOW?
New research from the European Molecular Biology Lab in Germany has found that the human brain's cerebral cortex – the seat of higher thought – is very similar to a clump of neurons inside the head of the simple marine ragworm, showing us what our ancestors' cerebral cortex may have resembled 600 million years ago.[217]

Paleozoic Era.

THEN THINGS BLEW UP

The first period of the **Paleozoic Era** (meaning "ancient life"), the **Cambrian Period (545-500 MYA** [Million Years Ago]**)**, was a time of intense geological upheaval. Multiple collisions between the earth's crustal plates gave rise to the first supercontinent, known as Gondwanaland. Gondwanaland was basically a primordial Africa, as most continental land was clustered in the southern hemisphere at this time, where Africa is today. The ice was melting, and the sea level was high, creating many large, warm, shallow seas ideal for thriving life. The sea levels fluctuated, because temperature fluctuations caused pulses of expansion and contraction in the South Pole ice cap.

Together, this contributed to what's known as the **Cambrian Explosion**, a rapid development of most major groups of animals having "hard parts" (such as skeletons), and a dramatic diversification of species, especially among life in the oceans. But life on land was growing as well. The first known footprints on land date to 530 million years ago, indicating that early animals may have hit the land even before plants!

The subsequent **Ordovician Period (500-438 MYA)** saw the spread of graptolites, trilobites, and primitive fish (the first vertebrates). The first primitive plants moved onto land, having evolved from the algae living along the edges of lakes. They were accompanied by fungi, which may have aided the colonization of land through symbiosis. Gondwanaland, later to become the southern part of the super continent, Pangaea, drifted over the South Pole, triggering another Ice Age.

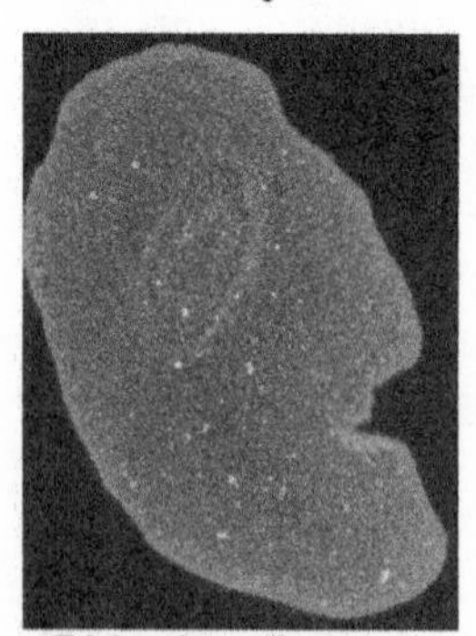

Trichoplax adhaerens

After four billion years, our atmosphere finally stabilized, ushering in the **Silurian Period (438-410 MYA)** and the first air-breathing animal, the scorpion. Meanwhile, large-scale glacial melting caused sea levels to rise again. In the now-growing oceans, the first jawed fish developed. Vascular plants (which can conduct water) evolved and spread on land, but didn't have stems or leaves yet.

Ferns and seed plants, including trees, appeared in the **Devonian Period (410-355 MYA).** These were the first plantlife with roots. And ferns are thus one of oldest complex lifeforms on the planet. No wonder then, that the Adinkra symbol, Aya, a fern, represents defiance, endurance, and resourcefulness. Primitive sharks, wingless insects and arachnids (spiders) appear in the fossil record for the first time.

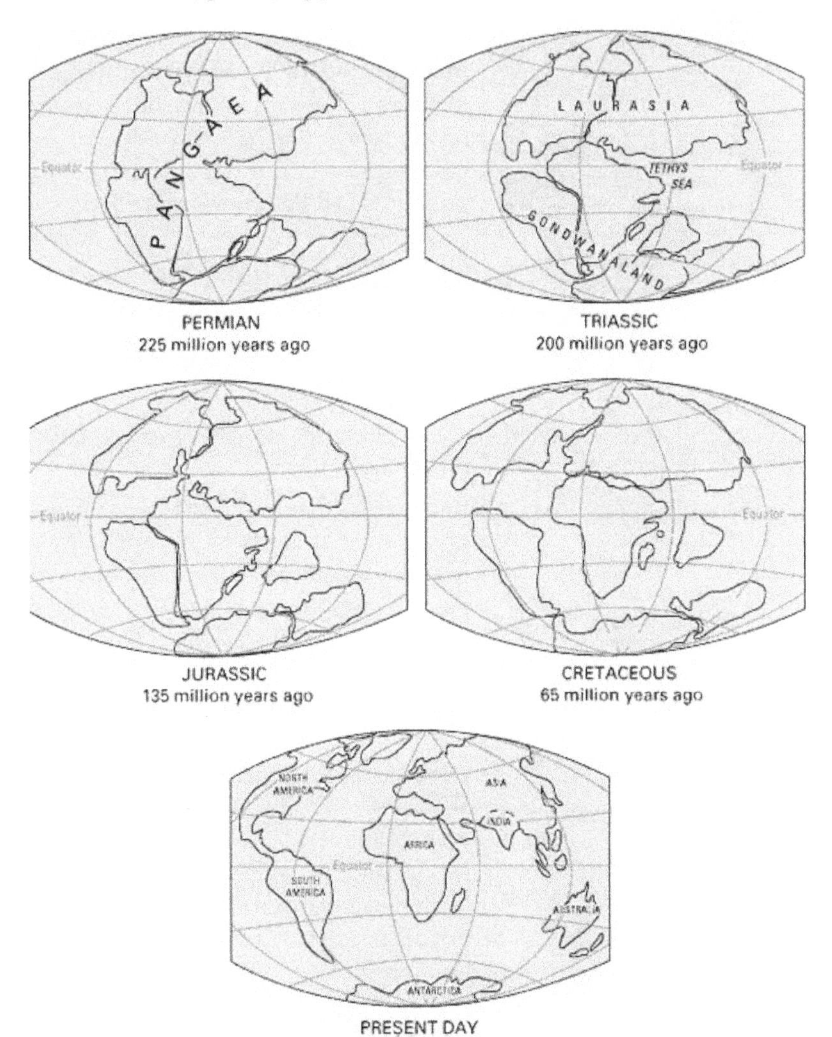

PERMIAN
225 million years ago

TRIASSIC
200 million years ago

JURASSIC
135 million years ago

CRETACEOUS
65 million years ago

PRESENT DAY

This is also the first appearance of *Ventastega curonica*, the earliest known four-legged animal (beginning the evolution of sea animals to land animals). **The Carboniferous ("coal-bearing") Period (355-290 MYA)** which followed produced the great swamps from which we

DID YOU KNOW?
Interested in what your ancestors looked like during the Cambrian Period? Look up a picture of *Trichoplax adhaerens*, the world's simplest known living animal. Trichoplax may be a direct descendant of man's earliest multi-cellular ancestors 550 million years ago. Genome sequencing reveals that, despite its simplicity, it has over 11,500 genes, 80% of which are shared with more evolved animals (like us). Trichoplax has no neurons, but "has many genes that are associated with neural function in more complex animals. It lacks a nervous system, but it still is able to respond to environmental stimuli. It has genes, such as ion channels and receptors, which we associate with neuronal functions, but no neurons have ever been reported..."[218] In other words, it thought, without a brain. Trichoplax shares over 80% of its introns – or "non-coding DNA" – with humans. Meaning it might just be your 66,000th cousin. Humans even share a method of sensing oxygen with Trichoplax. This sense emerged with the development of multi-cellular organisms, when the rise in atmospheric oxygen levels around 550 million years ago made it possible for multi-cellular organisms to exist.[219] This is the origin our relationship with oxygen.

get our present-day coal. It's also the first time the Earth was covered with tall plants and trees.

IT WAS FUN WHILE IT LASTED, BUT THEN THEY HAD TO GO

The Paleozoic Era ended with the **Permian Period (290-250 MYA),** which is important for several reasons. At this phase, all of the Earth's land areas became welded into a single landmass that geologists call Pangaea. This process created major changes in the world's topography. For example, in the North American region, the Appalachian Mountains were formed by this union. During this time, the first mammal-like animals appear, along with winged insects, beetles, weevils, and countless other species.

But Earth's greatest known extinction occurs at the end of this period. We're not sure whether it was caused by multiple impacts from space or massive volcanism, but the Permian-Triassic extinction, also known as the "Great Dying," destroyed nearly everything living on the planet. Yet, as we noted earlier, such events are multifaceted. While the Permian-Triassic extinction event eliminated over 95% of marine species and 80% of life on land (and led to the eventual separation of the continents), this "clearing of the slate" led to later diversification, particularly among land animals.*

SOON AS WE CLEARED HOUSE, ANOTHER PARTY STARTED

The **Mesozoic Era** (meaning "middle life") begins when the great Permian extinction ushered in the **Triassic Period (250-205 MYA).**

---

* See "Extinction Level Events."

Although life on land took 30 million years to completely recover, the Triassic saw an explosion of new species (including 800 species of cockroaches, termites, bees, and a dragonfly-type insect that was about 29 inches long, as well as a variety of plantlife, including the ancestors of modern ferns). It was the beginning of the age of dinosaurs and marine reptiles such as Plesiosaurus. The Triassic ended with a minor extinction as Earth's polar ice caps disappeared and Pangaea began to break up, leading to the reappearance of Gondwanaland, the southern supercontinent and Laurasia, the northern supercontinent.

This leads us to the **Jurassic Period (205-135 MYA)** where we find evidence of the appearance of more dinosaurs and reptiles, as well as the moth, fly, beetle, grasshopper, lobster, and shrimp. There are no "true" birds yet, but the first protobirds, such as Archaeopteryx, have emerged. Primitive crocodiles appeared alongside other early mammals including primitive kangaroos. Meanwhile, Gondwanaland (the southern supercontinent) drifted apart.

The **Cretaceous Period (135-65 MYA)** is when new dinosaurs such as Triceratops and Tyrannosaurus rex emerge in the fossil record. "True" birds emerged, flying between thriving species of deciduous trees and flowering plants. The first primates, including marsupials, also appeared, but were in a constant fight for survival against the larger dinosaurs. With the dinosaurs around, there was no way primates could thrive and expand.

A Few Words on the Dinosaurs

C'BS ALIFE ALLAH

You may not know any, but there are dinosaur-denyers. They believe that dinosaurs didn't exist. This idea is partly related to the rejection of evolution theory, but it all goes to one essential root (consciously or unconsciously): Young Earth Creationism, which states that the Abrahamic God created the Earth between 5,700 and 10,000 years ago. Thus the main reason that dinosaurs were rejected was simply because they were too old and threw off the Biblical timeline (which is held to be true by various groups of Jews, Christians and Muslims). This byproduct of Young Earth Creationism has made its way into the Black conscious community because Black consciousness tends to retain more religious elements than white New Age movements. However, some people have started producing "reasons" for why dinosaurs aren't real. Here are some of my thoughts:

- Most of the time it's just laughable. I mean, what would be the purpose of a worldwide conspiracy to make people believe in some big lizards?! Occam's Razor, people.

- Those who say that they are just "whale bones" have never seen a whale bone. They don't look anything alike.
- And for those who are threatened by the idea of a massive creature predating modern man (as if that displaces our legacy), how can you accept the existence of whales but not dinosaurs?
- There are plenty of slabs of rock where the fossils left behind allow you to see the whole form of the dinosaur.
- There is plenty of archived video footage of digs (check YouTube) as well as sites featuring "citizen science" live streaming of digs, where you can actually watch bones being dug up. In fact, depending on where you live, you might be able to dig some up yourself! Of course, you might need some training, because you're more likely to have chicken bones in your backyard than Velociraptor.
- The whole world is powered by fossil fuels. Fossil fuels are the result of mega-flora and mega-fauna decaying and being compressed within the Earth over millions of years. We can work backwards, using fossil fuels to estimate what was on the Earth long ago. Without dinosaurs, we'd be much shorter on fuel than we are now.
- The idea of what dinosaurs looked like and how they lived has certainly shifted a great deal over the past century. But changing interpretations doesn't equal "totally made up."

Finally, dinosaurs are no threat to your grandeur. You were here before, during, and after dinosaurs, in some form or fashion. After all, you were here when everything was so tiny you'd need a microscope to see it as well. And don't forget, had it not been for dinosaurs, we'd have missed several key stages in the development of the Earth's ecosystem. Most of those steps, including the extinction of the dinosaurs themselves, were necessary for man to thrive today.

The Rise of Mammals

SUPREME UNDERSTANDING

THE DINOSAURS HAD TO GO TOO

The dinosaurs weren't here to stay. Like the Permian Period, the Cretaceous Period also ended violently when the Earth was struck by several large objects from space. The Chicxulub impact in the Gulf of Mexico involved a meteor about six miles wide destroying thousands of square miles of land, setting off one of the largest "megatsunamis" ever, and sending up a layer of soot and dust that blanketed the Western skies. But Chicxulub was not alone. The Shiva impact off the coast of India involved a meteor at least 25 four times its size! And throughout the world, there are dozens of other impact craters from this period. This heavenly barrage killed off the great dinosaurs, as well as most other species on land and in the sea. There were few survivors. Yet, obviously, the ancestors of man survived. This point in time is known as the K-T Boundary – signifying the extinction of the dinosaurs and beginning of the reign of mammals. As Paul Davies has said, "What was a catastrophe for one population can be an advantage for another."

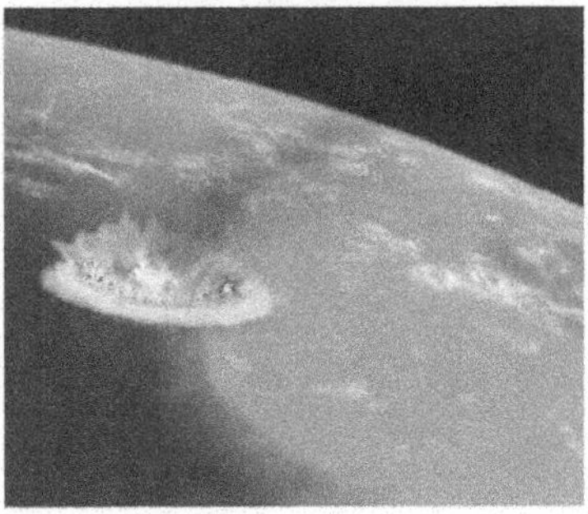

SORRY FOR THE WAIT, BUT YOUR TABLE IS READY NOW

Now it was time for mammals to take center stage. The **Cenozoic Era** (meaning "recent life") extends from 65 million years ago to the present day, and is divided into the Tertiary and Quaternary periods. The **Tertiary Period (65-1.8 MYA)** begins with the **Paleocene Epoch (65-55 MYA)**, which offered a cooler climate, allowing the polar ice caps to regenerate, which in turn allowed for more favorable weather patterns. Meanwhile, in the absence of predator dinosaurs and other large carnivores, small, early mammals diversified and spread across the continents. With the southern continents already separate, now North America separated from Europe and established its connection with South America.

In the **Eocene Epoch (55-34 MYA),** the oldest known fossils of most modern mammals appear. All were still small, including the early rhinoceros, camel, rodent, and monkey. In the **Oligocene Epoch (34-23 MYA),** the first elephants with trunks emerged, along with early horses and many of the grasses that now cover Earth's plains. These grasslands, along with the warmer climate of the **Miocene Epoch (23-5.33 MYA)** facilitated the rapid evolution of hoofed animals and larger primates like Sahelanthropus tchadensis. The Tertiary Period ends with the **Pliocene Epoch (5.33-1.8 MYA),** which is when we find the earliest recognized evidence of the first hominids.

The **Quaternary Period (1.8 MYA-present)** extends from that point to the modern day, widely considered the span of the "age of man." But if you understand the past several eras as developmental stages leading up to the emergence of modern man, it becomes clear that the story of man began long before this period. Of course, the story of hominid development is, in its own right, significant enough to deserve its own chapter.

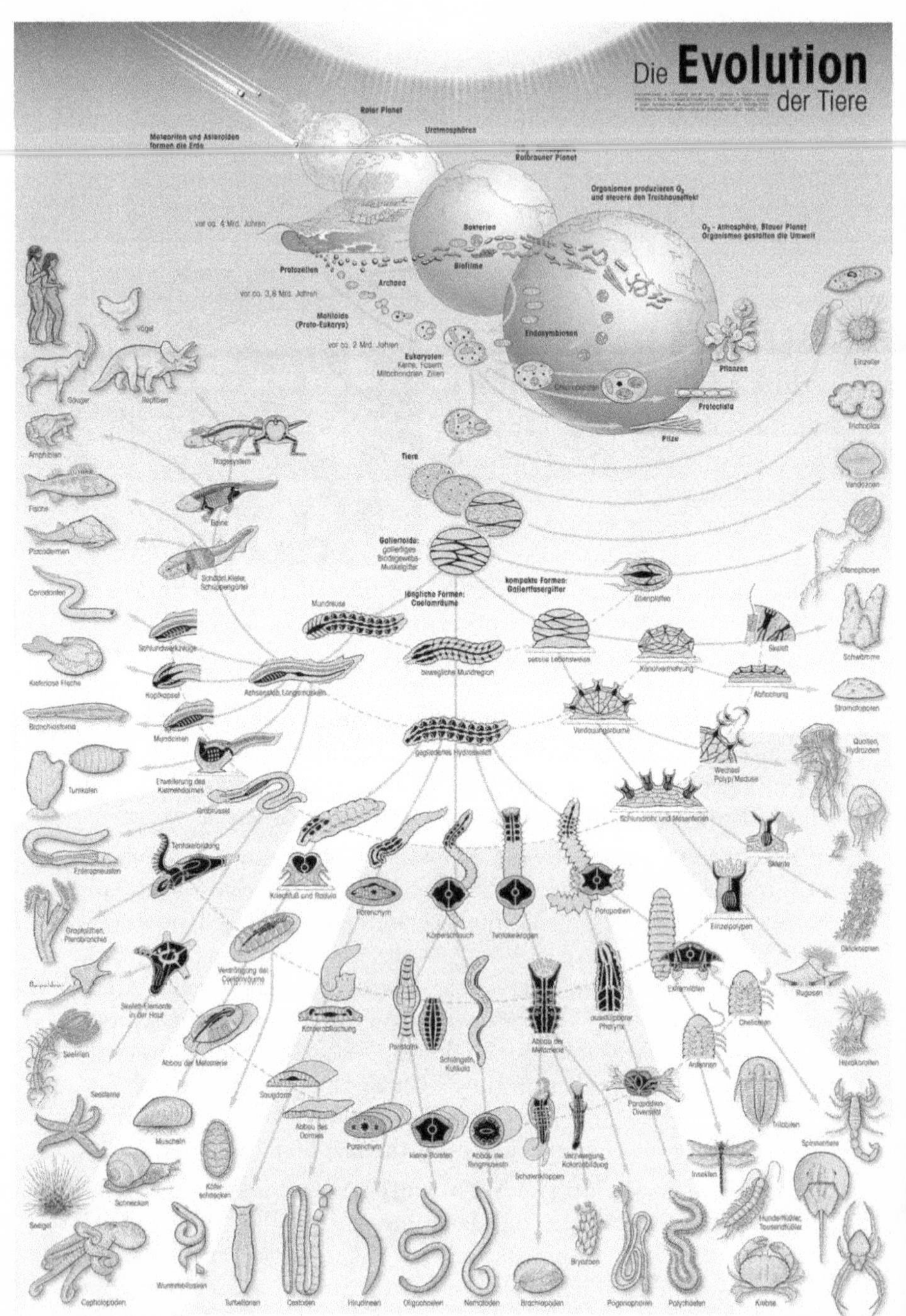

We chose this chart because it offers both a great overview of evolution as well as a challenging activity. This German chart isn't available in English, but thanks to the Internet, that should no longer impede your ability to use this or any other resource. Simply visit translate.google.com to get the English version of anything you see. There are also several free translation apps for smartphones, including one that talks to you.

The Original Man

THE ORIGIN OF ALL

"If I were president of one of these black colleges, I'd hock the campus if I had to, to send a bunch of students off digging in Africa for more, more and more proof of the black race's historical greatness." – Malcolm X

We've already shown how life came to be on this planet, and the laws we can observe by the course such developments take. We mentioned that the development of life on Earth resembled the development of man in the womb, but stopped short of completing the story with the development of modern man. And most readers are comfortable with the idea of cyanobacteria giving rise to multi-celled organisms which then gave rise to sponges, and so on. It's only when we reach the issue of man himself that we find people saying the rules of evolution don't apply to us. But considering how much misinformation is out there, we don't blame them. Again, that's why this book exists.

We're going to clear up some of the myths and misconceptions about the development of humans, showing how man is older than Homo sapiens, that Homo sapiens and all his ancestors were Black, and how today's Black man emerged as the heir to the throne of the Original Man. Most of us know that the Black man is the first man, but do we know the story behind this fact, or what it means for us today? And what about the Black Woman? We'll cover all that, along with a few discussions on the mysteries of our DNA, global extinctions, and several other issues that hold key insights into both our past and our current predicament.

Throughout this chapter, you'll see that not just cognitive ability, but consciousness has accumulated over time. Thus it is fitting that our species' current classification, Homo sapiens, literally means "Man, the Knower." Yet mainstream science says man only became fully conscious within the past 40,000 years, in direct association with the time he ventured into Europe. That's no surprise. We're going to demonstrate how man was quite conscious over a million years ago. To do so, we're going to have to show you who Man really is.

WHAT YOU'LL LEARN

- ❐ Why man is more than Homo sapiens.
- ❐ How sex and gender came to be on this planet.
- ❐ Why the Black woman is paralleled in nature by the Earth.
- ❐ Whether man came from monkeys or monkeys came from man.
- ❐ How man's ancestors directed their own evolution over several million years, and the decisions it took to do so.
- ❐ What any man must to do to survive and become successful.
- ❐ What it takes to survive a global extinction.

WHAT IS MAN?

SUPREME UNDERSTANDING

If you've noticed, in every chapter, we have to define the words we're using to establish what we mean, but also what we don't mean by those words. Without defining our parameters, ideas can be misinterpreted and carried to extremes that we didn't intend. So, before we discuss the history of the Original Man, let's define "man."

C'BS has already explained that, in essence, man means "mind." If this was a dictionary, his definition would be definition number one. Definition number two is the one I'll be using for this chapter, because we'll be dealing with biology, anatomy, and anthropology…which are "hard" sciences that require concrete subjects to study, rather than the abstract concepts we can explore when we go beyond those sciences. So, for the purposes of this chapter, when we say "man" we're often referring to the anthropologists' distinction of man as *Homo sapiens*, but we're also going to reject some of that limitation and extend the use of "man" to include our ancestors as far back as we can trace ourselves walking upright and doing "human stuff." In some cases, this is LONG before more scientists accept as the origin of "modern man" or *Homo sapiens sapiens.*

COULD MAN HAVE BEGUN TRILLIONS OF YEARS AGO?

Well, it goes back to how you define "man." If you're talking about "man" in the sense of the Mind, then we don't even have a starting point for that. As we've discussed earlier, the Mind was "here" when there was no "here" and when there was no time to begin with. If you consider the organizing properties of the universe as the origin of man, and you realize that man is the personification of those organizing principles, then you can trace "man" back to whenever you want. However, you wouldn't believe that "man" was hopping around butt-

> DID YOU KNOW?
> There are at least 5 million undescribed species on the planet; and some estimates say the figure may be as high as 50 million. Only about 1.9 million species have been given scientific names. Vast numbers of bacteria, insects, and creatures of the deep sea remain undescribed.[220] If you find one and identify it, you could name it what you want! Scientists have named a horse fly after Beyonce, a type of lichen after Pres. Obama, and a slime mold beetle after G.W. Bush.

naked when the planet was still a hot ball of molten rock, back when the ozone layer hadn't even formed. At that time, our ancestors were busy building the ozone layer so "modern man" could take shape, but those ancestors were single-celled organisms! Yet, if single-celled organisms today are any indication, we know that the organizing properties of the universe were present in them as well.

So, we won't find man on the planet until the planet was ready for man. Yet the origins of man are not the origins of "man" (as in male) alone, because it takes two to tango. Just as we can trace back the origins of sex to the relationship between photons and particles, we'll see a similar process unfold when we look at the origins of sex on Earth.

THE BLACK MAN IS THE ORIGINAL MAN

C'BS ALIFE ALLAH

"We have found the Black complexion or something relating to it whenever we have approached the origin of nations…In short all the…deities were black. They remained as they were first…in very ancient times." – Godfrey Higgins, Anacalypsis

Wherever you find elements of blackness, you are taking a journey to the primal nature of the universe. In acknowledging the Black Man as the Original Man we are stating that the biological genetic framework of the Original Man is a dynamic system which develops and passes down through time. Though there may be a primal man (or "first man") to trace back to, it doesn't take away from the modern Black Man being the "Original" Man, in that he is the heir apparent to the functional dominant biological blueprint. Seeing things this way, we will not spend years, months and days looking for "the ultimate ancestor," we will realize that we ARE the ancestors, whose genetic material has been encapsulated and renewed through time and space.

When we're looking at the Original Man, we're looking a huge superset of all possible values. Looking at the travels and journeys of the Original Man across the planet, you see one who has within him the full spectrum of dominant and recessive genetic traits. Though topical melanin is utilized as a major marker for our classification of "Original

Man" it is just one relevant marker, not the only one.* The Original Man is not merely identified by his activated dominant genes, but by his ability to sequence and re-sequence the full array of genes from dominant to recessive in new patterns and arrangements. This is evident by the variation of the Original Man across the planet.†

Futher, the variation of the Original Man is not simply "reacting" to the environment. Instead, he is drawing from his own genetic potential to rise to the occasion. If he didn't have the potential to evolve himself then he would perish. His whole genetic makeup is geared to move in tune with his environment because the environment is an extension of him. This is why we "interbred" with various other lineages of "mankind" before venturing into new regions of the globe. This was all about the optimization of the Original Man. For example, interbreeding with Neanderthals and Denisovans boosted the immune systems of those humans who ventured into Europe and the Pacific Islands (who wouldn't have fared as well without those synthesized traits).[221] This has continued into recent times. Ever wonder why enslaved West Africans were better equipped than Europeans or Native Americans to survive the diseases and conditions they encountered in the Americas? As we noted in *The Hood Health Hand*

Books like *The African Exchange: Toward a Biological History of Black People* by Kenneth F. Kiple and *Black Superman: A Cultural and Biological History of the People Who Became the World's Greatest Athletes* by Patrick Desmond Cooper actually explain how diseases like malaria actually strengthened West African populations and paved the way for Black physical dominance in the Western world.

In fact, many Native Americans believed Africans had "Great Medicine" in their bodies because they were virtually immune to the European diseases killing off the native populations. How much of this was coincidence, and how much of this was pre-adaptation?

There is a wealth of material in our genetic lineage that emerged independent of any stimuli in our environment. We are not just the products of our environment. We are the producers of our environ-

---

* Full classification would involve weighing phenotypical (surface level) traits such as skin hue, eye color, hair texture, as well as genotypical (genetic level) traits.

† Some may take offense at the classification of all "People of Color" as "Original" and not limited strictly to the "Negroid" phenotype/genotype. We assert that every aspect of the Original Man must be weighed equally. Thus we are looking at an active system who may have some genes turned on and others off. It is only when certain genes are ABSENT that we set up a separate classification system of man. In the foundational eras of most Original people throughout the world, we find the dominant genes active and thus darker skin.

DID YOU KNOW?
The Tiktaalika, a walking fish discovered in 2004, has limb functions very similar to land mammals. It was eventually recognized that this ancient fish had the genetic code for limbs and fingers long before they scrambled onto dry land.222 Many other prehistoric creatures possess early traits that would only later become fully fleshed out in the Original Man.

ment, and there is a whole world inside of us that chooses different ways to express itself, without waiting on the "green light" from the outside. In other words, we choose our own evolution, not simply to respond to environmental changes, but in anticipation of them.

This accounts for the adaptations of the Original Man who chose to explore the islands of the Pacific being different from the adaptations of the Original Man who chose to explore the mountain peaks of the Himalayas…which are both different from the Original Man who sojourned into the jungles of South America. In many cases, these changes weren't reactionary developments. Instead, such pre-adaptations draw upon the original genetic template, functionally proactively rather than reactively. We explore some of the background to this topic in "Another Look at Evolution" in the Appendix.

For now, it's only important that you understand that the story of life is not simply a story of adaptation, but also of pre-adaptation. Much of our success in transitioning through over four billions years of development has been due to us not simply choosing the right adaptive strategy for the present circumstances, but having the foresight to know what changes to make for future circumstances.

As we note in "An Introduction to Superorganisms" and "Another Look at Evolution," life accumulates consciousness through both aggregation (coming together) and across generations of evolution. Know what else has increased in concentration over the course of

ACTIVITY: QUESTION YOURSELF

Active learning (where you participate in the process) is more effective than passive learning (where you just take in information). So, throughout this book, we'll have activities for you to do. For starters however, this is not simply something you'll do once, but throughout the course of this book and anytime you encounter new information. This is another set of questions designed to help you verify and understand new information of any kind.

How could you explain this information to a 6 year old?
How can you use or apply this information?
Where can you look for more information on this subject?
What are the sciences and research methods used to draw these conclusions?
What are other topics worth researching after reading about subject?
What are some logical conclusions you can make with this evidence?
What are some illogical conclusions people can make if they don't understand this evidence?

evolution? Melanin. The Black man, the end result of all this consciously directed evolution and aggregation, not only embodies the highest form of living mathematics (Mind/Consciousness), but also the greatest concentration of both topical (akin) melanin and neuromelanin. This is no coincidence. As we note in "In the Heavens Above," melanin is the key chemical component that organizes organisms and turns energy into life. There is no coincidence to why the Original man is a Black man, and thus why the Black man is the Original man.

Creation of the Personal Body

DIVINE RULER EQUALITY ALLAH

There's no such thing as the beginning of man. There is a beginning, however, to man's personal body. That is, the physical form that man carries himself in. This personal body is grown from the energy of the sun's radiation and born through water and carbon. Whether we're talking about the development of man within the womb of a woman or the evolution of humans on this planet, the formation of man's body always occurs in stages.

This begins with transferring solar energy into the womb of the earth's soil to manifest unicellular organisms which building and add on to become more complex organizations of life. The original source and cause of such organization is from Allah, which simply means the original Mind and intellect; the original intent and purpose of my self.

When we refer to Man (or mind/mentality), we are really referring to the structure by which matter condenses (gravitation, electromagnetism) into a logical order whether it is organic or inorganic. A person is a sublime condensation of organic matter, but the "driver" of this vehicle is not the matter but the Mind.

The Mind exists beyond space/time, as it is just a reference for the underlying order and structure from the microscopic (quantum) to the mesoscopic (middle range) to the macroscopic (galaxies). This structure is seen (or activated) in space/time as a male principle (radiation) and female principle (matter).

As noted earlier, the present day, personal body of man came about through evolution. Through the process and eventual form, it shows a mathematical evolution. People get "confused" about evolution because they don't understand that it was a process of selecting and rejecting biophysical elements to be placed in our present personal body. If you look at the morphology of our bodies as we progress from single cell to the infant you will see the bodies of other animals that resemble those various stages. The same process of selection and rejection that occurred on the Earth is revealed in the womb, where elements are constantly added and eliminated. All of these parallels are generated by the same structure, which we call the Mind.

We also know that, during pregnancy, the baby lives entirely in water until it is born. In our planet's evolutionary history, we know that organisms living in water predated land animals. Man emerged with a body that is successful at taking advantage of all mediums on our planet: water, land, or air.

Dinosaurs were not intelligent enough to survive the climatic change, whereas our ancestors did. Yet the Homo sapien form, as with any forms that preceded it, is only a tool or body. MAN gives Homo sapien the direction. The Original Man is not just "the body," he is MAN. Thus, instead of searching for the correct ancestor, we know the common ancestor of ape and human, of plant and human, of bacteria and human, is MAN (the Mind). The body Homo sapien is simply a concentration of Mentality into command centers, in a form best suited for today's environment and circumstances. So when we say MAN, we mean the highest order of organization or mathematics, not merely the physical form this mind operates. Thus, this man is the man that was always here, before apes, and before dinosaurs.

GENDERATION (PART TWO)

C'BS ALIFE ALLAH

ARE YOU SEXIST?

Most of us are sexist, meaning we adhere to sexism. I'll explain. Any word with an "ism" as its suffix simply means a classification system, based on whatever the root word is. So yes, if you utilize the racial classification system, you are racist. If you recognize that there are male and female sexes, then you are a sexist. So, yes you're probably sexist.

Now, just because you utilize the racial classification system doesn't mean that you are an advocate of white supremacy. Just because you acknowledge male and female doesn't mean that you are a misogynist or chauvinist. Being a sexist, at its root, is observing and comprehending that there is a difference between man and woman.

Acknowledgement of these differences is what leads to a productive dynamic in society and in interpersonal relationships. Therefore sexism is seeing that there is a most beneficial norm for civilization that is measured from the qualities of gender. And yes, gender roles may shift somewhat over time and space, yet no society ignored collectively defining what said roles were to be. Many confuse this with "stereotyping." The difference is subtle, yet it is there.

NON-BIOLOGICAL GENDER VS. BIOLOGICAL GENDER

In the first article on "Genderation," we introduced the notion that there is a non-biological system of gender typing, which we can relate to the biological system of gender typing. There are many places where they overlap and a few places where they may seem at odds with each other. Just because we may relate one non-biological (or inorganic) gender notion to its biological counterpart doesn't mean that the biological counterpart lacks the other polarity of gender. For instance, we related Man to "energy" and Woman to "matter," yet it would be silly to say that a man's body isn't matter. It means that qualities that resonate with men and women appear in these objects and/or processes, but not necessarily in a rigid format.

THE EVOLUTION OF SEX

As Supreme Understanding noted in "The Origin of Life" there was a point in time where all of the original organisms on the planet were asexual beings.* This asexual process can be detected in microorganisms and plants, (as well as some worms, fish, and other animals) who are, on the tree of life, our distant cousins. If you observe the transitions in the morphology (form) of the human embryo, you can see the relationship (and lineage) for yourself. The evolution of biological sexual reproduction seems to parallel the process of cell division in the embryo, before sex is differentiated (notably, at a point when the embryo resembles other organisms that have gender).

Asexual reproduction also parallels the inorganic process of nuclear fission (splitting), while sexual reproduction parallels nuclear fusion (merging). Notably, nuclear fusion happens in the Sun, while nuclear fission happens within natural reactors in the Earth. Through the interaction between the Sun (or other stars) and the Earth, we can experience a progression along the periodic table of elements, where subatomic particles fuse to form elements from the lighter to heavier

* This means "without sex." It doesn't mean that they were the male and female sexes combined, which would be hermaphroditism. This was the point before even the creation of biological sexual genders.

elements, while fission produces the heaviest elements (like uranium, plutonium, and neptunium, which the Sun cannot provide).

Returning to the biological level, we know that single-celled organisms united to form multi-cellular organisms. This allowed for them to be more adaptable to the environment and to express more of the complex permutations of their biology that are encapsulated on the quantum level. In fact there is evidence that early bacterial and viral sex developed as a defense and repair mechanism from the early epochs of the oxygen-deficient Earth being bombarded with ultraviolet rays.[223]

THE EVOLUTION OF GENDER

But the emergence of sex (and biological gender) didn't happen overnight. Gender has been around as long as the universe has been binary, but it took a long progression from asexual to sexual reproduction to manifest it as male and female members of a species. At first, life reproduced strictly by asexual reproduction (splitting or budding) where we literally cloned ourselves (Not as people! But as micro-organisms!). At some point, organisms began to merge with each other, and there was a partial exchange of genetic material. Then there arose a phase where some organisms were able to reproduce asexually and sexually (through mergers). From that batch arose some that were able to reproduce only sexually.* This type of reproduction allowed for even greater permutations of the genetic material that was developing on Earth.

So it appears that life gained its foothold on Earth by unifying resources at the cellular level. Once these were consolidated mechanisms, they used sexual reproduction (mergers) to expand those resources in new and inventive ways. Organisms could select mates that would confer the most survival advantage to the offspring, allowing for developments that could not be accomplished in ten times as much time through cloning alone.† This allowed life to move beyond

* Though this may appear to be the simple end point, it is not. Sexual reproduction started with no dedicated sexes (as in plants and some lower animal forms). Organisms also developed that could become either male or female instead of having set reproductive forms. Finally, some organisms evolved to the point where the sexes are physically very different (known as sexual dimorophism), as in primates. Yet the difference are even greater in other creatures, such as pinnipeds (seals and sea lions) and ungulates (deer, cows, goats).

† After all, cloning only allows new traits to emerge by replication error or genetic mutation, which requires much more time before an advantageous change emerges

its former boundary, which, at that time, was the ocean. As we'll see in Volume Two, coming together is what got us free.

Leaving the ocean is synonymous to leaving the womb. Prior to the development of life adapting to land, many (if not most) organisms were born as complete organisms in the ocean and used the ocean and its materials to grow and expand, like a primordial amniotic sac. When we left the oceans and waterways, we took the water with us in the form of the female gender. Thus the female gender was made from the original organism, the same way that the Earth is made form the Sun, the same way that particles come from photons: The original organism breaks off a part of itself. This occurred so that the female of the species could be a receptor for the seed from the male of the species. There were no receptor organisms before this, as the seed was emitted intact. We see the precedents of the sperm and egg in the development of the cell within this ocean, as well as the evolution of life that follows.

THE SPERM AND THE EGG

The sperm is a mobile organism. The egg is a static interface and substrate that feeds that organism and allows it to grow and develop from its simple stage to the more complex stage.* We're not discounting the importance of either part of the equation, just identifying the role of one as the activator (the sperm) and the other as the receptor (egg). The feminine element is vital in the growth and development of life based on the underlying structure of the universe (the Mind). Without the materialization of the feminine principle in the early moments of the Universe, everything would have quickly rushed towards a steady state of energetic stagnation, known as heat death. In the cosmological sense, the feminine principle exists to postpone the "heat death" of the physical universe. In the anthropological sense, if there were only males, then we would "blow ourselves up" in similar fashion. Thus it follows that our ancestors – regardless of how they described the various personalities of the lower pantheon – nearly always described the source of the Universe as a "He" and the material used to fashion the Universe as a "Her." From ancient India to the Americas, myths describe the Creator – upon becoming aware of his own existence (in a void or vacuum) – immediately creating a feminine "consort" with whom to engender the universe and begin creation. On an individual-

---

(and an infinitely longer period of time if it's all "random").

* Keep in mind that we're not just talking about people here. We can see the sperm and egg in ferns and mosses (which were on the planet over 400 million years ago).

level biological scale, this same principle plays out as the female increases the survival rate of the species by offering an intermediate home for the fetus until it is able to breathe on its own in the world at large.

In recent years, some have argued that because mitochondria* can be traced back further in the female that she is the "precursor" of the male gender.† Mitochondria doesn't prove that females were on the planet first; it only proves that people inherent mitochondria through the female.[224] Since the male sex has, as one of its polarity-level traits, the tendency to evolve rapidly to meet the demands of the environment, it's not as easy to trace "stable" aspects of the male genetic lineage.[225] Again, this parallels non-biological gender typing, where matter (female) is more static with less flux, while energy (male) is more dynamic with less stability.

GENDER ROLES

"Man means mind and woman means the womb of mind" – Imam W.D. Mohammed

The social constructs of "Man" and "Woman" refer to the gender roles of the physical forms (which fulfill the potential of their genetic foundations) and of the gender typing of the mental forms. If we were to isolate the etymological roots, we know that "man" is rooted

A QUICK NOTE ON GENDER RELATIONS

Does the Original Man being the "personification" of the Mind and the Original Woman being the personification of Matter mean that Woman does not have a Mind? No more than you think we mean that Woman being Matter means that Man is not made of Matter! When Five Percenters describe their women as the Earth, the do so with reverence, knowing just how important the Earth is. Being the Earth does not equate to being treated like "dirt." At least it never should. Yet because of Western domination, many of us associate concepts that identify Man as the active principle (Yang) and Woman as the receptive principle (Yin) with concepts of "better" and "worse." If you look at the Yin/Yang symbol (known as the *Taijitu*), both portions contain a piece of the other, and neither side is bigger (or better) than the other. Some of us, in a rush to remedy rampant misogyny have crusaded against what makes sense (despite how people use or misuse it), in favor of lofty ideals that neither make sense, nor can be applied to anyone's benefit, but which "sound good." It does people no good to elevate their sense of self with ideas you can't substantiate. For example, I've never heard a reasonable explanation for what makes a "Goddess" any different from (or better than) the Mother Earth. After all, every divine being that Western anthropologists would label a "goddess" has been a representation of the Earth, Nature, Fertility, Matter, or the Womb (of Space), all of which are encapsulated within the concept of the Earth. Yet, paired next to "God," "Earth" just doesn't "sound good enough" for some of us. I bet you wouldn't say that to the Earth's face though. She'd destroy you.

in mind (which we are utilizing as a place marker for structure). Woman would be rooted in "womb of the mind" or "wife of the mind," which would be a place marker for substrate. In this sense, neither is fully independent of the other. Further, only biological principles have birth records. In the non-biological realms of gender, the singularity (man) and the field (woman) don't have a "beginning." Yet, in the order of operations, masculine principles would precede feminine principles. Mind precedes Matter because he initiates and gives structure. This is reflected on the biological level, as seen in the dominant gender determination coming from the male in the species, as the XY vs. XX. To be clear about this point, the Original woman is not the "opposite" of the Original Man. If that was the case, it would be YX vs. XY not XX vs. XY. She is not the counterpoint. The relationship between the Original Man and Woman is one of complement. It is the balance between the both that promotes productive activity. It is the harmony between activation and reception that keeps life and matter moving forward.

Did Man come from Monkeys?

SUPREME UNDERSTANDING

Okay, let's get something clear: Scientists don't actually claim that man came from monkeys. Really. They don't.

That's a myth spread by religious people, to make the idea of evolution seem more repulsive. And for Black people in particular, the idea of coming from monkeys seems especially heinous, considering how much abuse Blacks have endured at the hands of whites who called us monkeys, gorillas, and apes. C'BS once noted to me, "If we'd been told that we descend from lions, would we be so turned off by the idea? Or would we brag that we come from the mighty lion?" Personally, C'BS and I think we'd be yellin "Lion Power" or somethin right about now. But the monkey thing leaves a sour taste in our mouths, and I understand the resistance. But the truth is, early white scientists KNEW that the Black man was the Original man, so they made the "monkey connection" something we would hate, leaving us totally disinterested in hearing anything about our origins in this world (except for religious mythology, and you see how THAT went for us!).

Either way, there's no point in debating whether monkeys are actually quite respectable, because the scientific community only suggests that man and monkeys share a common ancestor. This ancestor is often called a primate, which doesn't bother me much, since I understand the root word of primate, primal, primordial, and primitive to be *prime*, which means FIRST. That's what's important to me. But I understand that most people still find the connection discomforting, so we'll use another term sometimes used for our common ancestors: Hominids. This basically means "like man." The question is, how much like man and how much unlike?

While we certainly agree than the development of man was clearly a "development" involving several stages of transformation, what is unclear is exactly how far back we'd have to look to find man when he was not yet man. This is because there's so many gaps in the fossil record, scientists are really guessing at how it all happened. Because most scientists still see modern Europeans as the highest stage in human development, working backwards means they're looking for something primitive and monkey-like in behavior and appearance. But just as a human is not "primitive" when it's a baby, there's no reason to view early man as some savage half-beast. The evidence doesn't support that view, either. In fact, remains of early 'true' primates still have not been uncovered in Africa.[226] But that doesn't stop mainstream science from reconstructing some imaginary ancestor of humans and monkeys, looking completely like a monkey, even when there's no fossils of this so-called ancestor.

Instead, the fossil record suggests that "evolution" may have been happening in two directions, with some groups of our ancestors becoming more like modern man, and other groups becoming more like modern monkeys. In fact, scientists are constantly finding hominid remains that rewrite the traditional timeline entirely. Whenever they find a "cousin" on our family tree that they didn't know about (or imagine), they have to revisit the whole idea of whe the earliest ancestor must have been. For example, a fragment of a pinkie finger excavated from a deep cavern in southern Siberia may point to a new species of ancient human.[227] The 40,000-year-old bone yielded DNA markedly different from that of modern humans or Neanderthals, challenging the current view of how our ancestors migrated out of Africa. The mitochondrial DNA of the unknown female hominid, whom they nicknamed "X Woman," differed from present-day human DNA at nearly 400 positions, twice the difference measured between human and Neanderthal DNA. The genetic patterns indicate that X Woman, Neanderthals, and modern humans shared a common genetic ancestor about a million years ago.[228] One of the researchers suggested that X Woman may belong to a group of archaic humans who migrated out of Africa at a different time from Neanderthals or modern

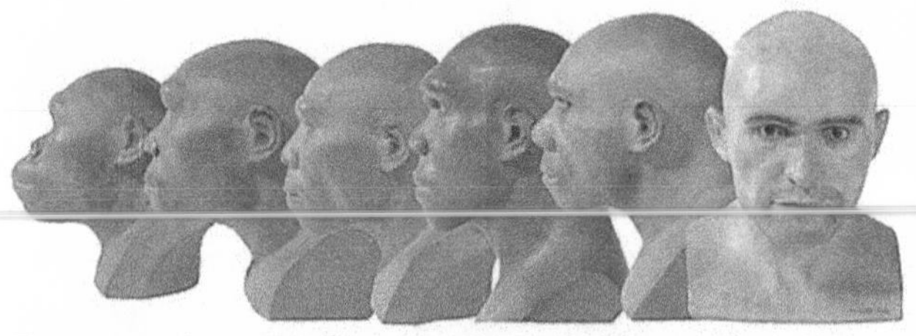

Reconstructions of (left to right) Australopithecus, Early Homo erectus (Java Man), Late Homo erectus (Peking Man), Homo heidelbergensis (Rhodesian Man), Neanderthals, and Early Homo sapiens (Cro-Magnon). Notice the "apex" of human development.

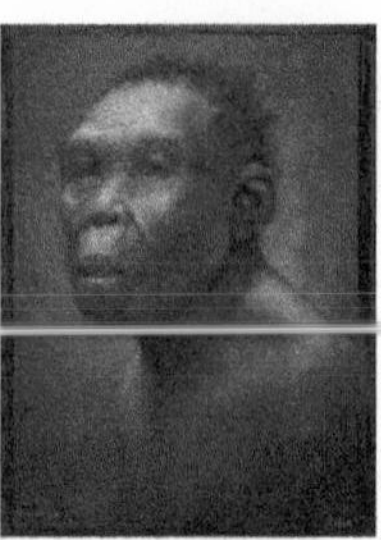

Artist rendering of Homo ergaster by Andrew Baker.

Alternate model of Homo ergaster by Viktor Deak

Homo heidelbergensis

Homo erectus

Bamboo raft made from stone tools to reproduce the voyage of Homo erectus from Timor to Australia

Homo sapiens idaltu

Homo sapiens sapiens

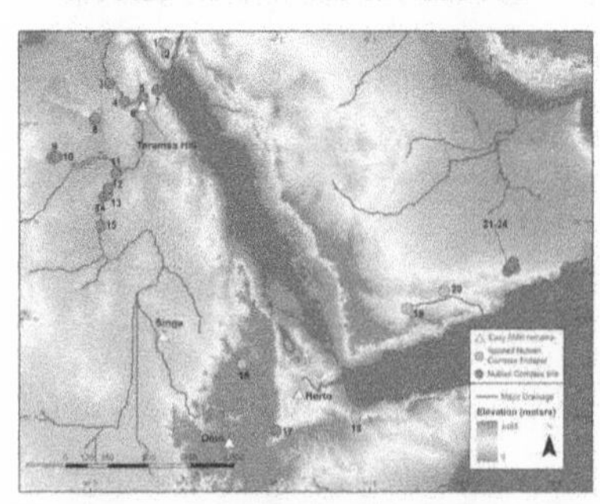

The "root of civilization," known as the Nubian Complex circa 125,000 BC, from which human populations dispersed to settle the globe. (above) The original people of the world (right)

DID YOU KNOW?
There's also anthropological evidence that the split between humans, chimpanzees, bonobos, and gorillas was recent enough for us to have some CULTURAL traits in common. For example, among chimpanzees and gorillas, there's evidence of tool use, precursors to language use, a preference for cooked food, agreed-upon social structure, percussion (as in drumming), and even instances of altered states of consciousness (as in shamanism).

humans.[229]

Let's keep looking. In 1981, *New Scientist* published a controversial pair of articles by John Gribbin and Jeremy Cherfas that argued:

- There are no fossils that are unequivocally ancestral to chimpanzees and gorillas but not to man;
- Therefore, the only good measure of the time when these three species split from one another is the comparison of genetic material;
- Genetic dating and serological techniques are unanimous in dating the chimp-gorilla-man split at about 5 million years ago.

While it was controversial enough to suggest that chimpanzees, gorillas, and humans diverged from a common ancestor only 5 million years ago (instead of the widely accepted 20 million years), the authors' next claim was even more shocking. Gribbin and Cherfas, after considerable fossil analysis, suggested that chimps, gorillas, and man descended from an ancestor that was more man-like than ape-like. That is, chimpanzees and gorillas are descended from man rather than vice versa.[230] Of course, this idea didn't go over well with the scientific community.

But that same view is finally getting some support from the mainstream. The recent discovery of a 4.4-million-year-old hominid named *Ardipithecus ramidus* (shortened to "Ardi") pretty much killed the idea of a chimplike missing link at the root of the human family tree. Fossil bones from Ardi and at least 35 other children and adults, dug up in the Afar desert in Ethiopia, suggest that our ancestors lived in lush woodlands and walked on two feet...before the hominid fossils that suggest we were hunched over in the grasslands. Okay, so even before Dinquenesh* and Australopithecus, we were more human than the science books have told us. And perhaps the more monkey-like Austrolopithecus wasn't an ancestor of modern man, but was one of the many populations that went in the "other" direction.[231] It makes sense, considering nobody's got the "missing link" to prove otherwise.

You may have heard about Mary-Claire King's 1973 finding that 99%

* Dinquenesh, meaning "wonderful thing" is the Ethiopian name for "Lucy," an important Australopithecus find.

of DNA between human beings and chimpanzees is identical.[232] But that's not entirely accurate. Later research modified that finding to about 94% commonality, with some of the difference occurring in non-coding DNA.[233] Even more recent research suggests that the commonality is 86% or less.[234] That's not so impressive, considering that humans have 60% identical DNA with fruit flies, 67% with mice, and 90% with cats.[235] This new data makes us wonder whether our common ancestor was more like chimpanzees or more like us.

We know it sounds a bit crazy to suggest that chimpanzees, bonobos, and gorillas (considered man's closest relatives) descended from hominids that were more like man than like them. But there's a solid scientific background to this perspective. Biologist Richard Dawkins, in his book *The Ancestor's Tale* proposes that chimpanzees and bonobos are descended from *Australopithecus afarensis*.[236] French zoologist François de Sarre also argued that some of the Australopithecines eventually evolved into gorillas and chimps. Yup, the same people that scientists say we descended from.

That is, Australopithecus (a "human-like" mammal, who many scientists say gave birth to the *Homo* lineage), gave birth to a species of mammals that became gorillas. Dawkins also argues that gorillas are descended from *Australopithecus robustus*, who lived about 2.7 million years ago, and descended from Australopithecus afarenses as well. Well, if monkeys descended from the closest thing to man before man, it would certainly explain why the fossil record has more evidence of early man than monkeys.

Another study published in the esteemed journal *PLoS ONE*, makes the argument that *Sahelanthropus tchadensis*, a likely ancestor of Australopithecus who lived 7 million years ago, could have been the hominid ancestor of apes and chimpanzees, citing a variety of evidence (such as anatomy and early bipedalism among others.[237] The study's author, anthropologist Aaron G. Filler, goes even further in his book *The Upright Ape: A New Origin of the Species*:

> The first "human" was probably Morotopithecus and probably lived 21 million years ago. The existing apes have a human ancestor.[238]

Other studies have corroborated Filler's findings, showing that early hominids didn't start out walking on their knuckles, and that apes actually developed this ability independently after splitting off from the last common ancestor with the human lineage.[239]

Some have argued that Australopithecus wasn't an ancestor of man at all, and was another descendant of the lineage that became man. Palaeontologist Yvette Deloison believes that, 15 million years ago, there

DID YOU KNOW?
The need to find a white origin for humanity has gotten so desperate that scientists – unable to find any "ancestral" fossils outside Africa for ANY hominid or pre-hominid for the past 39 million years of the fossil record – are now trying to suggest that "in the middle Eocene, 39 million years ago, there was a surprising diversity of anthropoids living in Africa, whereas few if any anthropoids are known from Africa before this time...this sudden appearance of such diversity suggests that these anthropoids probably colonized Africa from somewhere else. Without earlier fossil evidence in Africa, we're currently looking to Asia as the place where these animals first evolved."[243]
Translation: We, white people, hate the idea that we started out in Africa. We tried all kinds of theories but none of them worked. We kept ending up back in Africa. But we've FINALLY found a point where we can't find an ancestral fossil for these African ancestors of man, so we're excited to assume they must have come from outside Africa...and "colonized" it! Like we did!"

were 3 species of bipedal (walking upright on two feet) primates: one of them developed into hominids (Homo), another became the semi-bipedal, semi-arboreal (tree-dwelling) australopithecines, and the third developed into quadrupedal orangutans, gorillas, and chimpanzees.[240]

And let's go back even further. Surely, if we go back another 40 million years, we'll find an ancestor who was on all fours, looking more like a monkey than man, right? Wrong again. Recent studies of a well-preserved fossil dated to 47 million years ago reveals "an animal that had, among other things, opposable thumbs, similar to humans' and unlike those found on other modern mammals. It has fingernails instead of claws. And scientists say they believe there is evidence it was able to walk on its hind legs."[241] Researchers report that this extraordinary fossil could be a "stem group" from which higher primates evolved, meaning monkeys, apes, and humans all descended from something that looked surprisingly more human than expected.[242]

Does this mean that humans were around 700 million years ago, running from dinosaurs like the Flintstones? Or that there were humans around when life was forming in the Earth's oceans 4 billion years ago? No, we're not saying any of that. We're only saying that the traditional timeline is flawed, and the traditional reconstruction of the "ascent" of man is flawed as well. We're saying that the fossil record suggests that the common ancestor of man and monkeys was more like man than like monkeys.

So if monkeys could become man, than why do we still have monkeys? It appears that some of us went left, some went right, but that's the nature of evolution, which is not always an "upward bound" kinda process. Instead, much of evolution is about variation, and whether the variations that arise will be able to survive and thrive in their environments well enough to grow and have descendants. In the end, who survived, out of all these variants? Us.

WAS DARWIN RACIST?

SUPREME UNDERSTANDING

Charles Darwin wasn't the first scientist to propose a theory of evolution that sought to explain how modern life came to be. He was just the first to publish one that covered all the bases. Yet, even with the comprehensive scope of his *Origin of Species*, Darwin got some things right and some things wrong. What he clearly got right was that there's no way that a closed system (a biosphere like Earth which is not receiving a constant influx of foreign life) can have up to 5 billion species (only 1.9 of which have been identified), unless everything that is on the planet now, has always been here.[244] And we know that's not the case. The fossil record shows us what life was like during different periods in the Earth's 4.5 billion year history, and most of those creatures aren't even around anymore. Meanwhile, creatures that are familiar to us today are absent in the fossil records of ages past. Then again, they're not actually "absent" as their ancestral forms were there. So, in a way, we've always been here, but not always in the form.

The question Darwin sought to answer, was HOW? And this is where Darwin missed a few things. To his credit, Darwin was developing these theories over a century ago, long before scientists discovered DNA or any of the genetic markers we can now use to figure out what gave birth to what. But, let's also be mindful of the fact that Darwin and his contemporaries had an agenda. While they promoted the concept of natural selection, that is, the idea that nature will kill off those animals that are weak (survival of the fittest), while allowing those who naturally develop the best traits (those which are most adaptive to the environment's demands) to survive.

But how did these traits develop? Darwin argued that it was all natural mutations. While we know that mutations can certainly occur naturally, most natural mutations are deleterious (meaning they lose data, resulting in "mutants" which are usually "messed up" somehow). While some mutations can produce offspring that are better prepared to survive, there's other ways to produce these results. But Darwin glossed over these possibilities, and promoted the idea of "nature producing the best candidates for survival" for a reason. We know this because we can identify and date when Darwin got his "flash of insight":

> DID YOU KNOW?
> The process of heating and cooling steel (as is done with a samurai sword, for example) makes it stronger because of carbon. Due to the heating and cooling process, the carbon particles change position within the steel, making its internal structure stronger. Repeating the process increases the chemical change. A lot like us, carbon-based lifeforms forged by adversity and strengthened for the next challenge.

In October 1838, that is, fifteen months after I had begun my systematic inquiry, I happened to read for amusement Malthus on Population, and being well prepared to appreciate the struggle for existence which everywhere goes on from long-continued observation of the habits of animals and plants, it at once struck me that under these circumstances favourable variations would tend to be preserved, and unfavourable ones to be destroyed. The results of this would be the formation of a new species. Here, then I had at last got a theory by which to work.[245]

Malthus? As we explain in *The Hood Health Handbook, Volume Two*, it was the 1798 publication of Thomas Malthus's *Essay on the Principle of Population* that provided much of the foundation for the later eugenics movement seeking to sterilize or otherwise "eliminate" the "less desirable races" of the world. This line of thought carried itself all the way into recent developments such as Margaret Sanger's Planned Parenthood, the World Wildlife Fund's subtle 1984 campaign to decrease the human population in the Third World, and the UN Population Division's March 2009 investigation into how to best "accelerate fertility decline in the least developed countries."

It's true that Darwin wasn't advocating forced sterilization, but that's because the technology for sterilization wasn't developed until well after Darwin died. Darwin was certainly supportive of the early eugenics ideas promoted by his cousin Francis Galton (who actually coined the term "eugenics"), and he favored eugenic restrictions on marriage.[246] As he wrote in *The Descent of Man*:

> Thus the weak members of civilised societies propagate their kind. No one who has attended to the breeding of domestic animals will doubt that this must be highly injurious to the race of man. It is surprising how soon a want of care, or care wrongly directed, leads to the degeneration of a domestic race; but excepting in the case of man himself, hardly any one is so ignorant as to allow his worst animals to breed.[247]

And that's why the full **original** title of *The Origin of Species* was *On the Origin of Species by Means of Natural Selection, or the Preservation of Favoured Races in the Struggle for Life*. Darwin was arguing on behalf of white people, the "mutants" of a world populated by Black and brown people. Darwin and his colleagues knew that white skin was a new development on the planet, but preferred to argue that it was a mutation preferred by nature, his argument for natural selection thus relegating the dark races of the world to extinction. We know many of Darwin's critics had an agenda because they felt his ideas on evolution were a threat to the Creation story of the Bible. But let's not act as if Darwin, too, didn't have an agenda. Natural selection, minus its use in promoting white supremacy, isn't "wrong" in theory – it's just not the whole story.

CAN WE DIRECT OUR OWN EVOLUTION?

And we know that he knew better, because he noted that there were other models for the evolution of species, including sexual selection (when a species changes because it expresses a sexual preference that allows certain members to have more offspring than others), and direct selection (when humans consciously decide to "breed" a species towards a desired goal). While direct selection is common in plant and animal domestication,* Even Darwin noted that humans could take some credit for their own evolution because we, too, have directed the changes that have occurred in our own species.

But how long ago were we conscious of this process? We know that the earliest recorded evidence for purposeful animal breeding (wolves to dogs) dates back at least 70,000 years, but what about before then? While we know that the earliest humans didn't care much for the idea of changing the balance of nature, what do we say about the changes we produced in ourselves? Could we have consciously been modifying our own species to adapt and survive? As we'll see in Volume Three, anthropological studies of indigenous people have revealed that even the most remote tribes engage in purposeful cultural traditions designed to produce desired classes of people.

HOW FAR BACK DOES IT GO?

The further back we go, we can still find evidence that evolutionary decisions were made consciously. For example, some scientists suggest that our loss of body hair was a result of "sexual preference" rather than nature alone. Others have said that the Neanderthals became differentiated from other hominids (they were VERY different) because of conscious selection, rather than some accident of nature. Even further back, we'll see that the morphology (physical form) and survival rate of older ancestors (like Homo habilis) was a result of their own adaptive strategizing, meaning they decided how to evolve themselves!

---

* See "Nature vs. Nurture."

Let's go back much further then, to the dawn of early life on this planet. Could the consciousness in early life have played a role in their development towards later forms? Could cyanobacteria have directed their own evolution? We know that living things have a natural desire to survive, and to produce offspring. So, if survival requires modification, could life – even without a brain – choose to make the necessary changes to survive? Studies show that, yes, they can.

Even simple organisms like bacteria are capable of consciously making choices and changes necessary to not only their own survival, but the survival of the entire colony. In many cases, some organisms will even sacrifice themselves to allow the collective a greater chance at survival. In other cases, the collective will eliminate selfish individuals who do not reciprocate in this manner. This is among BACTERIA, not human colonies! Another mechanism behind conscious evolution involves pre-adaption via prediction. That is, organisms "bet" on how to produce their next generation to ensure the highest odds of survival and success.

PRE-ADAPTATION=PREDICTION=SINGLE-CELLED GAMBLING

Scientists have observed that bacteria living in ever-changing environments can "hedge their bets" on survival by trying different combinations of traits in their offspring. Technically known as "stochastic switching between phenotypic states" this ability may have been critical to the success of primitive forms of life.

Bet hedging "may have been among the earliest evolutionary solutions to life in variable environments," even preceding the ability to turn genes on and off. Bet hedging is well known in nature (e.g. disease-causing bacteria "randomly" produce different surface proteins, knowing that a few will escape immune system detection). But scientists didn't believe that nature "gambled" in this way. As science writer Brandon Keim notes: "After all, in any given instance, it's better to have the right surface protein. But it's not always possible to know what's right in advance, especially in highly variable environments."[248]

Now we know that not only do organisms apply such strategies, but this form of "predictive behavior" forms a template for later methods of human divination, which itself predates today's "hedge funds" that operate on the same principles. This mechanism could even explain the diversity we find in the ancestors of modern humans. Different morphologies for different environments, all diverging and merging to ultimately produce a man with all of those variations still within him, the Original Man.

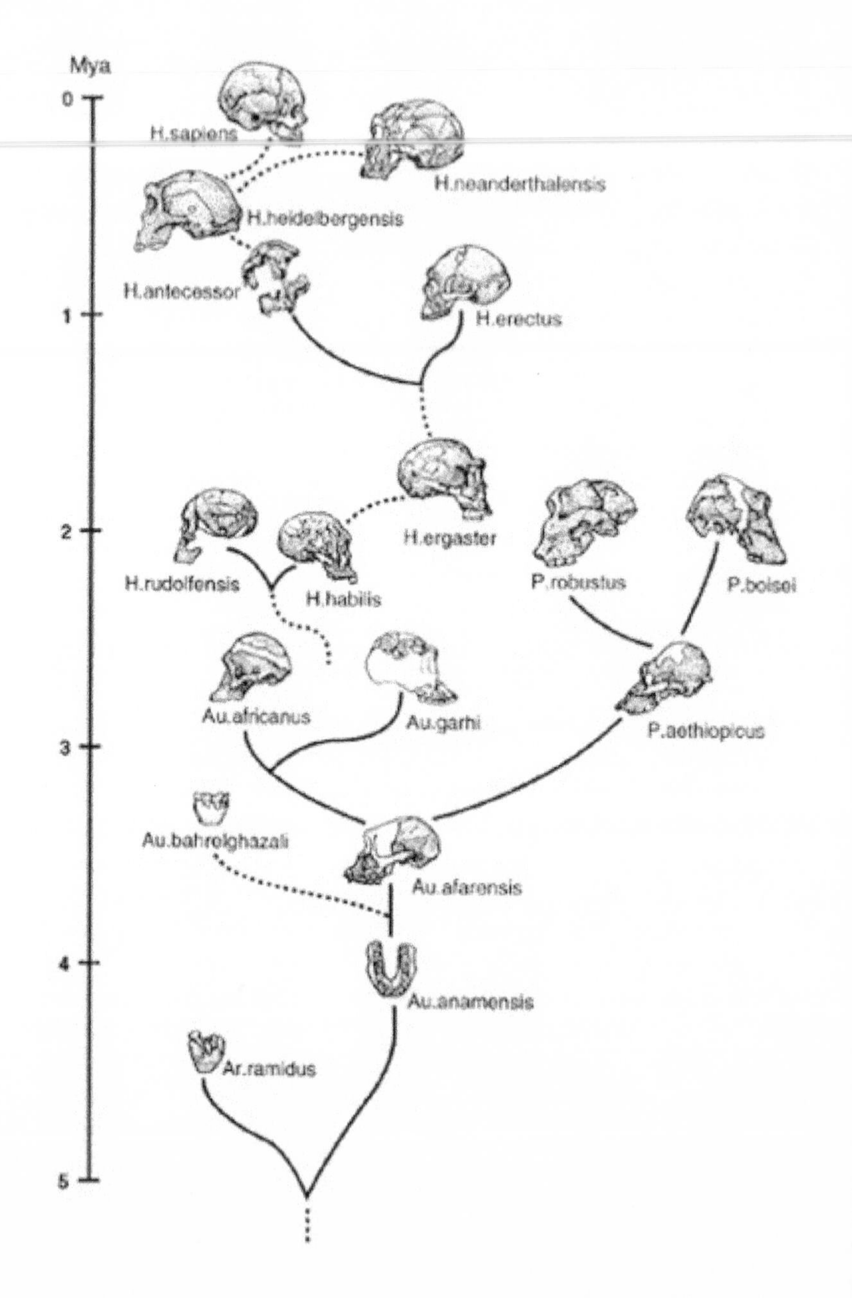

Hominid family tree from *From Lucy to Language* by Donald Johanson and Blake Edgar (1996).[249] Compare with the tree on the facing page.

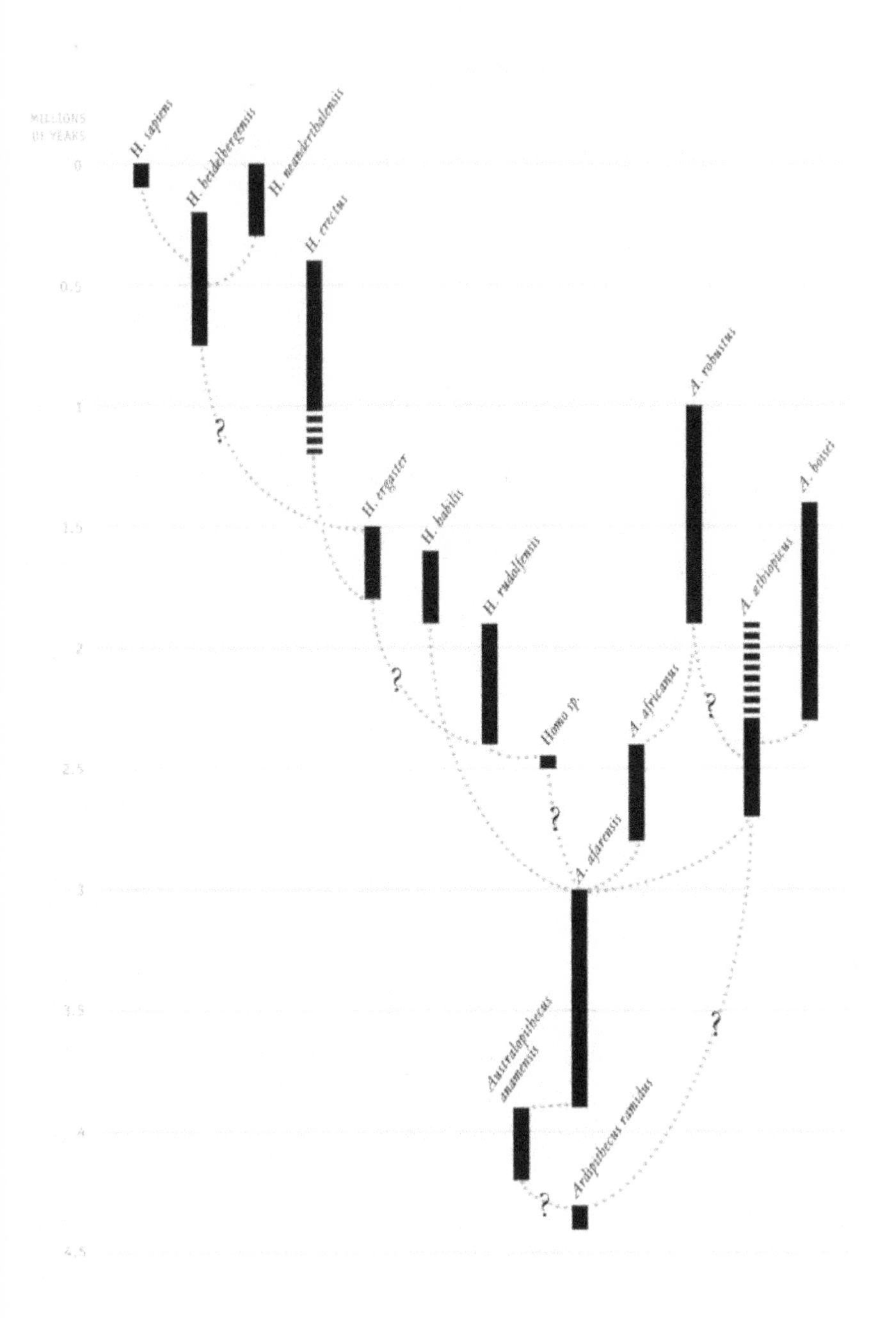

Hominid family tree from *Extinct Humans* by Ian Tattersall and Jeffrey Schwartz (2000).[250] One thing is clear: Much is unclear.

So let's recap. From about 7 million years ago, there were a LOT of folks running around Africa looking like the possible ancestors of man. Okay, technically, some of them may have doing just as much climbing as running, but you get the point. Despite the similarities, we can't identify ANY of these pre-hominid mammals as the direct ancestor of man. Why? First, because the fossil record is way too spotty, and there aren't enough transitional fossils to show who became what, and when. That's not to say that man circa 4 million years ago looked like Denzel Washington. We're just saying that scientists don't know if that man looked more like Kenyanthropus platyops or Australopithecus afarensis…or Denzel Washington.

VARIATION VS. EVOLUTION

The currently accepted fossil record for Hominids goes back about 7 million years ago, to *Sahelanthropus tchadensis*, a fossil nicknamed "Toumai" discovered in Chad. The next important find is from a full million years later, *Orrorin Tugenensis*, discovered in Kenya. From 5.5 to 4 million years ago Austrolopithecus anamensis lived in Kenya at the same time Ardipithecus ramidus lived in Ethiopia. From 3.8 to 3 million years ago, *Australopithecus afarensis* ran the streets of Ethiopia while *Kenyanthropus platyops* was kickin it in Kenya. From that point on *Australopithecus africanus* was the main player to be found in the fossil record, until *Homo habilis* emerges about 2 million years ago througout Kenya and Tanzania.

After Homo habilis, other early members of the Homo genus, i.e. *Homo ergaster, Homo erectus* and *Homo heidelbergensis*, migrated from Africa around 1.9 million years ago, possibly compelled by climate change or other factors. Whatever their reasons, these populations dispersed far and wide, with clear evidence of settlements in the Levant, southeast Europe, India, China, and Southeast Asia.* There is even some evidence – although it is only footprints – that Homo erectus made it to North America 1.3 million years ago. How? Considering the boat-building that would have been required for them to populate Southeast Asia at least 800,000 years ago, it's possible that these

* It's important to note that some scientists believe Homo erectus actually reached China about 2 million years ago, the same time Homo habilis shows up in the fossil record, which kinda throws a monkey wrench in this idea of a "linear progression," which we noted earlier.

DID YOU KNOW?
A 2011 study of prehistoric "human ability to manipulate fire and the landscape" in Africa suggests that we have known how to make fire for quite some time, but the use of fire was highly controlled. In fact, greater numbers of people didn't result in more instances of fire until the advent of plant cultivation and pastoralism (herding) called for the use of fire to transform the landscape. The authors report: "We show that substantial human impacts on burned area would only have started 4,000 years ago in open landscapes, whereas they could have altered fire regimes in closed/dissected landscapes by 40,000 years ago...The annual area burned in Africa probably peaked between 4 and 40,000 years ago." In fact, it seems our ancestors did more burning 4,000 than sub-Saharan Africans do in modern times.[254]

"primitive" ancestors navigated the Pacific Ocean to get there.

Talking about all these different "hominids" is where people get confused, because they think that anything that's not called "Homo sapiens" was some kind of monkey or mutant. So when we hear about Homo erectus, Homo habilis, or Homo ergaster, we picture something totally nonhuman. But scientists like Alan Thorne say that this is a false distinction, something like when "scholars" claimed that Africans and Europeans were different species because of the difference in our appearance. In fact, Homo erectus appears to have contributed some of his genetic legacy to the people of Asia, suggesting his community was still there where Homo sapiens came through 60,000 years ago.

HOMO ERECTUS: SOUNDS NASTY BUT IT WASN'T

So let's talk about Homo erectus. Thorne and others argue that Homo erectus wasn't another "type" of hominid, but just a "variant" of Homo sapiens that left Africa about 2 million years ago. Homo erectus left Africa after the Sahara (once again) became a desert. He settled throughout Europe (remains have been found in Georgia and Spain) and Asia (remains have been found throughout Indonesia, Vietnam, and China). And Homo erectus wasn't some upright monkey either. These guys were creating rafts that made the first sea voyages (currently known) over 850,000 years ago. Robert Bednarik's First Mariners Project proved that this journey could have been made with rafts made entirely with bamboo and stone tools (which says a lot about how advanced "primitive" people might have been).[252] Homo erectus made the first documented evidence uses of fire 1.5 to 1.7 million years ago in Kenya, South Africa, and China.[253]

However, we should note that (a) small groups don't make big fires,*

* Evidence includes thermally altered stone artifacts and circles of burned clay that have been dated to between 1.5 million and 1.7 million years ago. Anthropologists

> DID YOU KNOW?
> Homo erectus wasn't the only "extinct" population to stick around longer than we give him credit for. Another variant of early man, the Denisovans, contributed to the gene pool of modern Melanesians. And traditional accounts of indigenous people suggest that otherwise "extinct" populations may have survived in the isolated jungle and mountain areas of the world into recent times. Indigenous accounts describe people fitting the description of Neanderthals, Homo erectus, and even older populations. Could this be the "truth" behind the myth of Sasquatch and similar stories found throught the world?

and (b) ashes don't preserve well in the first place, so the use of fire may be much older than we have evidence to show. On the other hand, the rarity of evidence of fire use may be related to taboos on its use among early man. In fact, although Homo erectus brought the science of fire as far as China over a million years ago,[255] man-made fires weren't common among humans until about 15,000 years ago.

Does that mean we didn't use fire? No, we simply didn't MAKE fire. Most of our Paleolithic ancestors made use of natural bush fires, caused by lightning. Even today, the Andaman Islanders, the oldest population of Asia, never create their own fire, and instead carry natural fire (from lightning strikes) from site to site. They're by no means unable to master the technology required, as they know how to preserve a single lightning-struck fire for for months at a time using well thought-out techniques, including "preservation chambers" in the hollows of trees. It's not that they "can't" do it. They simply won't.

"I do as my father did, and his father before him, and the father of his father, and the first father of the Pygmies, who was God. He told us how to live, and His way is right." – African pygmy response to the question of why they refuse to adopt fire-making tools[256]

But for Homo erectus, the use of fire probably wasn't governed by taboos, nor was it used simply for warmth or protection from predators. Recent studies show that Homo erectus were the first to cook their food with controlled fires (possibly in hearths), to thaw out frozen meat,[257] and to process their food with tools.[258] Perhaps Homo erectus was able to get so far beyond Africa because of his mastery of fire and the range of survival options it afforded him.

It's also possible that mastery of fire turned our ancestors from solitary eaters into communal ones. With our changing environment creating a need for new food sources, we needed to cook with fire to deal with an increased dependence on meat. Those stone circles where we find the earliest evidence of Homo erectus' fires were probably communal gatherings around roasted food. This technological development may have led to new social developments. Thus, food and fire

---

who study hunters and gatherers say that fires made by these people are small and leave little evidence.

played integral roles in our development into a human community. You'll see this theme recur when we discuss another important period in our past, the Paleolithic. The lesson? Eat with your family and you'll grow closer.

Nonetheless, despite all of Homo erectus' abilities and growing social cohesion, they too went extinct.[259] By 150,000 years ago they were dead and gone. No descendants. End of story. And this happened to several populations. Because we know how and why it happened, the most notable people to go extinct were the Neanderthals.

THE NEANDERTHALS

SUPREME UNDERSTANDING

LIKE THE GEICO CAVEMEN BUT MEANER

Neanderthals were another variant of Homo sapiens (sometimes called Homo sapiens neandertalis) that began to "differentiate" from the ancestral human population around 300,000 years ago. This ancestral population may have been the so-called Homo heidelbergensis that left Africa about 600,000 years ago because of the Sahara becoming desert (yet again!).[260] Within about 180,000 years, they were significantly unlike the "anatomically modern" human population. They were bigger, stronger, and faster, with arms like Schwarzenegger and barrel chests. And they weren't dummies either. In fact, they had bigger brains than modern humans! And they developed more advanced (and deadly) weaponry than the population they descended from. Oddly, however, despite the difference in brain size, they didn't seem to have much of a culture, and appear to lack any traces of art or music in whatever their culture was. One thing is for sure: they did like bloodshed. These guys were brutal, and spent a lot of their time killing each other, or trying to kill us. We'll get to that in a minute.

ACTIVITY: MAKE YOUR OWN FLINT TOOL

How hard could it be to make those stone tools our ancestors used a million years ago? Those things just look like broken rocks, right? Well, that's the kind of thinking that leaves us with the impression that our ancestors were primitive and not brilliant. To see for yourself, why don't YOU try to (a) locate flint, and then (b) flake off blades from the piece of flint, leaving (c) a core you can use as a hammer. For extra points, try making a variety of tools from that one piece, as our ancestors did. For example, an Acheulean hand axe (1.5 million years ago), was a piece of flint with one side round for cupping in your hand and the other side with a sharp edge for cutting through wood or bone. Make that, and let us know how it goes! If primitive monkey men could do it, I'm sure we could all do it too!

Where did the Neanderthals come from? We'll be able to explore the topic in much more detail in Volume Four, but for now we can say that the process Helen M. Leach[261] and Peter Wilson[262] call "human domestication" appears to have played a role in the emergence of this unique "type" of mankind. Many paleoanthropologists have observed that early Neanderthals appeared to be more like later humans than later, "classic" Neanderthals[263] – you know, the kind they always show with the caveman features. So they once looked just like us, but then they became…something else.

How did they become so unlike us? The available evidence suggests that another early experiment in eugenics was at work. That is, certain traits were purposefully bred out of the human population tor produce the Neanderthal population. While there's no way to identify whether such a process was directed by conscious intent, or whether it was the work of larger, unconscious forces in the population, it is clear that it happened. One of the most telling pieces of evidence is a single genetic marker, MC1R. Recent studies have suggested that Neanderthals had red hair, based on a mutation in the MC1R protein. As paleoanthropologist John Hawks has written in summary of the available research:

> The pathways involving MC1R tend to influence the ratio of red [pheomelanin] versus black pigment [eumelanin], but they may be affected by other genetic changes…When MC1R is highly receptive, the melanocytes produce mostly eumelanin. This means that hair and skin will be in a range from light to brown-black, depending on the activity of other genes…Dark hair [among Neanderthals] seems unlikely – otherwise, why select for MC1R loss of function? I suspect by analogy with recent Europeans that Neandertals would have been under selection for pigmentation variants as well. The most common suggestion is sexual selection. But there's no reason for sexual selection in one region to remain the same over a long period of time. And it's rather unlikely that red hair was the target of this selection… More likely, the target of selection has been some other phenotypic effect of the recent alleles.[264]

In even plainer English, Neandertal genes show that they were "bred" to have red hair, but red hair – by itself – doesn't appear to be a good reason for sexual selection (whether natural or intentional), so it appears that some other pheonotypic effect (appearance change) was the goal. Thus, in addition to bigger arms, barrel-like chests, and other physical traits that were "bred out," Neanderthals may have also had a lighter skin pigment than the humans they emerged from.[265] But while the Neanderthals were becoming who they were, we too were becoming who we are.

DID YOU KNOW?
When rocks have been exposed to heat, they develop tiny crack marks as signs of heat exposure. After studying forty thousand pieces of flint tool artifacts, ranging from 1 to 2.5 million years old, from sites throughout Africa, Brian Ludwig of Rutgers University discovered that the artifacts he was studying started to show these small fractures only after about 1.6 million years. This finding suggests that by that time, Homo erectus was not only using campfires regularly, but also hunting and possibly "forging" his tools in fire.

WHY WE DON'T ALL LOOK ALIKE

SUPREME UNDERSTANDING

Before we get too deep into this subject, let's start with a quote from Albert Churchward's *Signs and Symbols of Primordial Man*:

> Inasmuch as we find the same hieroglyphics, signs, symbols, etc., in various parts of the world, also that the earliest race of beings were Negroid, it is but natural to believe that at one time these were universal, and that their birthplace must have had one common centre, which must have been Africa. Throughout the whole world, where the ossified remains of earliest man have been found, we find the same type, which is Negroid. This is proved by the shape of the bones, the flattened tibia, the shape of the skull, etc., and the long forearm with wide space between the two forearm bones (Ulna and Radius), the discussion of which would be outside this work.

What is Churchward saying? That no matter what you find, where you find it, or the trivial differences in the features you find, wherever you find the earliest people, they are Black. As African anthropologist S.O.Y. Keita would agree, the most telling indication of Blackness in these fossilized remains is not necessarily the facial structure, but the shape of the skull and bones in the limbs that reveal a "tropically adapted body." In other words, a body that is African.[266]

Now that we've got that out the way, let's talk diversity. We're gonna get pretty deep into this subject, so prepare for a long ride. The reason being, we have to work overtime to correct some of the misconceptions that abound in the current field of Black studies. Many of us are still working off of myths that were created by Europeans, like the myth of a "true Negro" type. We'll get into that, and a lot of other stuff related to exactly who and what is Black in Volume Two. But if you want the quick and easy version, here goes:

Look around you the next time you're in a barber shop or hair salon. Observe all the different facial structures. You will see sloping foreheads, round foreheads, and flat foreheads. You will see thin noses and wide noses. You will see people with strong West African features and light skin, alongside people with more European-looking features and dark skin. You will see a variety of hair textures. And if you happen to see a dark-skinned Indian or Arab while you're there, you'll see almost all the features you find in an East African, with the exception of their hair texture. And sometimes, an East African (such as an Ethiopian) will have finer hair texture than a South Indian with coarse or curly hair. This is all because Black people are the Original people of the planet Earth. The ancestors of today's Black people must have possessed all of the features you find in all people today, or at least a "primordial" (not yet developed, but containing the potential to develop) form of these features. And today's Black people also possess a variety of these features.

But, you're right, what defines Black people becomes much harder to tack down when you realize that there is not just one type of Black nose or hair texture. We could say that dark skin alone should be the deciding factor, but – if we were to do that – we would have to eliminate the San people of South Africa near the same point where we eliminated the people of Peru and China. So who and what is Black? As one of my favorite 5% 'plus lessons' said, "Blackness is color, culture, and consciousness."

> DID YOU KNOW?
> A 2010 study published in the *Journal of Human Evolution* attempted to do the activity on page 221. The authors reported that only experts with extensive training and experience would be able to predict and control the shape of the stone flakes, and that this skill alone "requires the acute exploration of the properties of the core and hammerstone to comply with the higher-order relationship among potential platform variables, kinetic energy of the hammerstone at impact, and flake dimension that reflects the constraints of conchoidal fracture." Well, okay then. The authors concluded that "given the difficulty and the nature of the skill, the evidence of precise control of conchoidal fracture in the Early Stone Age record may be indicative of the recurrence of a learning situation that allows the transmission of the skill, possibly through providing the opportunities for first-hand experience." In other words, we had "stone tool schools."

WHO ISN'T ORIGINAL?

C'BS ALIFE ALLAH

As we've already explained, the Black Man is the Original Man.* He is Black because he is Original, not simply the other way around. That is, being Black is a byproduct of being Original, like teeth being white is a byproduct of teeth being healthy.

A major part of being "Original" is

* Often, when white scholars acknowledge that the Black Man came first, they are following what I call the "Pre-Adamite" format where they can acknowledge the Original Man's antiquity without crowning him with full "human-hood." So the "Pre-Adamite Man" (the Black man) becomes the "closest to the apes" or a "proto-man," while the white man is the "pinnacle" or "end result" of human development.

that it contains quite a lot of potential for diversity. The Original Man can manifest in a variety of forms that are conducive to his circumstances, and these are not limited only to those with the darkest of skin. If this were true, many of us here in the U.S. would have to redefine our Blackness!

But everyone isn't the Original Man. Throughout this book, we group the Black, brown, and yellow-skinned indigenous or aboriginal people of the planet as Original, and this includes so-called "ethnic" people of many different shades and physical types. So why isn't every man the Original Man? Naturally, some of our readers are concerned about white people.

Our classification system excludes white people for several reasons:

Genetics: In terms of the genetic landscape, Caucasian people have a predominant quantity of recessive traits. This signals that they are at odds with the environment genetically. This is evidenced by their inability (and lack of adaptability) to coexist with the Sun, the primary source of energy for all other life on the planet (as evidenced by rates of skin cancer and other solar ailments). This is also evidenced by their worldwide negative birthrate and other evidence of "incompatibility" with nature. The idea that whiteness is a "natural" response to cold climate is unfounded, as we'll explore in Volume Four.

History: Historically, the European relationship with people of hue has been one of colonization, genocide, slavery, imperialism and destruction of the environment. Though all indigenous cultures did not live in an idyllic golden age before the coming of white people, these elements weren't consistent, nor were they typical, until the advent of white cultural domination. This has been the consistent relationship of European peoples with the world. So the impetus, or the fundamental defining feature and most salient objective, of his civilizations, has been that of a predatory or exploitative relationship with the rest of

ACTIVITY: DECIPHERING DOUBLE-TALK

Can you decipher scholarly-sounding doubletalk? Try this example from Sir Harry Hamilton Johnston's 1910 *The Negro in the New World.* Not only is it steeped in big words, but it's also from a different era, making it even worse. Here's the quote:

"In the external male and female genitalia the Negro sub-species has developed peculiarities which are divergencies from the common human type but are not simian features. It is not necessary to redescribe them here in detail, but it might be mentioned that the hypertrophy of the intromittent organ which is characteristic of male negroes (perhaps not male Bushmen) – with a corresponding exaggeration of the clitoris in the negress – is also met with in the Asiatic negro (Andamanese and North Solomon Islanders)..."

Now what is it really SAYING?

the world. That is, they have been at odds with Indigenous people since their appearance on the planet. This has all been within a short span of time, as their appearance upon the world stage came at a much later date than other people. The global evidence for this point, too, is documented in Volume Four.

Sociology: For those who are overly concerned with offending white people, let's be honest with ourselves: White people have recognized their "difference" from the world, long before we wrote this book. They created the social construct of whiteness to identify themselves as unlike the rest of the world of hue. Whiteness became a rallying point and common culture whose most salient trait is being unlike (and opposed to) the Original people of the planet. No other people have historically distinguished themselves from everyone else on the planet in this way.

So before anyone says we are being "racist" by excluding them, please consider the fact that Black and brown people have historically been the most tolerant and accepting people you'll ever encounter (sometimes to our detriment), and this premise of exclusion came from white people themselves. It is only us who are confused about where they stand. Now, certainly, some white individuals attempt to confront and resist these norms. Those who have attempted to do so in earnest have learned these lessons the hard way. White people who actively resist whiteness (and all its norms) are outcasted, disowned, and reviled by other members of their group. That is what defines the community, the collective identity, not the occasional resister who knows that "treason to whiteness is loyalty to humanity."* Since this is a book focused on the paths and processes of the Original Man, covering a period of time when white people had not yet emerged on the world stage, we won't get into much more depth on these issues. We've dedicated an entire volume (Volume Four) of *The Science of Self* series to answering any of the questions we've raised in this essay.

For now, let's return to the story of the Original Man.

Homo Sapiens

SUPREME UNDERSTANDING

How did we become what is currently considered "anatomically modern humans" or *Homo sapiens sapiens*? And how long ago did this

* The byline of Race Traitor, an online journal (written by white people) dedicated to tackling issues of race and privilege. www.racetraitor.org

happen? That's another one of those issues that mainstream science made a mess of. Until recently, the oldest discovered remains of modern humans were the skulls of two adults and a child found in the Afar region of eastern Ethiopia. Dated to be about 160,000 years old, the remains filled an important gap in the human fossil record and proved that modern *Homo sapiens sapiens* didn't arise independently in different parts of the world, as previously assumed.[267] Both the fossil record and mitochondrial DNA indicate that the ancestors of all modern humans can be traced back to present-day Africa, most likely in the region now covering eastern Ethiopia and southern Arabia. But did this population group develop only 160,000 years ago?

In 2005, two skulls originally excavated from near Kibish, Ethiopia, in 1967 (and then estimated to be about 130,000 years old) were retested using new dating methods and found to be at least 200,000 years old.[268] These skulls were now the oldest evidence of modern man, but they're just the oldest "evidence," not a way to determine the date for when people like this first walked the planet. The problem is that there are large gaps in the fossil record, so the next find could very easily be 100,000 years older. But as of right now, experts agree that there is little information about the fossil record of humans from 100,000 to 500,000 years ago.[269] Plus, the scientific community almost never agrees on dates for early human remains found throughout Africa, but considerable evidence suggests that modern humans are much older than what is suggested by the currently recognized timeline for human development. In *Forbidden Archeology: The Hidden History of the Human Race*, Michael Cremo and co-author Richard Thompson document nearly a thousand pages of evidence that goes against the accepted timeline.

Of course, Cremo and his work were reviled by the WST. They condemned Cremo's work on the premise that – despite his painstaking research, and the fact he is a member of the World Archeological Congress and the European Association of Archaeologists – he was also a member of a the Bhaktivedanta Institute, the scientific research branch of the International Society for Krishna Consciousness. As a "Hindu creationist" Cremo's research wasn't trustworthy, they said – despite the fact that Cremo and Thompson pulled all of their information from reputable scientific publications that came from the WST themselves.

Since then, new challenges to the origin date of modern humans have emerged. In 2011, the *American Journal of Physical Anthropology* published a study reporting on teeth resembling those of a human at the

Qesem Cave in Israel. The teeth were between 200,000 and 400,000 years old.[270] So, the story of human origins isn't written in stone, by any means.

What we do know, thanks to the advanced genetic knowledge we've developed since the 1980s, is that every human on the planet today – and we mean Black, white, Inuit, everyone – can be traced back, through their mtDNA, to a "mitochondrial Eve" who is estimated to have lived about 200,000 years ago. This doesn't mean there weren't humans around before this time, only that whoever was around before "Eve" hasn't left any surviving descendants through whom we can trace their DNA. So whether we date the earliest Homo sapiens at 200,000 or 500,000 years ago, we're not beginning the chronology of man at either of those dates. After all, we were engaged in "advanced" social activities long before then.[*] When we talk about "man" throughout this book, we're talking about Homo erectus, Homo sapiens, and a number of other variants that fall under the Homo designation.[†] When we use "modern humans" or "early humans," we're referring to man as he was in his most recent incarnation, as Homo sapiens.

What was life like for us 200,000 years ago? Through the work of archaeologists and paleoanthropologists, we know that we were hunters and gatherers living in tiny, separate bands. Now geneticists have found that these bands remained separate for over 100,000 years, developing unique genetic lineages and cultures, until environmental changes brought us all to the brink of extinction. A recent genetic study examined our evolution from 200,000 years ago to the point of our near-extinction 70,000 years ago, when the human population dropped to as little as 2,000. Paleoclimatological data suggests that our homeland, Eastern Africa, went through a severe series of droughts between 135,000 and 90,000 years ago. These droughts may have contributed, at first, to the population splits, but, by the end, they were leading to our near-extinction. That's right. We almost died off entirely. But, apparently, we realized the tribalism thing wasn't working to our advantage, and we did something different. We came together. This, you'll see, is a recurring lesson we keep learning and forgetting.

---

[*] See "Homo Erectus" in this volume and "The First Industries" in Volume Two.

[†] Again, however, those disctinctions aren't as real as scientists would like to think, since many of these different "variants" of man were able to interbreed, showing that they weren't different species. In fact, these folks were neighbors, competing against each for survival other (and sometimes cooperating). They were all early man.

According to Spencer Wells, director of the Genographic Project,* these tiny, separate bands, finally came together as…

> …a single pan-African population, reunited after as much as 100,000 years apart…Tiny bands of early humans, forced apart by harsh environmental conditions, coming back from the brink to reunite and populate the world. Truly an epic drama, written in our DNA.[271]

The same process of climatic change that was about to kill us off also helped open the doors for our growth and development. Once we survived the droughts that occurred between 135,000 and 90,000 years ago, related changes in the climate, occurring between 90,000 and 70,000 years ago, caused sea levels to drop dramatically and allowing our now rapidly-growing population to crossing the Red Sea into Asia. Of course, we still had Neanderthals blocking our path, so it took us a while before we could make a full excursion. But once we got into the Near East, free and clear, we populated the entire globe. When you think about that, you can almost look at the world as a video game that wouldn't allow us to progress to the next stage until we mastered the goals of the previous stage.[272]

THE STORY IN OUR BLOOD

SUPREME UNDERSTANDING

DNA is the language of life. DNA is a code encrypted into a 4-letter language of proteins that our system processes to produce our genotype and phenotype. Although DNA is microscopic, we have so many double-helix strands of DNA coiled up within our ten trillion cells that – if they were stretched out end to end – it could go all the way to the Sun and back 50 times![273]

Molecular biologists can use DNA to determine a lot about you. We all know from CSI that DNA can be extracted from our blood, spit, semen, skin, hair, or bone (just about any cell in our body!) and used to identify our ancestry (or "race"), our blood type, and even our genetic predisposal to certain diseases.

Evolutionary biologists can now use DNA signatures from human remains and modern populations to track migrations and population expansions, as well as admixture between different groups, and these

* The Genographic Project is an excellent resource for understanding the human journey. Find it online at https://genographic.nationalgeographic.com Another solid online resource is Steven Oppenheimer's Journey of Mankind at www.bradshawfoundation.com/stephenoppenheimer/index.php Neither resource is perfect, but they're some of the best models currently available.

studies can help us pinpoint the origins of various people, as well as the interactions they may have had over the past 70,000 years or so. Why 70,000? Because DNA studies have determined that all of our ancestors can be traced back to a population group that lived around that time. I'm talking about EVERYONE on the planet descended from a few folks living at that time. Surely other people, long before then, also had descendants, but – as we'll soon see – they either died without leaving descendants, or their descendants didn't make it into the present day. We are basically the survivors. And our survival story can be found in our genetic book of records, our DNA. Within our DNA, we find the story of our ancestors…and in many ways, we find our ancestors themselves.

THE GRAMMAR OF DNA

Our cells have this amazing ability to translate DNA's 4 nucleotides into sequences of 20 different amino acids in a polypeptide. It's downright linguistic, like taking 26 letters and a few punctuation marks to produce this book. The way it works is by using a code consisting of groups of three nucleotides – known as codons – to specify each of the different amino acids.[14] This set of rules, known as genetic code, is essentially universal among all living organisms. But it gets more interesting. Genetic code has 64 codons. But since it only needs to encode 20 amino acids, some of the codons code for the same amino acid, just as we have different words for the same concept. DNA also has punctuation. Some codons, called stop codons or nonsense codons, code no amino acids. (For example, the codon UGA is a stop codon.) These codons always occur at the end of the gene, informing the cell where the polypeptide chain ends. Other codons act like capital letters and tell the cell where a polypeptide "sentence" begins. For example, the codon GUG not only encodes the amino acid valine, it also specifies the starting point of the polypeptide chain. There is clearly a "universal language" in our genes. Our DNA tells our bodies what to do, writing our history in advance. And even today, we are the authors of these instructions, dictating the history of our descendants.

WHAT IS "JUNK" DNA?

"Junk" DNA is the term used to identify DNA that doesn't serve any

RECOMMENDED VIEWING: THE HUMAN FAMILY TREE

In this NatGeo documentary, geneticist Spencer Wells and National Geographic's Genographic Project trace the human journey from its African origins to the ends of the world, using genetics to illustrate the migrations that populated the Earth.

DID YOU KNOW?
There are approximately 12,500 genes in the human genome whose function remains undetermined. These represent about 50% of the total estimated number of human protein-encoding genes. And genes make up only about 2% of the entire genome.[276] According to Richard Dawkins the nucleus of each of our cells contains a digitally coded database larger, in information content, than all 30 volumes of the Encyclopedia Britannica put together.[277]

known purpose. Humans have quite a lot of it. The labels "junk" or "selfish" DNA are used to highlight the imperfection scientists expected from "random" natural processes and the idea of "junk" DNA became mainstream among evolutionary biologists. But throughout the living world, this non-coding, or "non-specific" DNA can make up anywhere from 30% to nearly 100% of an organism's total DNA content (genome).[274] Now scientists are beginning to see "junk" DNA in a new light, and this "trash" might be the Original man's treasure. For one thing, evolutionary biologists are growing to regard "junk" DNA as one of the most potent pieces of evidence for biological evolution. When different organisms share the same junk DNA, this clearly indicates that these organisms shared a common ancestor. Pseudogenes are one class of junk DNA. Evolutionary biologists consider pseudogenes to be the dead, useless remains of once-functional genes. Presumably, severe mutations destroyed the capacity of the cell's machinery to "read" and process the information contained in these genes. A recent article published in *Trends in Genetics*, points out that several research teams have reported that DNA sequences identified as pseudogenes or "junk" DNA play a critical role in regulating gene expression, and that up to 50% of pseudogenes in some genomes appear to be transcriptionally active, meaning these "non-coding" DNA are actually coding – although in a more indirect, "shadowy" mechanism.[275] What exactly are they coding?

Researchers have also found that non-coding DNA determines the size of a cell's nucleus, an especially important consideration in complex cells like ours.[278] Why? As overall cell volume increases, the nuclear volume, and hence DNA content, too, must increase to give the cell's nuclear contents room to communicate effectively with the cell's cytoplasm (stuff outside the nucleus). In other words, this "junk" DNA ensures that – no matter how large the "body" (cell) becomes, the "head" (nucleus) is large enough to reach everyone it needs to communicate with. Essentially, it maintains the ratio of 5% to 85% within the cell's system.[279]

EXTINCTION LEVEL EVENTS

As you'll see in much greater depth in future volumes, a lack of unity (or adaptive strategy) can spell doom. This is, in part, why 99% of every species that ever lived on Earth has now disappeared.

Although all humans today are descended from the population that left the Nubian Complex about 75,000 years ago, we know that it's just not logical to think that no one left the area before that. We know, of course, that Homo erectus ventured far beyond, reaching as far as southeast Asia. But, there's also evidence that there were humans occupying India before 75,000 BC. What happened to those people? Like so many other populations before (and after) them, they may have been wiped out.

In what is now Lake Toba in Sumatra, Indonesia, there was once a volcano – bigger than any volcano found on Earth today – which scientists now call the Toba supervolcano. When Toba erupted between 69,000 and 77,000 years ago, it killed off nearly everyone in southern Asia, from the Arabian Sea to the South China Sea. Yet there are remains showing that, somehow, some Indians survived. The Neanderthals survived as well. What this suggests is that – through some mechanism – extinction events can kill "selectively." That is, whether it's that some living organisms possessed the necessary traits to survive an extinction event, or that the events themselves were structured to only kill off certain living organisms, it is clear that some things are "meant" to survive while others are not. In effect, modern humans are the product of over 4 billion years of selective elimination. We're the "best" because we've survived the tests, so to speak. And scientists agree that what makes us "special" is not our speed, our strength, or our ability to destroy the Earth, but rather the fact that everything that has happened on the Earth up to this point – including mass extinctions – has led up to producing us as we are today. Scientists and religious people debate whether this is the "design" of the universe or simply a coincidental byproduct, but in my eyes it's clear: We made the process that made us. And here we are.

HELL ON EARTH - DINOSAUR FRICASSEE

Before Toba, there were many others. The most well-known is probably the one we associate with the end of the dinosaurs. This is known as the Cretaceous Extinction.

At the end of the Cretaceous Period 65 million years ago, most scientists contend that a massive asteroid struck the Earth, wiping out 3/4ths of all life on the planet. The impact likely covered the sky in clouds of ash and soot, blocking out the Sun and creating what is known as a "nuclear winter" scenario on Earth.[280] Plus, the shock of the asteroid impact would have triggered widespread volcanic eruptions AND destabilized the immense amounts of highly-flammable methane that laid frozen and dormant under oceanic sediments all over the world. The result? Massive wildfires would have swept across a dark planet, subjecting the dinosaurs to what could have resembled Hell on Earth. This could explain why the layer of soot that once blanketed the planet was so thick it contributed up to 10% of the carbon found in today's biomass.[281]

Ten million years after the dinosaurs were deep roasted by a global firestorm, another methane burp appears to have erupted from the oceans. This one didn't ignite, but slowly filled the atmosphere with enough greenhouse gas to cause yet another period of global warming.[282] The methane eruptions shifted some landmass around, while the temperature rise killed off many of the animals that once populated North America, freeing it up for settlement by the deer, horses, and canines that streamed across the now-open Bering Land Bridge.[283] Once again, the Earth cleaned house, replacing the old with the new.

And the asteroid that killed the dinosaurs may not have been the only one of its kind. There are massive craters all over the Earth, demonstrating clear evidence that our planet has been bombarded by interstellar matter since its origin. In fact, if it wasn't for the Earth being so beautifully covered in a garb of water and vegetation, we might see her resembling the equally battered but naked Moon, who is unable to conceal her scars. Yet many of the Earth's massive craters can still be seen with the naked eye. One of the largest, the Shiva Crater, is found in India. Throughout Africa alone, there are at least 14 impact craters larger than 2 miles in diameter. At nearly 200 miles across, the 2-billion-year-old Vredefort crater of South Africa is the largest verified impact crater on Earth.

When we think about events like the Toba explosion or the Shiva crater, we're not thinking there's any rhythm or reason to these natural disasters. But there's a growing number of scientists who have noted that such "extinction level events" appear to be more periodic than we think. One theory has it that comet showers, coming around about every 26 million years, have created this regular pulse beat in the history of life. But Roger Lewin's 1998 study, analyzing marine extinctions over a period of 600 million years (as revealed by the fossil

DID YOU KNOW?
According to biomecular engineer Richard Green, a migration left Africa between 300,000 and 400,000 years ago and quickly diverged, with one branch becoming the Neanderthals who spread into Europe and the other branch moving east and becoming Denisovans. When another group, the ancestors of modern humans, left Africa about 60,000 years ago, they first encountered Neanderthals, and then – after traveling further east – the Denisovans, leaving traces of Denisovan DNA in the genomes of humans who settled in Melanesia. Denisovans contributed about 5% of the genes of modern Melanesians.[287]

record) – suggests that while many extinctions are periodic, their occurrences somehow "cut across functional, physiological, and ecological lines." Yet despite them being unrelated to any detectable ecological activity, the "major pulses of extinction result from geographically pervasive environmental disturbances."[284] In other words, it's like the Earth cleanses itself quite regularly – like a menstrual cycle – so that new life can grow and old stuff can go.

In other words, an extinction in one thriving population can be fortuitous to another struggling population. For example, geological records from the end of the Permian period about 250 million years ago show that a mass extinction of marine and land species accompanied the death of massive forests and other plant life surrounding the eruption of a supervolcano in Siberia. This global extinction event – where more than two-thirds of reptile and amphibian families died off and 95% of oceans life went extinct – was probably caused by poisonous volcanic gas. Scientists believe that volcanic gases from the eruption depleted Earth's protective ozone layer, increased carbon dioxide in the atmosphere, and acidified the land and sea.[285] Yet, thanks to all the dead trees left by this extinction, an ancient fungus, Reduviasporonite, was able to thrive and spread across the planet by eating the rotting vegetation.[286] Life always finds a way. This fungus may have aided the colonization of land (by plants and animals) through symbiosis.

Know what else survived? The supervolcano's eruption triggered a huge spike in global warming that helped kill off whatever survived the acidic land and water. But some animals, like the relatives of Kombuisia Antarctica, survived by making their homes in Antarctica. The species was about the size of a house cat, burrowed in the ground, walked on the surface and lived in trees.[288] Thus, Antarctica, now seen as a frozen desert barren of life and activity, was once a haven for those creatures that would survive to become our ancestors. Just a reminder that what some of us consider the "third world" could some day soon become our "first world" – something the filmmakers of *The Day After Tomorrow* seem to have realized.

THE FIRST EXTINCTION EVENT EVER?

Perhaps there was even some life on our planet when the Moon was ejected from the Earth. Considering this happened about four billion years ago, any life would naturally have been rudimentary and unicellular (simple and single-celled), as they would have been the primordial ancestors of the planet's earliest organisms. Considering how far back we have evidence of life on this planet, it's reasonable to theorize that life could have existed then as well, with the majority of these organisms totally obliterated by the upheaval of the planet during the birth of the Moon. If there were survivors, it was these survivors who became us. And thus the story of life.

So let's recap. During the Permian Extinction, most life on this planet died. During the Cretaceous Extinction, it happened again. Each time, our ancestral lineage survived. Over the next several million years, millions of other species disappeared, leaving no descendants. Our ancestors survived. Despite his successes, Australopithecus went extinct. Homo habilis went extinct. Homo erectus went extinct. Neanderthals went extinct. Denisovans went extinct. Each time, it was our ancestors who survived. Of all the Homo sapiens who lived then, we are the descendants of only the thousand or so who made it when humanity itself was near-extinction 130,000 years ago. And despite the millions who died in the Maafa or via any of the many genocides the Western world has inflicted on Original people over the past 2,000 years, our ancestors are the ones who survived. We have survival written into our DNA, bottlenecked through literally dozens of near-extinction events where only the survivors were ones who could pass through the bottleneck and give their DNA to the next generation.

Want to know something scary though? Mass extinctions aren't just a feature of the distant past. In fact, there's mounting evidence we could be facing another mass extinction in the near future. Scientists have found that the modern extinction rate exceeds that of ancient events, even the one that killed off the dinosaurs. According to a summary of the research, a "flood of invasive species can trigger mass extinction events...By studying the similarities between modern times and that of the Late Devonian period, scientists say we could meet a similar fate."[289]

THE EASTER ISLAND MYSTERY

There are many historical instances of Original people falling victim to their own "advances" and subsequently going near extinct. Typically, this happens as a result of overconsumption, that is, when man takes

DID YOU KNOW?
What allowed the ancestors of man to beat out their competitors and survive? What allowed Australopithecus to remain while Tchadensis and others died out? What allowed Homo habilis to survive when Australopithecus died out? In both cases, it may have been the ability to try new things.[291] In evolutionary language, this is known by a simple word: adaptation. Keep in mind that adaptation is a strategy. You have to adapt enough to meet the challenge, but not so much that you're unable to keep it up. Sounds like life huh?

more from nature than he replaces. One example, cited by Jared Diamond in his book Collapse, was the mysterious demise of Easter Island, which was first credited to tribal warfare, and then later credited to the natives using up all of the trees that provided them sustenance. While there is some merit to this theory, it's never that simple. In fact, deforestation played a role, but the bulk of it was actually caused by rats. In 2005, University of Hawaii anthropologist Terry Hunt found evidence suggesting that Polynesian rats, which came on the boats that populated the island, played a key role in deforestation. While the natives consumed much of the island's 16 million palm trees, building the ropes and wooden devices used to transport and erect the massive stone heads that are associated with Easter Island, they made efforts to replant as well. However, they couldn't replant fast enough to compensate for the rats, which fed on the seeds, making reforestation nearly impossible. Before long, the island was nearly barren of palm trees, which were key to sustaining Easter's human population. As a result, population levels fell. Still, with no evidence of early warfare, the people of Easter Island weren't on their way to extinction.[290]

That is, until the Dutch came in the 1700s, bringing diseases like smallpox and murderous slave raids. "Older explanations essentially blamed the victims for their demise," says archaeologist Patricia McAnany of Boston University. "The island still represents a cautionary tale," she says, "but one of the dangers of invasive species."[292]

Where did these rats come from? In the traditional cultures of Hawaiians and Polynesians, rat was a common food. Yet, the Polynesian king was traditionally considered too sacred to touch or eat rat, suggesting that the Polynesians set taboos on eating rat (like the Shipibo people of Peru and Sirionó people of Bolivia – who may have had contact with the Polynesians) because they were – at some point in their past – aware of the rat's primary function, garbage-disposal, and hence its destructive potential to a region's food supply. Nonetheless, the Polynesians apparently "domesticated" rats as a food supply, much as early Europeans did with pigs. Well, so did the Polynesians. In fact, the Polynesians brought pigs with them to most of the islands they populated first, but not to later settlements like Easter Island, where they now only brought rat. Archaeological records of pig extinctions on Polynesian islands suggest that the Polynesians figured out that pigs were bad for business at some point.[293] But they didn't learn this lesson about the rat quickly enough, and Easter Island suffered total environmental collapse. Similar examples occurred when the Spanish introduced scavenging pigs into North America.

The botton line is that the majority of life to ever live on this planet neither survived nor left any descendants. Today's species are thought to represent less than 1% of all species that have ever lived.[294] What does it take to make it? As we'll see in greater depth in the next essay, it appears that the deciding factor is not random genetic mutation, but conscious adaptation. In other words, only a few are chosen. And the chosen few choose themselves.

REVIEW

As you read the previous chapter, you may have noted how the development of man followed the same mathematical process we discussed in a previous chapter. This process (1) began with the foundation of the first man, who then (2) diverged, spread, and attempted whatever was in his capacity, and then (3) came to realizations about what worked and what didn't. This led to (4) humans developing institutions, cultures, and traditions based on what was effective, as well as social groups that engaged in the same patterns of behavior. Many of these bands of people (5) grew considerably in size and industry due to their consolidation and shared culture, and developed into (6) egalitarian societies with an equal distribution of labor, ownership, prestige, and responsibility. However, this eventually gave way to (7) the rise of leadership and privilege, with some individuals esteemed over others, and therefore possessing more property, knowledge, and/or rank in the society. When this occurs, it leads to (8) a natural process of accumulation and loss, progress and problems, advance and sacrifice, power and poverty. In essence, the rise must come with a fall. This is all part of the natural order though, and even massive extinctions of entire population groups can sometimes be the necessary consequence of the trajectory of this course. That is the nature of how all processes are brought to (9) completion, so that another cycle can begin.

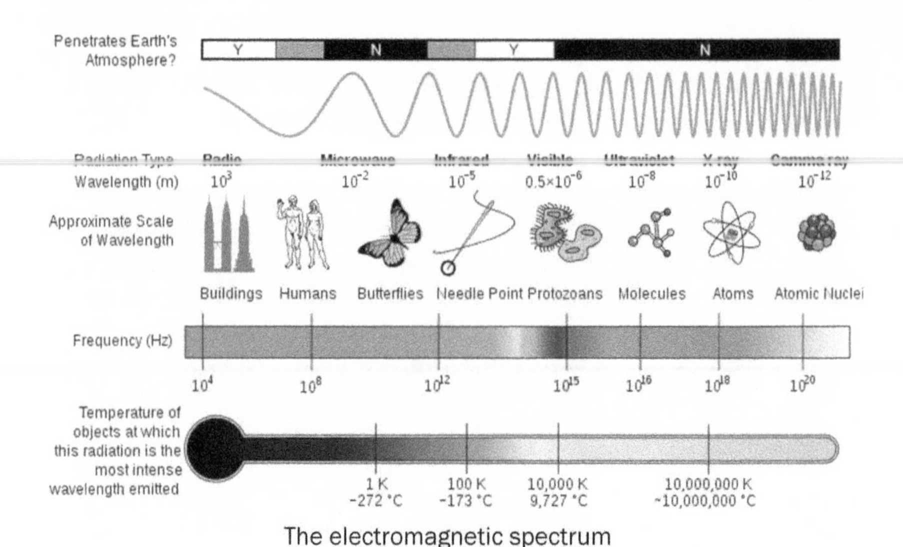

The electromagnetic spectrum

AGE OF EMBRYO (IN WEEKS)
FETAL PERIOD (IN WEEKS)
FULL TERM

1 2 3 4 5 6 7 8 9 16 20-36 38

dividing zygote, implantation and gastrulation

CNS, heart, eye, heart, limbs, eye, heart, ear, teeth, palate, ear, external genitalia, brain

CNS
heart
upper limbs
eyes
lower limbs
teeth
palate
external genitalia
ear

LOSS OF CONCEPTUS
MAJOR MORPHOLOGICAL ABNORMALITIES
FUNCTIONAL DEFECTS AND MINOR MORPHOLOGICAL ABNORMALITIES

TIMING OF AIR POLLUTION RISKS:

Interrupted placental development
Fetal growth restriction
Reduced weight gain
Early susceptibility to later preterm birth
Preterm birth
Heart defects

Note: Blue bars indicate time periods when major morphological abnormalities can occur, while light blue bars correspond to periods at risk for minor abnormalities and functional defects.

Milestones in fetal development

MAN AND NATURE

RELATIONSHIPS WITH THE EARTH

"Regard Heaven as your father, Earth as your Mother and all things as your Brothers and Sisters." – Native American Proverb

Our ancestors used many terms to refer to their relationship with nature. Yet, whenever we look at indigenous cultures or ancient Black and brown civilizations, we see the same reverence for the Earth and nature, despite whatever terms they used. Yet, these same ancestors typically described themselves as the "masters" of nature somehow.

Is there a contradiction here? What does it mean for man to have dominion over nature? To different groups of people, this idea has meant very different things. To Original people in collectivist societies, man was a steward of the Earth. He sustained through its providence, and in turn he took care of maintenance. This is a common concept among indigenous cultures throughout the world. On the other hand, cultures that have been heavily influenced by Europeans tend to see things differently. They picture man constantly at odds with his environment, in a timeless battle for domination and survival. While you can find traces of both streams of thought among Original people today, the "Original" perspective is definitely not the idea of beating the shit out of Mother Nature and making her our bitch. That's something the Europeans hit us with. And when they did so, we were part of that "dark jungle" they were attacking!

I learned of this relationship during my studies of the first degree of the NGE's Student Enrollment, which identifies the Original man as "...the maker, the owner, the cream of the planet Earth..." While some may only see wording of "owner" as a property claim, I looked at the other meanings of the word "own." To "own" something also means to take responsibility for it. And if the Earth is "ours," then it is certainly OUR responsibility. While this is certainly a pressing issue today (considering the rampant abuses of the environment that have occurred under Western rule), what was man's relationship to the Earth in the distant past?

To understand man's relationship with Earth, we first have to understand the true nature of the Earth itself.

WHAT YOU'LL LEARN

- ❒ Whether aliens had any role in human development.
- ❒ What a superorganism is, and what it can teach us about ourselves.
- ❒ How and why the Planet Earth is a living organism.
- ❒ Why evolution tends to produce cheaters, exploiters, and followers.
- ❒ The "nature" of the relationship we have with Nature.
- ❒ The role of Original people in maintaining the Earth.
- ❒ How the Original man literally made mountains, lakes, and rivers.
- ❒ The signs we left behind to remind ourselves of who and what we are.
- ❒ How our scientific exploration and experimentation led to the emergence of global Black civilization.

AN INTRODUCTION TO SUPERORGANISMS

ROBERT BAILEY & SUPREME UNDERSTANDING

The Earth is indeed alive. Yes, but to understand how and why, we have to start off small. We already explained what it means to have life and the science of living organisms. So here we'll explain how Earth, like us and countless other seemingly complex systems, is one huge living organism.

WHAT IS A SUPERORGANISM?

David Sloan Wilson and Eliot Saber define a superorganism as "a collection of single creatures that together possess the functional organization implicit in the formal definition of organism." Earlier, we learned what it means to be an organism. So a superorganism, in other words, is many coming together as one to perform functions they wouldn't be able do individually. Superorganisms are representative of a sophisticated level of organization and coordination, sort of like how your body is composed of billions of cells that can't do much on their own, but can do everything together. Kinda like Voltron. The same way, bacteria and other simple creatures (like slime molds and fungus) are capable of aggregating and moving collectively as one unit, and are able to make better decisions as a collective. Ants and bees can do the same, yet have more divisions of labor. You can find superorganisms on every scale of life, from some of the simplest organisms, such as bacteria and amoeba, to the insect kingdom, to more complicated examples, such as Earth's vegetation. In fact, as we'll see, superorganism theory is fundamental to the increasing complexity of life on Earth.

EXAMPLES OF SUPERORGANISMS

BACTERIA ARE EVERYWHERE

Bacteria are prokaryotes (single-celled organisms) that are found alone or clumped together forming a biofilm (sometimes with other micro-organisms). A biofilm is a superorganism composed of countless, specialized bacteria all working together.[295] Think that's weird and nasty? Well you've actually got a biofilm in your mouth right now. Dental plaque is a common example of a biofilm.

Most of us know that we are composed of millions of cells, including bone cells, muscle cells, blood cells, brain cells, etc. So on a cellular level, you have a network of cells communicating and interacting in various ways throughout your nervous system. Though we're more familiar with the electrical signals, our cells also communicate chemically and perhaps through mitogenic radiation.[296] This intra-cellular communication goes back billions of years, to the origins of multicellular life, when cells developed the ability to send signals to other cells, followed by the ability to send a signal to another cell on one side, but not the cells on its other side. As we noted in "The Story of Life," this is a foundational aspect of life itself!

And these intricate methods of communication help cells learn what is going on around them so that they can make decisions concerning metabolism, division, even whether to die or not, ensuring that they thrive together as a superorganism, ultimately allowing for us to function. That's already superorganism status by itself! But wait, there's more.

What's lesser known is that we, as human beings, are in fact complex systems, with bacteria being an integral part of us. And as they are often quite a bit smaller than human cells, bacterial cells can actually

RECOMMENDED VIEWING: I AM: THE DOCUMENTARY

Tom Shadyac interviews scientists, religious leaders, environmentalists and philosophers – including Desmond Tutu, Noam Chomsky, Lynne McTaggart, Elisabet Sahtouris, Howard Zinn, and Thom Hartmann. The film asks questions like: "What is the Nature of Man?" "What's Wrong with the World?" and "What Can We Do About It?" It raises many good points. For exmple, all kinds of lifeforms (fish, birds, insects, and mammals) value cooperation and even vote democratically, and this emphasis on cooperation is found among all aboriginal/indigenous people (who see an excessive focus on competition as a mental illness), but not among...well you know. Shadyac also discusses the neurological roots of empathy (more on that in Volume Four), the electromagnetic field of the heart, extra-sensory perception, and several other topics covered in this text.

outnumber the "human" cells in the body itself. And these are not simple mindless organisms. They collectively possess at least 100 times as many genes as the mere 20,000 or so in the human genome, and have at least 20 different niches in our systems (not just in our mouths and stomachs). For example, the group of bacteria that reside on your inner-forearm are different from the group a few inches away on your inner elbow, with both sets specialized for their own purposes.[297]

And these bacteria aren't all "bad germs." Many bacteria work WITH your body, serving to break down foreign substances and aid the processes happening within your body. In *The Hood Health Handbook,* we talk about the good and bad bacteria in your stomach and intestines. The good kind helps digestion and prevents disease, and that's why people take probiotics (including yogurt) to increase those bacterial cultures! In fact, the way that bacterial cells (and some viruses) work together with our own cells makes man himself (as an individual) a superorganism. But as a group? We'll get back to that.

FUNGI TAKE OVER FORESTS

Fungi are some of the longest and oldest superorganisms on the planet today. Some have been found covering more than covering 1,500 acres and at least 1,000 years old.[298] Yes, that means an organism (really a "superorganism") as big as 1,500 football fields. And these guys were here waaaay back in the day, making them effectively bigger than anything else moving on earth. Take that, dinosaurs.

AMOEBAS BECOME SLIME MOLDS

Amoebas are very small eukaryotes that can't do much of anything by themselves. But when given enough food, certain species of amoeba reproduce asexually via mitosis and cytokinesis (two processes of cell division) until they aggregate into a "slime mold" that sends out streamers and sort of flows along the surface looking for food. These slime molds can demonstrate enough intelligence to solve mazes.

ANTS BECOME COLONIES

Ant colonies are able to form some of the most sophisticated and interesting superorganisms. You can look at a single ant as the equivalent of a cell in our body: It's short-lived (depending on the species, between 1 and 10% of the entire worker population of a colony die each day), yet very specialized. Colonies of ants have their workers, soldiers, administrators, etc., like our body has its different groups of cells forming our organs, each performing their special duties. The queen stays pumping out children something like our gonads (ovaries or testes). Some would argue that the nests of some ants correspond to a

DID YOU KNOW?
Agriculture is neither a "leap forward" in civilization, nor is it solely a human enterprise. Not only do Attine ants and Ambrosia beetles engage in sophisticated acts of farming, but scientists have also found agriculture among the social amoeba Dictyostelium discoideum. D. Discoideum will save some of its food and harvest it on to its body in order to plant some in a new location, to produce a new crop. This way they'll always have food. This gives a different starting point for the origins of agriculture.[300]

body, some nests being so large that they are comparable to the skeletons of whales. To communicate across such large colonies – some can house 8 million ants – takes sophisticated systems equivalent to that of the human nervous system. Together, the ants form a little brain, following specific rules that allow them to operate the way they do, with thousands working together on one accord for a common cause.

In his review of *The Superior Civilization: The Beauty, Elegance, and Strangeness of Insect Societies* by Bert Hölldobler and Edward O. Wilson, Tim Flannery writes:

> Nothing in the brain of a worker ant represents a blueprint of the social order…and there is no overseer or "brain caste" that carries such a master plan in its head. Instead, the ants have discovered how to create strength from weakness, by pooling their individually limited capacities into a collective decision-making system that bears an uncanny resemblance to our own democratic processes.[299]

Some ants have learned to herd and milk bugs; there are even ants that take slaves, ants that lay their eggs in the nests of foreign ants and there are even Attine ants that have discovered agriculture.

Attine leafcutter ants also "sing" – which helps them cut leaves by imparting vibrations to the mandible that is cutting the leaf – while they work and are able to cry for help by a form of stridulation (a sound created by the rubbing together of body parts).One species of leafcutter ant from South America, can contain nearly two thousand individual chambers, some with a capacity of fifty liters, and can involve the excavation of forty tons of earth and extend over hundreds of square feet. The leptanilline are possibly the most primitive and smallest ants out there. In packs they prey on venomous centipede much larger than themselves. And then there's the lethal army we find in Africa: Siafu ants can literally move as one deadly swarm to overtake a large animal…or a human.* None of these feats would any of the ants have been able to accomplish individually.

If you put a hundred army ants on a flat surface, they'll walk around in circles until they die from exhaustion. That won't happen with a col-

* For more on them, check out *How to Hustle and Win, Part Two.*

ony of a million army ants, however. The colony acts as a sophisticated superorganism, carrying out raids, keeping nest temperatures constant to within a degree, and doing a bunch of other stuff unimaginable to a single ant or even a hundred.

Nigel R. Franks speculates:

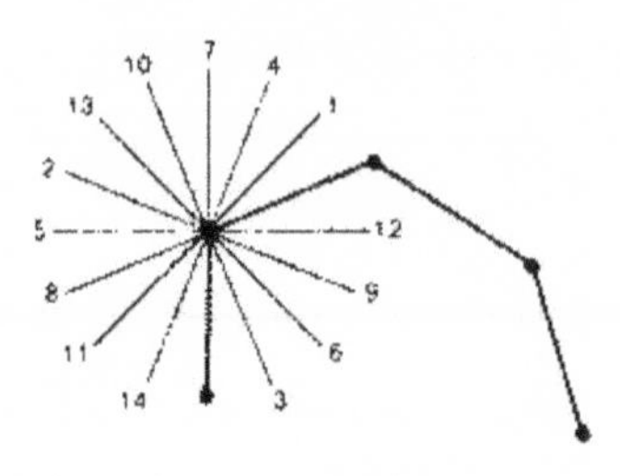

> It seems that intelligence, natural or artificial, is an emergent property of collective communication. Human consciousness itself may be an epiphenomenon of extraordinary processing power. Although experts prefer to avoid simplistic definitions of intelligence, it seems clear that all intelligence involves the rational manipulation of symbolic information. This is exactly what happens when army ants pass information from individual to individual through the 'writing' and 'reading' of symbols, often in the form of chemical messengers or trail pheromones, which act as stimuli for changing behavior patterns.[301]

Franks describes two remarkable capabilities of an army ant colony:

- Precise time-keeping, seen in the colony's nomadic phase of 15 days (during which larvae are growing) and the 20-day stationary phase (during which pupae develop). The queen's egg-laying also sticks to this schedule. Raids into the rain forest occur in both phases.
- Satellite navigation during raids, as in "multiple satellites" (ants) triangulating locations and directions as a unit. During its 20-day stationary phase, an army ant colony scatters about 14 foraging raids exactly 123° apart, as diagrammed. (The heavy line indicates the colony's path during the nomadic phase.) This scattering allows time for new prey to enter the previously raided areas.

What's amazing about this geometric precision is that each ant doesn't have enough information (on its own) to figure out where the next 123° raid should go. After all, army ants have only a single facet in each eye (rather than a compound eye like other insects). Franks says:

> The mystery is how the colony can navigate with each of its workers having such rudimentary eyesight. In my wildest dreams, I imagine that the whole swarm behaves like a huge compound eye, with each of the ants in the swarm front contributing two lenses to a 10- or 20-m wide 'eye' with hundreds of thousands of facets.[302]

THE BIRDS AND THE BEES

No we're not going to talk about that. In fact, we don't need a lot more information to explain how superorganism theory applies to the beehive. As Lee Alan Dugatkin writes in *Cheating Monkeys and Citizen Bees: The Nature of Cooperation in Animals and Humans*:

> If you doubt that honeybee colonies are at least reasonable conten-

DID YOU KNOW?
Agriculture is neither a "leap forward" in civilization, nor is it solely a human enterprise. Not only do Attine ants and Ambrosia beetles engage in sophisticated acts of farming, but scientists have also found agriculture among the social amoeba Dictyostelium discoideum. D. Discoideum will save some of its food and harvest it on to its body in order to plant some in a new location, to produce a new crop. This way they'll always have food. This gives a different starting point for the origins of agriculture.[304]

dors for the title of superorganism, just walk over and shake a nest of bees. The response will be swift, painful, and very, very organized. Antipredator behavior in honeybees is extremely complex and mind-bogglingly cooperative. There are distinct "defender castes" and caste determination is genetic. Colony defenders can be further subdivided into two groups: defenders (who attack very dangerous predators) and guards (who guard the nest from inter and intra-specific parasites and may attack less dangerous threats to the nest).[303]

Even flocks of birds flying in V-formation and schools of fish demonstrate the "instant" communication that appears to occur in some superorganisms. That is, there appears to be a "hive mind" or "swarm intelligence" that doesn't require a lapse in time between when one members sends the signal and when another receives it. This is rarer than cases where there's clearly a chemical signal that follows a pathway, but it has been seen in several species of animals that move in large groups but all "know" when and how to turn without any other cue.

WHERE DO SUPERORGANISMS COME FROM?

It goes back to the laws of nature. Individual organisms can grow and establish "power" on their own, but it is only through aggregation, co-operation, symbiosis, and distribution of resources and information (which we can collectively identify as the principle of "equality") that they are able to exceed the limitations of their previous standing in life and achieve greater standing or dominion over their environment. This brings us back to the principles we found in the ant colonies.

This phenomenon is found in the non-biological world of particles and chemicals that precedes the living world as aggregation. Aggregation is both common and consistent in inorganic nature. For example, nanocrystals are microscopic clumps of atoms about 1-10 nanometers long. These "lifeless" nanocrystals are also able to come together, forming bigger structures.

There's evidence that the tendency to consciously self-organize is pre-programmed into all life and matter. Elsewhere in this book, we talk about the way atomic particles "respond" to observers, as if they "know" they are being watched or measured and respond accordingly. But even particles are even more conscious when they come together.

According to Steve Davis:

> The origin of life itself required cooperation between complex molecules as they began performing the functions that we now associate with life, i.e., metabolism, homeostasis and reproduction. Cooperation therefore preceded evolution...The first life forms that survived while others failed were groups of complex molecules. The eukaryotic cell that became the basis of all modern life forms was/is a grouping of bacteria. An organism is a group.

So is a grove of trees an organism? Is the Earth an organism? Is the Earth alive? Is the universe alive? In a Nature article titled "Live Universes," M.G. Bjomerud suggests that the criteria for what constitutes a living organism should vary based on scale:

> There is no reason to expect that super-organisms would meet criteria based on observations of individual organisms. Isn't it time to consider the possibility that the boundary between life and non-life may be diffuse, non-stationary over time, and dependent on scale?[305]

FAMILY PLANNING 2 BILLION YEARS AGO

The "equality" demanded in the process of coming together as one does not, however, come without its sacrifices. In a summary chock-full of jewels (for those who can connect the microscopic to the macroscopic), science writer Michael White notes:

> Single-celled microbes function very well as individuals. Some of that individuality has to be given up for the greater good when cells hitch their evolutionary fates together as one multicellular organism. A key example of conflict resolution is the evolution of genetic limits on cell division: to have a coherent, multicellular body plan, individual cells can't just divide with abandon, the way bacteria do. When cells escape these genetic controls on division in humans, you get cancer. [306]

The principle of "coming together for the greater good" is critical to the origins of complex life. As we've noted, the first multicellular organisms (from which we descend) were born from two single-celled organisms coming together (scientifically known as endosymbiosis). White describes how a species of single-celled green algae became multicellular. First the algae cells began forming aggregates of cells. This collective appears to have grown increasingly conscious with aggregation:

> [T]he rate of cell division began to be controlled genetically. Unlike single-celled organisms, which reproduce whenever the surrounding environment is right, the new multicellular algae began controlling exactly how many daughter cells they produce. This is a critical step towards establishing a multi-cellular body-plan with genetically controlled dimensions. After this step, the cells basically started to become more organized, working collectively as a group, and ulti-

DID YOU KNOW?
Superorganisms can be huge. Some species of fungus can be over 1,000 years old and over 1,000 acres in size.[310] But when it comes to massive superorganisms, plants hold the record: "A grass clone, Holcus mollis, has been found with a diameter of 900 metres and an age of over 1000 years. A clone of box-huckleberry has been found with a diameter of 2000 metres and an age of 13,000 years. The big granddaddy is, however, an aspen (Populus fremaloides) covering 81 hectares and over 10,000 years old."[311]

mately developing specialized functions in a sort of division of labor.[307]

In other words, instead of continuing to reproduce freely, our ancestors began planning their families. They then began assigning responsibilities to their children, which became jobs. After this step, the cells started to become more organized and worked as a group. White continues:

> The researchers suggested that these major events coincided with the inventions of new ways for resolving conflicts among individual cells in the organism: in other words, formerly independent cells had to learn how to be civilized.[308]

As we examine the story of life, we see that several properties common to the major transitions in evolution are in fact, characteristics of superorganisms:

- Smaller entities have often come about together to form larger entities. (seen in chromosomes, eukaryotes, colonies, communities)
- As part of a larger entity, smaller entities often become differentiated. (seen in DNA/protein, organelles, anisogamy, tissues/organs, castes/social order)
- The smaller entities are often unable to replicate without the larger entity. (seen in organelles, tissues, castes)
- The smaller entities can sometimes disrupt or exploit the development of the larger entity. (seen in Meiotic drive phenomenon, parthenogenesis, cancer, coup d'état)
- New ways of transmitting information have arisen. (seen in DNA/protein, cell heredity, epigenesis, universal grammar, mathematics.)[309]

INTERDEPENDENCE: THE CIRCLE OF LIFE

As we work our way even further up the food chain, we see more division of labor, more complexity, yet the same principles. That is, competition is necessary, but cooperation is the evolutionary accelerator.

Mutualism (a relationship across different species in which both members benefit) can be found all around if you look closely. Not only will plants activate their own defenses, but they'll enlist help from the enemies of their enemies. "When spider mites attack lima bean plants, the plants release a chemical SOS that attracts another mite that preys on the spider mite. Mechanically damaged plants do not produce the cues; most likely, only elicitors in the saliva of the insect can trigger the plant to produce the right molecules." Scientists say that a plant talking to their bodyguard is likely to be a characteristic of most, if not all plant species.[312] Again, this isn't a feature confined to the plant kingdom; you'll find acts like these in the world over. In *Influence: The Psychology of Persuasion*, Robert Cialdini explores a mutual relationship between large grouper fish and the small "cleaner" fish that picks food from its teeth. As you can imagine, the grouper eats smaller fish, but not the cleaner. The cleaner does a little dance to say "I'm here to help," and the grouper sits still with its mouth open. The grouper gets a free dental cleaning (which prevents parasites and fungal infections), and the cleaner gets a free meal.

We can find the same scenario among some species of caterpillars, which cannot survive from predators unless they are protected by ants. The caterpillars send out "vibrational signals" across leaves and twigs to call their ant homies for back up. The ants, in turn, enjoy secretions produced by the caterpillars.[313]

In *The Superior Civilization*, Tim Flannery explains how deeply intertwined Attine ants (the ants who practice agriculture) have become with the fungus they harvest. The ants can't live without the fungus and the fungus can't live without the ants. This is known as symbiosis (a close and long-term relationship between different biological species) and it may take thousands of years of evolution to develop such interdependence.

COOPERATION, COMPETITION, AND ALTRUISM

The principle of cooperation and competition are critical to the story of life itself. In fact, they ARE the story of life itself, and cooperation more so than competition. To be successful as a population, nature re-

A QUICK NOTE ON ETHICS VS. KNOWLEDGE

In nearly all indigenous cosmologies, the notions of knowledge and wisdom always go hand in hand with notions of ethics. One who is a master of knowledge, or who is highly successful in life, but who is not "good," is considered less than one who has very little knowledge or wealth. This ethos can be found in Taoist, Buddhist, Native American, African, and Aboriginal traditions. Hoarding knowledge or wealth were seen as crimes against humanity. If either was accumulated, it was to be shared equally. In Western ideology, this relationship was reversed to value individual dominance (whether financial or intellectual) over any emphasis on ethical or prosocial behavior. Among Five Percenters, the notion of being a "poor righteous teacher" exemplifies an urban resurgence of indigenous values. "Poor" is said to mean not necessarily one who is in poverty, but one who "transcends materialism," while "righteous" is defined as doing what is for the greatest good of all parties – that is, what is not merely "right" but "just." We'll explore the value systems of Original people in depth in Volume Three.

DID YOU KNOW?
You've heard the saying, "You don't s*** where you eat." Turns out there's an ecological reason for that behavior, which can be seen throughout all nature, from single-celled slime molds to indigenous people. In 1983, Nature reported that this "system" for nutrient dispersal and fertilization occurred both on land and in the water:
"It has long been recognized that the movement of grazing animals from one terrestrial ecosystem to another, feeding in one and defecating in the other, may result in significant movement of certain [chemical] elements between them (i.e. the ecosystems). What has now been made evident, in work on the coral reefs of the Virgin Islands, is that a similar process takes place in aquatic ecosystems."[314]
Science Frontiers research William Corliss commented, "This process may emphasize the fine-tuning of the Gaia Hypothesis, in which life-as-a-whole operates in ways that make the planet-as-a-whole more productive of life."

quires that we cooperate more than we compete. But before we can cooperate, we must compete! That is the theoretical model proposed by the leading specialists in the field of superorganism theory, foremost of whom is Bert Hölldobler of Arizona State University's School of Life Sciences and Center for Social Dynamics and Complexity.

According to Hölldobler, the path to colonial superorganism is first paved by the maximization of the inclusive fitness of each individual of the society. He suggests this fitness arises by competition between individuals in the same society, which diminishes as the colonial society becomes larger, better organized and contains better division of labor and ultimately, cohesiveness.

This cohesive in one colony leads to increased competition with other colonies of the same species. Hölldobler notes that the competition between societies soon becomes a major force reinforcing the evolutionary process. Such a nested tug-of-war model, he says, might also be applied "equally well to the analysis of the evolution of other animal societies" and give insight into the evolution of cooperation in humans and our hominid ancestors, in addition to such things as collectives of cells and the formation of bacterial films.[315]

According to Steve Davis, cooperation and altruism go hand in hand:

> Cooperation is the principle underlying survival, and therefore underlying evolution. Cooperation can be seen at all levels of life, between molecules, within cells, between cells, between organs, between organisms within groups, and between groups. And because cooperation within groups has assisted individual organisms in survival, reproduction, and just plain old comfort, it has brought about the further development of altruism. Altruistic acts occur when individuals see themselves as belonging to a greater entity.

So let's recap. Cooperation. Communication. Hard work. Precision. It seems we could learn a lot by watching ants. Of course, nothing's perfect. Ponerine ant colonies are made up of predatory ants who are more capitalist than collectivist, and their colonies tend to be small and so competitive that ponerine colonies can't be considered a superorganism. In other colonies, some ants get forced into low-status jobs and are prevented from moving up by others in the colony. In some cases, the ants who are able to prevent upward mobility are not contributors themselves, but maintain standing through the "appearance" of work. Yeah, it gets deep. We'll get back to this dynamic in nature. It's known as exploitation or "cheating."

What else can go wrong? Worker ants often slack off once the queen is removed, and the female ants fight to become the dominant babymama. Yup. Plenty of other conflict and competition goes on in a colony. However, as Science writer Brandon Keirn notes in "Thoughts on Ants, Altruism and the Future of Humanity":

> Assuming there isn't some limit on cooperation, it's conceivable that between-colony competition in ants could be replaced by between-colony unity, just as group competition in people can be replaced by mutually beneficial relationships.[316]

Some of these principles can be applied to our societies as well. Businesses are using models based on insect behavior to increase their effectiveness. For example, analysts at Phoenix Sky Harbor Airport are even using a software program based on the theory of swarm intelligence – the idea that a colony of ants works better than one alone.[317]

MAN AS A SUPERORGANISM?

I know we've talked about the "internal" ecology of man as a superorganism, but let's move up in scale. What about man himself? And we mean HUMANITY. Is man a superorganism?

That's a tricky one. As we noted in "Man is the 7 in the Center of the Universe" and elsewhere throughout this book, man is something of an anomaly. Clive Backster has said:

> We get two different kinds of bacteria very much in synch with each other. We get plants responding to our intent. We get plants responding to the death of other creatures. All my work, which consists of file drawers full of this kind of very high quality anecdotal data, has shown time and again that these creatures – bacteria, plants, and so on – are all fantastically tuned in to each other. Now, as you get to humans, this capability gets lost.

Jeffrey M. Dickemann, Professor of Anthropology Emeritus at Sonoma State University, has contended that man's tendency to compete more than he cooperates is what keeps him short of such a title:

> Tim Flannery is so taken with the ants, and especially southern fire

> ants, that he is tempted to think that humans are becoming a superorganism, which he somehow sees as a salvation from our destructive current path. But the human species is precisely not a superorganism: its Darwinian success is precisely due to that fact...Competition, not only between states but between cities, communities, and families, at all levels of social organization, distinguishes us (and other mammals) from the ants, who have laid aside competition at these lower levels in favor of unquestioning collaboration. [318]

Flannery disagrees:

> Like us, ants have been shaped by Darwinian evolution, and if left unrestrained by predators and disease, they too would doubtless exhaust their resource base...The trend of human development over the last ten thousand years has been toward ever larger collaborative units, which entails economic specialization of the individual, enforcement of the rule of law (which promotes peace within the social unit), and ever greater interconnectedness between individuals. All of this is concordant with the superorganism concept...As Dr. Dickemann says, we shall soon see.[319]

But we may not need to peer into the future to see a time when man becomes a superorganism. Perhaps man was once much closer than he is now. For example, in *Cheating Monkeys and Citizen Bees*, Dugatkin compares the selfless cooperation of bees to human communities like the Canadian Hutterites, a Christian community resembling the Amish, who practice a way of life that Dugatkin applauds as the closest thing to superorganism status on Earth. Of course, the principles adopted by the Hutterites are the same principles indigenous societies have carried for thousands of years!

We can even find genetically determined castes of warriors (such as those found in the beehive) in many of these indigenous societies. Even when the members of these warrior castes descend from different clans or lineages, they would often do like the sô of East Africa and take a blood oath that bound them as blood brothers, united against threats to the society's survival. It would then be taboo for any of the sô to kill each other, regardless of what one of them said or did to the other, because "it would be like killing yourself." This pact was called avusô (literally, tomb of the sô).[320]

When you look at the story of man, you will see that man HAS repeatedly come together to move as one, and that has been where progress and success came. The same with the early phases of life itself, where prokaryotes came together to become eukaryotes, or where the life that swam the seas formed communities to populate the Earth. Perhaps this is why so few superorganisms are found on Earth, compared to the nearly 20,000 species that fit the description on land?[321]

Perhaps it was the act of coming together that allowed us to make that transition, just as the act of coming together allowed us to settle the globe as early humans?[322] But then what happened?

EXPLOITATION AND CORRUPTION

While collectivism is definitely the key to "advance" for any living organism, we can't expect 100% of all stakeholders to be on board. After all, collective identity requires sacrifice, and some individuals won't be interested in that. That's part of nature as well. In any culture, man-made or ecological, there are degrees of freedom that allow for divergence. Sometimes this divergence turns out to be for the greater good. Other times, it represents a threat to this good.

Thus we see selfishness and "individualism" among even the simplest life forms. The roots of this behavior are found beyond the genetic level, at the blueprint that allows for new developments to occur through such divergence. But when scientists like Richard Dawkins promote the idea of a "selfish gene," or a biological predisposition to selfishness, we have very little evidence from nature that directly supports this. In fact, most selfish organisms die off much faster than those who cooperate, and history reveals the same about human societies. Then again, perhaps Dawkins, like Darwin and countless others, is promoting a model that supports a scientific basis for European norms, as selfishness and individualism are mainstays of Western culture, and almost universally abhorred in indigenous culture everywhere else in the world. Davis continues:

> Selfishness is an insignificant factor in evolution because it is an exaggerated and therefore less common form of natural self-interest. Self-interest can be explained as the assertiveness necessary for survival. The procuring of food for example is a manifestation of this assertiveness. Selfishness on the other hand is a learned behaviour found among organisms that are nurtured by parents. It is discarded as the organism matures.

If only that were true. We have countless examples that adults do not always "discard" selfishness, even though scientists describe excessive selfishness among adults as evidence of intellectual immaturity or sociopathic deviance. Yet, perhaps selfishness is indeed a learned behavior that – in environments like the one fostered by Western, capitalist culture – is rewarded enough to make selfishness at least appear to be worth retaining. We know that selfishness and individualism are trademarks of European culture, yet the evidence may be more than philosophical. Studies suggest that, as organisms grow, they often develop a class of "cheaters" who seek to reap the rewards produced by altruism, without being altruistic themselves. We can see examples of this wherever there are collectives, including among bees, ants, and even slime mold. What produces this trait? We know that living organisms display selfishness when they are concerned about their own survival over that of the group, particularly if they have reason to see themselves as separate from the colony somehow.

As biologist Dr. Bill Hughes noted:

> When studying social insects like ants and bees, it's often the cooperative aspect of their society that first stands out. However, when you look more deeply, you can see there is conflict and cheating – and obviously human society is also a prime example of this. It was thought that ants were an exception, but our genetic analysis has shown that their society is also rife with corruption – and royal corruption at that![323]

In 2010, Amdam wrote an article for the journal *Advanced Cancer Research* entitled "Order, Disorder, Death: Lessons from a Superorganism," where he used the superoganism analogy to explain the nature of disease:

> Bee colonies have some of the same challenges that we have as singular organisms. They need to defend themselves against pathogens, to maintain homeostasis. They have to protect themselves against the multiplication of units or cells that are not a positive contribution to the organism as a whole. Cancer is a multiplication of cells that are not following the rules any more; you die from that. This can happen to bees as well.
>
> There's a type of bee that can insert itself into other colonies and produce copies of itself. After a while the social continuum of that colony breaks down because of this worker that's not performing social functions. They're just using resources and burning calories but not contributing anything. After a while, the population grows so large that the whole society breaks down. That's how bees can die of something similar to cancer...[But] the parasite bee can only survive as long as these "cancerous" individuals are not recognized by the colony. The moment they're recognized as something else, they're killed. That's how a cancer survives, too. But sometimes, a cancer suddenly disappears, suddenly regresses, and it might be because the immune system suddenly realized that these cells are not us.
>
> These invasions of clonal bees happened in the 1950's and in the 1970's in South Africa, where there's now an infestation, and in both cases the clone disappeared. It's unclear how it happened – but it's clear...that a mechanism arrived by which they could recognize the intruder and kill them. This is what our immune cells do, too. They feel our way through the body, look at surfaces, and decide – is this me, or something else? And if it's something else, they kill it. And the moment you can get an organism to recognize a cancer as non-self, that's when you have the opportunity to win.[324]

It's really deep, if you think about it. If you don't, I guess it's boring.

"We shake down acorns and pine nuts. We don't chop down the trees. We only use dead wood. But the white people plow up the ground, pull down the trees, kill everything...the white people pay no attention...How can the spirit of the Earth like the white man?...everywhere the white man has touched it, it is sore." – Wintu Woman, 19th century

Because the Earth's ecology is composed of a variety of superorganisms, Earth itself can be considered a super-superorganism itself. Without all the fancy words, we're simply saying that the Earth is alive. It is this understanding that forms the foundation of the Gaia theory. As just as the human body will work as one unit to expel parasitic organisms, and just as groves of acacia trees will work in solidarity to kill off greedy antelope, and just as a colony of bees will target and exile its unproductive workers, the Earth has a hell of a way of letting people know when they've gone too far abusing her resources.

THE EARTH IS ALIVE

ROBERT BAILEY & SUPREME UNDERSTANDING

THE EARTH AS A SUPER-SUPERORGANISM

Plants die so man can eat them. Man dies so plants can eat them. The circle of life! This is all part of the way the Carbon Cycle, Nitrogen Cycle, and – to a lesser extent – the Phosphorus Cycles of the Earth's ecosystem operate. Meanwhile, other systems such as the water cycle help regulate and contribute to the maintenance of the Earth. Nutrient dispersal is another example of an ecological system that encompasses different species. The movement of animals eating in one ecosystem and dumping their load in another helps ensure that nutrients and minerals are distributed properly. This process occurs on land and in

RECOMMENDED VIEWING: DISCOVERY ATLAS 4D

This visually stunning documentary details the interconnectedness of the Earth and its inhabitants. It begins with a focus on the formation and history of the Rift Valley of East Africa. It touches on several familiar themes we get into throughout this book, from the way natural systems are interconnected in subtle, yet significant ways, to the way man shapes his environment while his environment helps shape him. The first installment covers a variety of topics from origins of the salt trade to the role of the dung/scarab beetle in maintaining the ecosystem of the grasslands. One highlight: the acacia (which we talk about elsewhere in the book) evolved spiky thorns to defend against herbivores. The Massai thus used the acacia to create the first "barbed wire" fences for their cattle pens. The documentary then talks about how the European "invaders" took over East Africa and killed off the native Massai with smallpox. Talk about familiar themes.

the water.[325] The Earth even has a regular "cleansing" cycle to flush old life out and allow new life to take root. As True Wise notes in "The Black Woman is the Earth," all of these cycles and systems replicate the internal dynamics of the Original Woman. This brings us to the Gaia hypothesis.

THE EARTH IS ALIVE

The Gaia hypothesis asserts that "all of living matter on Earth functions as an enormous homeostatic superorganism that actively modifies its planetary environment to produce the environmental conditions necessary for our survival." In other words, like us, Earth is a superorganism, and like our bodies, Earth takes care of itself. The Earth itself replicates both the diversity of our body's different systems and the way they work together to carry out the processes fundamental to all living organisms (metabolism, maintaining homeostasis, the capacity to grow, responding to stimuli, reproduction, communication, and many others).

James E. Lovelock explains that the Gaia hypothesis postulates the Earth is "a self-regulating system comprising the biota and their environment, with the capacity to maintain the climate and the chemical composition at a steady state favorable for life." With Lynn Margulis, Lovelock developed the Gaia hypothesis in 1974. Margulis was an author of *Micro-Cosmos* and an advocate of evolution-via-endosymbiosis; that is, diverse organisms uniting to create new species.[326]

Before Lovelock and Margulis elaborated the Gaia hypothesis, an early prototype of this theory was found in the work of Vladimir Ivanovich Vernadsky, a Russian mineralogist and geochemist. Vernadsky is considered one of the founders of geochemistry, biogeochemistry, and of radiogeology. In Vernadsky's theory of the Earth's development, the noosphere is the third stage in the earth's development, after the geosphere (inanimate matter) and the biosphere (biological life). Just as the emergence of life fundamentally transformed the geosphere, the emergence of human cognition fundamentally transformed the biosphere. In this theory, the principles of both life and cognition are essential features of the Earth's evolution, and must have been implicit in the earth all along. This systemic and geological analysis of living systems doesn't contradict Darwin's theory of natural selection, but adds onto it by looking at its context within a set of broader principles, rather than only looking at species-level changes.

THE BLACK WOMAN IS THE EARTH

TRUE WISE ALLAH

How can we possibly know the Nature of a person?

"Yeah I can see what were ARE doing, but what are we supposed to be doing? What are we designed to do? What is our NATURE?"

With so much socialization, media brainwashing, mass cultural manipulation, etc. how can we possibly claim to know what is natural vs. what is simply learned? How do we identify someone's 'nature', and then differentiate that from that person's cultural programming. This is a very serious and legitimate question. To avoid inventing explanations arbitrarily, it's important to consider what has already been discovered or at least studied on this topic.

The topic of 'nature vs. nurture' has been studied for a long time. Essentially, this question can be boiled down to a question: "do we act the way we do because we are born that way? Or is it because we are taught to act this way?" Years of behavioral science research has explored the issue of where human behavior comes from. After decades of study, the scientific consensus is that human behavior is explained by both biological and environmental (social) factors. Neither nature nor nurture by itself can fully explain human behavior; it's a mix of the two sides. Not only do both sides contribute to human behavior, but the two sides also interact with each other dynamically, which makes the issue even more complex. For example, sexuality is obviously based on biological factors; however, the details of exactly how that sexual attraction is focused or applied is influenced by societal factors (e.g. people have sexual arousal in any culture, but what is considered attractive will vary according to culture). Environmental

A QUICK NOTE ON PREHISTORIC GENDER RELATIONS

As we'll explore in great depth in Volume Three, in hunter-gatherer/forager societies, men were responsible for most of the hunting, while women were responsible for gathering wild plants and firewood.[327] But sexual division of labor in the Paleolithic was much more flexible than it became in the Neolithic. In present-day hunter-gatherer societies like the Hadza of East Africa or the Aborigines of Australia, men may participate in gathering plants, firewood and insects, while women captured small game animals and assisted the men in driving herds of large game animals (such as woolly mammoths and deer) off cliffs. Archeological evidence from burial sites suggests that women could achieve high status in Paleolithic communities, and it is likely that both sexes participated in decision making.[328] With the adoption of agriculture however, it appears that women's status in society declined in many societies, as they were expected to do more demanding work than women in hunter-gatherer societies, while men – who were doing less hunting – had less to do.[329]

factors can even change the DNA expression itself. If anything can be concluded about the nature vs. nurture debate, it's that these two 'sides' are very complex and they interact dynamically. However, this does not mean that nothing can be concluded about one's nature. Determining one's nature is all about using the right method.

The right method to use here is the scientific method. Scientific method basically mean: 'doing everything you can to not be biased'! In a laboratory, this is relatively easy because most things in the laboratory can be controlled to some extent. For example, if you want to study what hydrogen does when it mixes with oxygen, all you have to do is control what chemicals and substances are and aren't allowed in the test tubes, beakers, etc. You control the amounts of things, their temperatures, their purity, etc. But the issue of human nature is more difficult. First of all, we aren't studying something that can be concluded in 5 minutes or even in 5 years. We need to observe across hundreds or thousands of years. Also, we can't control anything directly. So the specific method we need to use here is one that combines the scientific method with the study of history: compared examples.

Comparing historical examples is the scientific approach to studying history. It is important to compare examples and not just "give" examples. For instance, someone can say "slavery was not a white crime perpetuated against blacks, because everyone had slavery, and even the Africans enslaved each other." That's what it means to "give" an example. However to understand the truth about slavery one would have to actually 'compare' African slavery to New World (American) slavery, and observe how they compare. Since the goal is to remove bias and find the truth, we must compare historical examples as if we were dealing with experiments. As another example, someone could cite the existence of African warrior queens and then conclude that Africans did not make gender distinctions when it came to military and warfare. However, the scientific approach would be to compare a variety of African militaries, over many different places, over many different times, at many different stages of each civilization. Only then could we start forming conclusions about African gender distinctions in military. By simply 'giving' and example, one can seem to "prove" anything. However, if we want to be responsible and find the unbiased truth, we need to compare examples and treat history like a huge set of experiments. When we approach difficult questions such as "What is the Nature of the Black Woman" we need to take the most rigorous approach possible to avoid being biased.

DID YOU KNOW?
The modern word "Earth" is derived from the Old English *Eorthe*, which comes from the personified Earth deity of Nordic Europe, *Jord*. It should be noted that *Jord* is pronounced like *Ard*, the Arabic word for Earth. This connection may stem back to the occupation of northern Europe by Afro-Asiatic people, who settled Europe and influenced local language and customs about 4 to 5,000 years ago.

WHAT IS THE NATURE OF THE BLACK WOMAN?

Black people have had many different societies in many different places and under many different conditions. Although it is wrong and inaccurate to say African civilization has been all the same, it is certainly possible to find elements common to the majority of African civilizations. Some scholars (such as Dr. John Mbiti) have successfully done this. Their method is to study a wide variety of African cultures, and then compare the examples for commonalities. From this method the Black community has been managed to derive sets of principles, philosophies and behavior patterns that can at least somewhat applied to much of Africa and more generally the Diaspora of Original People. From studying this Diaspora, along with the Woman's consistent role within this Diaspora, we can begin to understand the Nature of the Black Woman.

Black civilization's progress may be viewed in terms of the contributions from both men and women. In other words, we can look at all of Black civilization (collectively) and extract out the adult female role as our basis for describing Black Womanhood in history. The Black Man and Woman have always had gender roles; in fact, most scholars of African history maintain that Black people have always had prominent labor-division arrangements based on gender. The modern perspective on gender roles tends to brand it as sexism and oppression of women. However, when we look at historical trends in African gender roles, we find that they are usually founded on a desire for labor efficiency rather than oppression. In fact, many researchers have argued that oppression is actually less likely when men and women are not in direct competition against each other. Some scholars have taken the importance of gender roles even further by demonstrated that unraveling these roles played a very key role in dismantling the cultural strength of Black people. In essence, gender roles have been important in Black history, and they can reveal the specific roles typical of Black Women.

Determining the roles is only half the battle. To answer the question: "what is the Black Woman's Nature," we must not only focus on gender roles, because some of these roles may have been decided arbitrarily. Therefore our discussion of the Black Woman's Nature will be based on two things: her historical roles and her biological predispositions. In other words, what has the Black Woman done, and how does that match what she is biologically 'built' to do.

THE EARTH CONCEPT AS A SCIENTIFIC MODEL

In any scientific model (a framework for understanding things), it is necessary for that model to describe a certain set of phenomenon. A model is a simple view of a complex reality. For example, the "billiard ball model of gas" takes what we all know about billiard balls (that they bounce around on a table and gradually spread apart) and applies this to the reality that it's trying to describe (that gas molecules bounce around and gradually spread apart). Some models serve as weak analogies, which mean that their application is very limited. For example, "the big bang" (whether that theory is correct or not) may be considered a weak analogy, because the idea of a "bang" doesn't tell us much detailed information about the creation of the universe. On the other hand, a model may serve as a strong analogy. Such models are rare in science because the purpose is to paint a picture of reality rather than to fully describe it. In this writing, it will be argued that the 'Earth' is a very strong analogy for the Nature of the Black Woman.

Over the course of this essay, the Black Woman's Nature will be described in tandem with our home planet's Nature. Both of entities (the Woman and the Planet) will be referred to as "Earth." Earth here is seen as a unifying concept, a principle serving as a strong analogy describing who the Black Woman is and what her nature is.

The Mother of Civilization is not a light title, and its gravity should be appreciated. The Mother of Civilization means the female component to the entire creation process of all the greatness we witness in humanity. This is title belongs to the Black Woman. As an analogy, what else in the universe can be considered the mother of civilization or more generally the mother of life? A planet, and as of now the only planet that is known to produce life is the planet we live on. Whether or not other planets in the galaxy can produce life does not take away from the fact that our planet (Earth) will forever be our own reference point for life producing planets. Wherever humans may end up in the galaxy or universe, the first planet, our original mother will always be Earth; our first woman the Original woman will always be the Black Woman, whose title is Earth.

"Beneath the clouds lives the Earth-Mother from whom is derived the Water of Life, who at her bosom feeds plants, animals and human" – Algonquian legend

For this essay we will also focus on six aspects of the Black Woman's Nature. This is consistent with the symbolism of six representing balance, harmony and equality. This is not religious, numerological or superstitious symbolism, but rather mathematical and scientific symbolism. Six is known as a "perfect number" in mathematics which is rare. Perfect numbers are numbers that add up to themselves. This may be seen as a top-to-bottom, bottom-to-top balanced structure on a basic numerical level. When circles of the same size are placed together (this can be demonstrated with coins) are placed together, it takes six of them to create a perfectly round formation that fits another circle of the same size in the middle. Hexagons are a six sided shape that appear strangely not only on this planet, but other planets such as Saturn. The explanation for this is very complex, but put simply it relates to spinning and complex geometry. These hexagons are manifested all over nature. For instance, snowflakes form six armed figures. Bees – among the most mathematically astute animals on the planet – choose hexagons as their shape of choice when building honey combs. They do this because hexagons maximize space while minimizing materials. This is called efficiency. There are more mathematical and scientific illustrations for six being a very special number. Here, six will be used to represent a universal principal of balance, harmony and equality.

Six core aspects of the Planet and the Original Woman will be focused on in our explanation of the Earth concept. What we view as the "planet Earth" is actually a biosphere or a place where life exists. "Earth" is simply the name of a concept; a concept shared perfectly by our planet and the Original woman. These six aspects of this principal will form the scientific foundation for this ideological connection.

THE ENERGY CYCLE – RESILIENCE

The energy cycle of our planet begins with the sun. Energy starts at the sun and travels approximately 93,000,000 miles at 186,000 miles per second and strikes our planet's atmosphere. Roughly 30% of this energy is reflected away back into space and never used by the planet. 20% of the energy gets absorbed on its way to the planet's surface. 50% of the energy strikes our planets land surface and water surface. How much energy is reflected away or absorbed, or hits the ground and water is very relevant. This is how our planet interacts with energy; and it is the reason why life is possible on our planet and impossible (or less likely) on other plants. It is a perfect balance for life. Some planets reflect too much energy away and therefore don't have enough energy to born any life systems. Other planets absorb and take too much energy, and therefore have too intense an environment to born any life systems. Our planet is very special in her ability to use energy.

All living beings process energy. Our planet is a living being that processes energy for her life systems. This can be proven when we evaluate the energy that arrives at our planet and then leaves our planet. There are two things that make the processing apparent: (1) the time lapse, and (2) the difference in the energy's wavelength. The time lapse is long and the difference in the wavelength is big. This means that the energy came, was cycled around and used in a living system, and then let off later.

Energy cycles through the biosphere fueling its ability to continue. This cycling of energy gives the biosphere its ability to adapt to change, its ability to evolve, its ability sustain life despite anything that threatens life. Specific species may die out, however the entire living system as a whole continues as long as the source (the sun) provides energy. The sun throws out gigantic amounts of energy constantly with a very stable energy system. It would take a very powerful force to threaten the sun's energy cycle. The earth's energy cycle on the other hand, is very easily affected (as evidenced by the constant climate change throughout the ages). However, the planet deals with these threats by adapting. The sun's energy system may be described as "strong" while our planet's energy system may be described as "resilient."

The Black Woman, like our planet is also extremely resilient. First, this backed up by widespread historical records of adaptation in the face of the most horrific ultra-harsh circumstances. The Black people as a whole have

DID YOU KNOW?
Vladimir Ivanovich Vernadsky and Pierre Teilhard de Chardin independently formulated very similar theories describing the evolution of the Universe. Three levels are described:
(1) Cosmogenesis or the formation of inanimate matter (the Physiosphere of Wilber), culminating in the Lithosphere, Atmosphere, Hydrosphere, etc., or collectively, the Geosphere. Here progress is ruled by structure and mechanical laws, and matter is primarily of the nature of non-consciousness.
(2) This is followed by Biogenesis and the origin of life or the Biosphere, where there is a greater degree of complexity and consciousness, ecology comes into play, and progress and development is the result of Darwinian mechanisms of evolution.
(3) Finally there is human evolution and the rise of thought, and a further leap in complexity, resulting in the birth of the Noosphere. Just as the biosphere transformed the geosphere, so the noosphere (human intervention) transformed the biosphere. Here the evolution of human society (socialization) is ruled by psychological, economic, informational and communicative processes.
For Teilhard, there is a further stage, one of metaphysical evolution, the Christing of the collective noosphere, in which humanity converges in a single divinization he calls the Omega Point.

Woman

been very good at survival. It is argued here that the "strength" aspect of this survival is likely to come mostly from the men, while the "resilience" aspect of this survival is likely to come mostly from the women.

Based on the extensive medical and biological documentation males demonstrate more strength while females demonstrate more resilience. Females are hit more severely by disease and are also hospitalized more often (Givens, 1979). While getting hit harder by disease, women live much longer than males (National Center for Health Statistics). This is an example of men being less affected in the short-term (strength) and women being less affected in the long-term (resilience). When it comes to stress and coping, similar patterns show in the medical research. Females of all ages are much more efficient in dealing with stress and less damaged by it (Frankenhaeuser, Dunne, Lundberg, 1979; Collins, Frankenhaeuser, 1978; Frankenhaeuser et al, 1978). All of the studies indicate that females have this advantage because of an ability to flexibly, adaptively and resiliently respond to negative aspects of the environment. One widely observed phenomenon (inside and outside of scientific study) is that women cry more often than men; which also relates directly to resilience.

Crying is a shock-absorber against stress. The same chemicals that are found in high-concentration within depressed people are also found in emotional tears. Emotional tears are different from other types of tears. Some tears only exist to keep the eyes moist; these are called 'basal tears'. Other tears only come in reaction to offensive fumes; these are 'reflex tears'. "Psychic tears" is the term given to the kind of tears that come from emotions. Psychic tears contain corticotrophin (ACTH), which is associated with depression and also the production of androgens (masculine hormones). ACTH causes the production of cortisol which is known as "the stress hormone." Therefore, when people cry, they are unloading hormones that cause stress. Women cry more than men on average, therefore, women are emotionally more resilient than men.

A woman should never feel abnormal about crying a lot. Crying is a resilient response to stress. She also should not force herself to be "strong" or manifest the properties of brute strength. Women who attempt the behavior pattern of male strength usually end up more stressed. A woman's natural comfort zone is to express herself freely

RECOMMENDED VIEWING: HOW THE EARTH CHANGED HISTORY

This BBC documentary uses some breathtaking HD camerawork to illustrate the history of the four elements (Earth, Wind, Fire, and Water) on man's history, as well as man's history of shaping the Earth. This series digs into geology, geography and climatology to tell some fascinating stories.

as the emotional being she is. Her man should also learn to accept her emotionality and view it as a normal and healthy process of self-healing. When both genders understand their own nature, peace is easier to achieve. Strength is the ability to take a hit; Resilience is the ability to bounce back from a hit.

THE CARBON CYCLE - PHYSICAL BIRTH

All life on our planet is based on carbon. This means that the carbon molecule forms the basis for all known living things. The reason carbon can do this is because of its bonding properties. More specifically, carbon has four valence bonds and these bonds are not too strong nor are they too weak. This allows molecules to be easily created out of carbon, but also be dismantled when necessary. Carbon is also built to react well with other molecules. This makes carbon perfect for being a basis for all life. The carbon cycle is the process whereby carbon circulates throughout the planet allowing life to be born. Carbon exists in the environment, and then it is absorbed within our planets plant-life (which is apart of the biosphere). When carbon passes from the environment to the plant-life it becomes "biologically fixed carbon." This means it exists now as the tissue of a living organism. Life is therefore "born" and expanded as carbon becomes fixed into the living systems. Carbon flows throughout the planet, and life is born from her carbon cycle. Biochemically, the carbon cycle is like a 'great womb' from where all life on the planet is born.

It is a well known fact among anthropologists, geneticists, linguists, archeologists, and historians in general that Black people are the Original humans. Such a statement is not racism, but rather pure scholarship. Black people are Original, not only in the sense of 'arriving first' but also establishing human civilization first. Of all the life to exist on our planet for billions of years, never did anything (of any kingdom of life) reach the intellectual magnitude of humans during the foundational building of human civilization. It is the Black Woman who was the mother of this civilization. She is "great womb" just as our planet is the "great womb" of advanced civilization.

DID YOU KNOW?
Much has been written about "Black Madonnas" in Europe,[330] which are dark-skinned representations of Mary the mother of Christ holding dark-skinned representations of Christ. These representations are said to derive from pre-Christian European goddesses of fertility or representations of the goddess Auset (Isis) and her son Heru (Horus). What has been often overlooked is that the darkness of the fertility goddesses is not simply related to fertility but to the fact that they were representations of Black woman. The two concepts were synonymous in the ancient world.

The carbon cycle also forms the basis for our planet's fertility cycle. The planet is most fertile when carbon levels are high and least fertile when carbon levels are low. Carbon has daily cycles and also yearly cycles. As our planet turns on a daily basis, she exposes herself to the suns energy, which in turn initiates the carbon cycle to born life. The part of the planet facing the sun at any given time is the most fertile part of the planet. This is because photosynthesis is stronger when the sun's intense energy is able to reach our planet's plant-life. This spinning which happens at approximately 1,037 1/3 miles per hour creates a 24 hour fertility cycle. On a yearly basis, our planet's exposure to the sun shifts back and forth between the northern hemisphere and the southern hemisphere. The angle of how the sun's energy hits our planet is what causes seasonal changes. This is a yearly fertility cycle which is directly linked to the carbon cycle. The changing of the carbon cycle affects the fertility of our planet on a daily basis and also a yearly basis.

Women also have fertility cycles. The ability and desire to conceive (especially if she exercises a holistic nutritional foundation) changes throughout the day, month and the year. For Black Women in particular the quality of the actual pregnancy and the prenatal development process is heightened. For example, studies reveal that Black Children at birth are neurologically superior to white children (Falkner, Tanner, 1978). Black infants are also born with superior motor intelligence (Scott et al, 1955; Bayley, 1965). The amazing fertile capacity and developmental potential of our planet's Carbon Cycle is symbolic to the Black Woman's womb.

THE NITROGEN CYCLE - MENTAL FERTILITY

Most of our planet's atmosphere, approximately 80% of it, is some form of nitrogen. A large portion of this nitrogen is in the form 'atmosphere nitrogen'. Atmospheric nitrogen (N2) is useless to living things. In order for nitrogen to be used by life, it must first become "fixed," which means changed to a different form. Most of this fixing is done by bacteria, thereby making nitrogen available for life. This nitrogen is used by our planet for fertility and also to provide the foundation for genetic material.

All life on our planet is able to exist and repeat itself because of deoxyribonucleic acid (DNA). Without genetic codes, life would not have a blueprint to be passed on. This genetic code is extremely complex, and a display of intelligence. It is not an intelligence decided by some far-off mystery god; rather it is an intelligence that is self-created through a slow yet continuous and dynamic process of evolving life. DNA is essentially a physical display of intelligence. Our planet may be thought of as having a 'mind'; a mind that continues on and on.

Our planet has a 'mind' in the since that she changes intelligently over time. These changes are not only reactive (in response to threats to ensure survival) but also proactive (growing and elevating the life systems to new levels). The intelligence of our planet's biosphere can be thought of as a mind, because it takes in complex information and uses it to create things; in our planet's case, she uses this information to advance life through multi-generational changes in DNA. It is the availability and distribution of nitrogen that makes DNA possible. Therefore, the nitrogen cycle can be thought of as the basis for our planet's system of applied intelligence. It is 'mental fertility' in the sense that it is intelligent (mental) and also available for new and continued mental growth (fertility).

Human culture can be compared to this continuing life-maintaining mind. Culture is a complex code of life, just as DNA is a complex code of life. Culture attempts to survive; DNA's purpose is attempting and managing to survive. Culture continues over long periods of time; the same is true of DNA. The Black Woman is the oldest domestic culture bearer in human existence, just as DNA is the oldest continuing system of complex code on the planet. The Black Woman's ability to apply her intelligence in this multigenerational manner for the longest (domestically) of all humans, is similar to the Nitrogen Cycle's support of the DNA genetic code.

THE OXYGEN CYCLE - NURTURING

All life requires nourishment. Nourishment is when a living organism receives what it needs to sustain itself. Oxidation-reduction is a dual process that makes it possible for living things to get nourishment. Oxidation is when oxygen gives an electron to another molecule; reduction is when oxygen takes an electron from another molecule. This dual process is the basis of metabolism. Metabolism is how living things are able to be nourished. Metabolism is made possible by the availability of oxygen on our planet. Not all planets have much oxygen available, which is one of the many reasons why other planets do not have life (or at least none that we know of). Oxygen is also necessary for healing and cellular repair to take place. It is the oxygen cycle that makes this possible. Our planet is very special in her superior ability to nourish life.

The Black Woman nourishes the child in three stages. The first stage of nourishment is through the umbilical cord, and of course only a woman can do this. The second stage of nourishment happens through the breasts, which is made possible by the woman's mammary glands. The third stage of nourishment happens through mother's (primarily in most cultures) preparation of food. Years of behavioral science research have demonstrated women display a superior level of empathy. This nature of woman is very useful to the family's need for emotional and psychological healing; which is empathy. Oxytocin is known as the "attachment hormone." It is released in high volume during child labor, and also during breastfeeding. This plays a role in women being highly emotionally attached to their children. To some degree, this causes women to be emotionally vulnerable. For instance, oxytocin is released in females during sexual activity. This makes women biochemically less capable of being satisfied with casual, non-commitment-based sex. Functionally, this facilitates the natural tendency profound attachment to men with whom they've been sexually involved. All of these phenomena play a role in why the woman cares so much about, and is so dedicated to the well-being of her family. The Black Woman is biologically constructed to be incredibly good at nurturing and healing.

THE MINERAL CYCLE - NATURALLY GIFTED COMMUNICATION

Minerals move from the inside (liquid core) of the planet to the outside (solid rocks), often from volcanoes. The minerals on the surface of the planet are exposed to constant weathering which causes the particles to disperse throughout the atmosphere, the oceans, and even back to the core of the planet. At the core of the planet where minerals exist mostly as hot liquids, the minerals are also moving around. The core of the planet is made up mostly of ferromagnetic metals such as iron and nickel. Ferromagnetic means that a material can be magnetized. The creation of a magnetic field through the movement of our planet's core is called a "dynamo process." The metal moves around and works-up a magnetic field called the magnetosphere. Compared to all the other rocky and mineral-based planets, our planet has the largest magnetosphere. The 'field lines' of the magnetosphere have been measured to extend 36,000 miles into space. Our planet is a giant magnet.

The sun is constantly sending out particles that are very high-speed and high-temperature. These particles are called solar wind. This solar wind has an electrical charge, and so it interacts with our planet's magnetosphere. Normally, the wind blows around the outside of the magnetosphere following its contours. Occasionally, the path of the solar wind breaches our planet's magnetic barrier. This causes the electrically charged particles to be guided inside our planet's atmosphere leading it to touch the ionosphere. When the electric particles interact with the ionosphere they create auroras, which are lights that appear in the sky. These auroras are a geomagnetic storm which is an intimate interaction between the sun and our planet. Each of these storms reveals information about things going on around our planet. The Earth is revealing information about herself. Historically this has brought about everything ranging from scientific curiosity to feelings of spiritual upliftment to humanity. In essence, the mineral cycle and the magnetosphere provide a beautiful form of communication from the Earth.

The female verbal advantage over males is extremely well documented. This is not an effect of socialization, as it appears even during infancy and persists throughout life (Suess, P., 2000). This makes sense considering the nature of traditional domestic roles in women. Domestic work in indigenous societies tends to involve multiple mothers, and the work is collaborative, detailed, and dependent on communication. They need to balance and interpret a wide array of information coming from the other women, the children and their husbands. This communication in this kind of setting needs to be very fluid, clear, constant, effective, efficient and accurate. Hundreds of thousands of years of this activity make the Black Woman perhaps the largest developer of human language. Her propensity for talented communication is as modern as it is ancient. Just as the mineral cycle enables our planet to be 'gifted' in its ability to create signals, the Black Woman is a naturally gifted communicator.

THE WATER CYCLE - EQUILIBRIUM

Our planet is very linked to her outside environment. She is linked in the sense of having a heightened sensitivity and reactivity with her outside environment. In fact, she is able to be more sensitive and react more to outside energy more dynamically than any other known planet. The reason for this is because of the water cycle. The water cycle happens as water molecules change back and forth between liquid and gas states. Through evaporation, condensation and precipitation, weather is created. This weather represents the sensitivity of our planet external (solar and lunar) forces.

The Black Woman is known for her sensitivity, intuition and perception to details. Just as our planet is highly sensitive to outside energy and forces, the Black Woman is also sensitive. One example of this sensitivity is hormonal sensitivity. Women who live in close proximity and are constantly around each other will find that their menstrual cycles begin to synchronize. Women are not only emotional, but also 'collectively emotional' in that they share and synchronize on hormonal and emotional experiences. She is also intuitive. Intuition is the ability to access implicit knowledge (that is knowledge which exists, yet is difficult to verbalize). Women have a proportionately larger corpus callosum (Gyldensted, 1977; De Lacoste-Utamsing, Holloway 1982). The corpus callosum is the bundle of nerves connecting the left and right hemispheres of the brain. The increased level of communication between the left and right hemispheres of the brain causes a more fluid interaction among the different faculties of the brain. All of these biological factors play a role in the Black Woman's ability to sense and perceive information in the manner she does.

Our planet's water cycle is unique and depends on more than just the mere presence of water. Water exists on other planets and even the moon; however, these entities do not have water cycles capable of supporting life. One of the reasons for this is the distance from the sun. All other factors being equal, if our planet was too close to the sun all the water would evaporate; too far and the water would freeze. However, distance is not the only important factor in the ability to have a water cycle; gravity is also important. For example, the moon and the Earth are the same distance from the sun, yet the moon's water does not produce a water cycle like the one on our planet. The Earth's gravity is 6 times that of the moon, therefore it is able to hold an atmosphere. This aspect of the water cycle is important, because it gives our planet the ability to spread life and life supporting nutrients across her whole biosphere. Water's life-supporting properties are based on the fact that water is polar and has a high dielectric constant – giving it the ability to dissolve nutrients and spread them. All of these factors converge to give the water cycle its property and principle of 'equilibrium'.

The Black Woman's body is also very good at cleansing itself, balancing itself, and maintaining a healthy equilibrium. The woman's body is a life center not only for herself, but also for any potential child that grows in the womb. Her brain also displays the quality of equilibrium, as it has less lateralization (McGlone, 1980). Lateralization is when each side of the brain (hemisphere) is highly specialized. Women have less lateralization which means that each side is less specialized for one thing; the brain tasks are spread more evenly across both hemispheres. Just as our planet is very skilled in its ability to dynamically distribute resources, the Black Woman's body operates in the same manner. This principle – Equality – is perfectly descriptive of this nature shared by both our planet and the Original woman, the mother of civilization.

AN INTRODUCTION TO ECOLOGY

ROBERT BAILEY

Ecology (from the Greek word *Oikos,* meaning "home" and *ology*, meaning "the study of") is the branch of biology that studies the distribution and abundance of organisms as well as the interactions that determine such distributions and abundances; It also covers the cycles of energy and matter. The scope of ecology covers everything from organisms and populations of organisms (species) to communities of populations and onto planetary scales. Ecologists seek to understand the influence organisms and their environment exert on each other. They find the answers to questions such as why and how populations came to be and why others became extinct, as well as explanations for fluctuations in population numbers and the composition and organization of ecological communities. All that talk about being "top of the food chain" is rooted in ecology. An ecologist may obtain their findings through a combination of observation, research, lab and field studies. Through their studies we come to new understandings by which we can preserve the earth and maintain the balance of life.

"Traditions are the products of generations of intelligent reflection tested in the rigorous laboratory of survival. That they have endured is proof to their power." – E. Hunn

While the name may be new, this ecological understanding of the interdependence of life and accompanying holistic practices have been around for millennia. TEK (traditional ecological knowledge) shouldn't come as a surprise to us, as it was common thought and practice to early hunter-gatherer societies in order to sustain life. Early societies actually had limits on what they did and didn't do, sometimes even without government regulations and some still maintain these traditions. For example, Fikret Burkes wrote in his book *Sacred Ecology: Traditional Ecological Knowledge and Resource Management,* that while fishing, the Cree Indians did not overfish and had community-based resource management.[331] Greek philosophers such as Hippocrates and Aristotle are said to be among the first to record their observations on natural history and so history of ecology usually begins with Greek origins. Modern ecology came out of natural history. With numerous contributions from different thinkers such as Charles Darwin, James Hutton and Jean-Baptiste Lamarck, it developed into a "more rigorous" science in the late 19th century.

The First Environmentalists

SUPREME UNDERSTANDING

MAN'S PLACE IN NATURE

Man is inextricably tied to all other life on this planet. Yet while man is certainly a living being, his consciousness (particularly his level of self-awareness) puts him outside the parametes of the complex microorganism scientists call Gaia. Basically, man is the "cream of the planet" because he is organically made from the same elements as all living things, and embodies the functions of any other living organism, but he also rises above the role of simply "living" by thinking about how he is living, and how this affects all other life. This is a capacity no other living creature possesses (and some would say most modern men lack this trait as well!).

How long has man had this awareness? There's no way to point to a time when he didn't, just as there's no way to point to a time when man was not man. As a result, we have countless indigenous traditions (most of them having no said birth record) that exist solely to preserve nature's balance. Since the dawn of human history, man has been sustaining himself while preserving the ecology of his home.

THE WAY THE ECOSYSTEM WORKS

Ecosystems are made up of four major categories of organisms:

- Primary producers (plants), which convert sunlight and inorganic materials into biomass and stored chemical energy, form the ecosystem's base.
- Primary consumers (herbivores) feed on plants.
- Carnivores, located at the ecosystem's top level, consume the herbivores.
- Decomposers convert the ecosystem's remains into inorganic materials used by the primary producers.

And thus Hakuna Matata, the circle of life. But it's not all "don't worry, be happy." Maintaining the ecosystem's stability is serious, and requires a means to regulate the levels of each category of organisms. The amount of sunlight, nutrients, moisture, and temperature regulates the abundance of primary producers. Herbivores also affect plant levels through consumption. If not checked, exploding herbivore numbers will cause an ecosystem to collapse by over-consuming the primary producers. So there exist natural mechanisms to regulate herbivore levels.

One is top-down, where carnivores control herbivore numbers. The other is bottom-up, where plant defenses control herbivore levels.* These large animals provide what John Terborgh has referred to as a "stabilizing function."[332] Animals like black caiman, jaguars and harpy eagles maintain the remarkable diversity of tropical forests through "indirect effects," a term referring to "the propagation of perturbations through one or more trophic levels in an ecosystem, so that consequences are felt in organisms that may seem far removed, both ecologically and taxonomically, from the subjects of the perturbation."[333]

And then there's man. Throughout man's history, our diet has shifted, with man going from being a vegan raw foodist to an omnivore.[334] Throughout the world, most indigenous people eat a diet that is 75% fruits, vegetables, seeds, nuts, lentils, and tubers and 25% animal matter. Of course, the Western diet is quite a bit different from the rest of the world, thanks to the European legacy of consuming massive quantiies of animal flesh. Still, even Europeans were not exclusive carnivores, and the closest any man has come to being "carnivorous" were the Neanderthals, apex predators who ate mostly meat and some human flesh.

But perhaps our historical dietary shifts weren't simply adaptive. That is, perhaps we weren't simply adapting to consume whatever was most available in the local ecosystem (although this "flexibility" seems to have ensured that our ancestors survived while others did not). Perhaps we were actually regulating the local ecosystem through our consumption habits.

At the very least, we know that our lifestyles were conducive to the maintenance of the Earth – beneficial even – rather than exploitative and destructive. In fact, we were the first environmentalists. In one of our many balancing acts, we maintained a balance between sustaining ourselves and the balance of nature.

A recent study by Jon Erlandson of the University of Oregon suggests that the Chumash people who settled Channel Island (near Santa Barbara, California) pioneered sustainable fishery management over 12,000 years ago. When certain areas became depleted, they simply moved to another, effectively imposing a "no-take zone" in the old fishing grounds. And when harvests dwindled throughout the region, they switched to hunting and eating otters until shellfish numbers recovered.[335] Of course, this is only the first study to confirm sustainable

* See "The Secret Life of Plants"

DID YOU KNOW?
One common aspect of many indigenous traditions is the idea of animal guardianship, which according to Kent H. Redford and colleagues, "places animal spirits in the role of protecting wildlife against human abuse. For example, Hitchcock (1962) describes the role of the tabanid fly in regulating the fishing of the Montagnais of eastern Canada. The fly is considered overlord of salmon and cod, and hovers over the fishermen whenever the fish are taken from the river "in order to see how his subjects are being treated." Occasionally, the fly will bite the fisher as a punishment for wastefulness; the [sick] man would expect poor fishing for a time as a further chastisement. Guardian animals may also punish those who disturb a forest or other ecosystem unnecessarily. Certain communities in the Peruvian Amazon fear and respect the "animal-demon" *shapshico*, who shoots a tiny dart causing illness and hysteria 'if you cut down a tree in his [rainforest] garden.'"[336]

fishing in the past. We know humans in South Africa were hunting the seas 120,000 years ago, and evidence from Israel suggests that fishing goes back at least 800,000 years.

The indigenous history of environmentalism is timeless, and detailed in considerable depth in books like *Original Instructions: Indigenous Teachings for a Sustainable Future*, edited by Melissa K. Nelson, *Indigenous Traditions and Ecology: The Interbeing of Cosmology and Community*, edited by John A. Grim, and *Sacred Ecology*, by Fikret Berkes. These books describe how Original people have interwoven the tenets of sustainable resource management into their cultural traditions, taboos, and mythology, long before Europeans wrecked the Earth and then cashed in on the idea of environmentalism. Ironic that the same people who made the Earth go grey to get their green are now the ones pushing "going green" to get more green.

Derrick Jensen, the author of *Endgame, A Language Older than Words, Walking on Water*, and about ten others, goes a bit further. One of the strongest non-indigenous advocates of indigenous sustainability that I am aware of, Jensen argues that it is white-owned corporations and governments who ruin this planet, and that the masses of white people passively support this exploitation, which primarily hurts Original people (but will soon wipe us ALL out).

How bad is it? Pretty bad. Even the American Museum of Natural History and the United Nations agree that "we are in the midst of a mass extinction of living things, and that this dramatic loss of species poses a major threat to human existence in the next century."[337] J. N. Michael, editor of *The Biodiversity Crisis* notes, "We are in the middle of a sixth major mass extinction…The last great extinction event occurred at the end of the Cretaceous period, about 65 million years ago, when an estimated two-thirds of all species, including all the dinosaur groups except the birds, were obliterated."[338] In the same book, H. Raven states, "Over the next few decades, we could lose about 50,000 species per year, a rate 20,000 times the [average natural] rate. By the year 2100, perhaps two-thirds of the Earth's current species will have disappeared or be on the way to extinction."[339] These species, as we noted above, are all connected to the maintenance of our ecosystem. It's fine-tuned, and if you think losing a few species doesn't matter, imagine taking just a few wires out of your car and seeing how things go down the road.

But indigenous people had the solutions. In fact, they still do. In *Decolonizing Nature: Strategies for Conservation in a Post-Colonial Era,* Marcia Langton, who has done considerable work on preserving the knowledge of the Australian aborigines, has noted: "The slowness of the advances, where there are any at all, in recognizing the contribution of indigenous peoples to maintaining biological diversity may contribute to the collapse of faunal and floral species that have been sustained by these groups for much of human history."[340]

Asked where examples of "radically sustainable living" – the type of change needed to "save" the Earth now – could be found, environmentalist A. C. Keefer responded:

> Right where you are! Look to the contemporary or historical practices of the traditional indigenous peoples who live now, or had lived, for long periods of time in your area. That so many indigenous groups have acted completely sustainably (at least before encountering overwhelming foreign negative influence and disruption) is no mistake or

ACTIVITY: GETTING BACK TO YOUR OLD SELF

"So I be ghost from my projects, I take my pen and pad/ For the weekend hittin L's while I'm sleeping/ A two day stay, you may say I need the time alone/ To relax my dome, no phone, left the nine at home/ You see the streets had me stressed something terrible/ F***ing with the corners have a nigga up in Bellevue" – Nas, "One Love"

Take a drive to the nearest nature preserve or state park. Catch a bus if you have to. Find a spot so "natural" that you can't see anything manmade no matter which direction you face. Try to find a spot near some water. Take your shoes off and let your bare feet connect with the Earth. And just breathe. You can try some of the exercises described in The Science of Breath. Now ask yourself, what really matters? I bet your value system will be - at least momentarily - affected. When you get back in touch with nature, you're getting back in tune with your primordial self. You know, the one who lived a GOOD life without anything the West gave us (and with everything it took from us). Just try it and you might see life differently, if only for a moment. (Note: Don't get lost in the woods. Turn your phone OFF and put it away where you won't keep reaching for it, but DON'T leave it in the car. I've been lost in the woods before and it's not fun. If you've got GPS on your phone, this is a situation where it could be very handy. In fact, there are apps for most smartphones that show you where the natural trails are in your area and how to hike them.)

> the product of ignorance. That is, this sustainability cannot be attributed to a "lack of enterprise or creativity" or "inability to innovate," but rather is the result of a cultivated, intimate awareness and of and reverence for ecosystems, and of deliberate vigilance against excessive waste and environmental harm. Find out also if at any point indigenous and non-indigenous peoples forged decent relationships and respectful lifestyles with one another (though it can't be said to have happened very often) and let the activities practiced by that alliance inform your decisions.[341]

But here's the problem. White environmentalists, green advocates, and animal rights advocates aren't fighting for you! Derrick Jensen gets into some of this in his books, but I make the point a little harder in How to Hustle and Win, when I explain that white people are saving the planet for themselves, at YOUR expense. Consider how they'll dedicate all their efforts (and money) to saving a forest creature or plant while allowing the local population to die of famine or preventable disease. Sometimes, they'll even villainize local indigenous people, blaming them for the extinction of local species, when it was these indigenous people who PRESERVED the ecosystem until the colonists wreaked their havoc.

In *A Critical Evaluation of Conservation and Development in Sub-Saharan Africa* – a $700 set of books – Paul DeGeorges and Brian Reilly document considerable evidence from indigenous societies in Africa, tracing the preservation of local ecology as far back as the earliest man. Yes, this scientific book documents the ecological lifestyle of our ancestors over the past three million years. DeGeorges and Reilly explain that their goal is…

> …to explain the relatively sophisticated systems that already existed/exist and are still for the most part being ignored as a means of integrating traditional value/management systems into modern conservation concepts. It might be better argued that colonialism and independence have impeded the evolution of these traditional management systems, ignored them and in some cases exterminated them, but they did exist prior to colonialism. In most cases, they are still present today, though often in a deteriorated state.[342]

What are some of the means by which we kept the Earth straight, even as we hunted, gathered, and shaped the landscape with our settlements? DeGeorges and Reilly continue:

> Ultimately, by means of culture, taboos based on religion and traditional value systems, local rules and regulations enshrined in farming practices, land tenure systems and folklore, the indigenous Africans protected and conserved ecosystems and their associated biodiversity. In addition, territoriality, mobility/migration, fire, harvest regulations (e.g., no harvesting of pregnant female or young), seasons, hunting

> guilds, etc. resulted in biological and sociological controls. Habitat manipulation was also (e.g., fire) employed to design landscapes needed to suit the various lifestyles of rural Africans.
>
> Rules and regulations regarding access to natural resources were precise and codified, although not written down, and had been enforced since time immemorial. The result was sustainable exploitation of natural resources. This had been accomplished with no ecological purpose (e.g., maintenance of biodiversity) in mind as we understand it, but out of a sheer instinct for self-preservation. Conservation of game animals and fish was necessary in order to provide for survival in the future.[343]

When did it change? I'll let them tell you.

> While Africans may have controlled and manipulated the landscape to the benefit of livestock and agriculture, there seemed to have been a place for wildlife as a supplemental resource, while Europeans, with their modern firearms, devastated the game, bringing a number of species to extinction.[344]

Pre-colonial Africans even had guilds responsible for resource management:

> In pre-colonial Africa, human populations were low relative to the resource base. Wildlife, forestry, fishery, pasture and agricultural lands were extensively managed. Due to the high ratio of resources to people, it was unnecessary to practice intensive management. Most resources tended to be linked to a territory owned by the ancestors, access to which was ultimately in the hands of the king/chief and his headmen who were elders within the community. Most resources were managed as common property for the good of the community, although there was some commercial use. Over a large part of Africa, hunting was considered a profession, requiring innate skills and acceptance by elders into a guild where a three to six year apprenticeship was undertaken. The apprentice would spend most of his life under a "master hunter" until he became a "master," who provided him with direction, and taught him in the secrets of the hunt linked to the ancestors and spirit world. Without such knowledge, one's life and that of one's family would be in great danger. Thus limited access to these guilds helped control hunting pressures, assuring sustainable use of wildlife for the greater good of the community. In most cases, individuals within the lineages were given responsibility for the management of aquatic and terrestrial resources for the good of the community. Non-lineage migrants had to request permission within a given lineage's territory to access its resources. Intercropping, manuring and fallow were used to maintain agricultural production. In many high yielding areas where irrigation was practiced, over-population was experienced, even in pre-colonial times.[345]

In *What About the Wild animals? Wild Animal Species in Community Forestry in the Tropics*, the authors connect animism, totemism, taboos, and even shamanism to indigenous plant and wildlife management. They also identify traditional practices that fall under the category of "soft management:

> Increasingly, researchers are beginning to realize that the distinction between wild and domestic species (of animals or plants) is not as clear-cut as once thought. Humans have been manipulating wild species for millennia and careful research has shown many cases in which indigenous people engage in "partial domestication," which could also be referred to as "soft management."
>
> Most of the literature documenting this type of practice is concerned with plants. Alcorn (1981) points out that "agriculture is only one type of plant management, and domestication is only one of the processes to which people submit plants." Soft management practices for plants include slashing, neglecting, sparing, protecting, transplanting or planting. The result of these activities is such that what may appear to be undisturbed forest to the uninitiated observer, can in fact be a vegetation formation in which species composition and distribution are largely a product of human action. Balée (1989) refers to these types of plant communities as "cultural artifacts."
>
> There are many other examples of soft management of wild animal species. The indigenous peoples of the Amazon refrain from cutting wild fruit-bearing trees in gardens in order to increase populations of game animals. Farmers have been known to deliberately plant more crops than are needed in order to provide food for game animals. Other soft management practices include rotation of hunting zones, restraint from killing females, taboos and seasonal movements by hunters.[346]

But at some point in the transition from the Paleolithic to the Neolithic – the time when so many other things changed for us – we start seeing some of the early failures that precipitated (but came nowhere close to the scale of) the destruction brought on after European contact.

For example, another recent study found that over-harvesting of North American marine life grew exponentially worse after the colonial era, but began with Stone Age populations, who fished (and hunted) heavier than the people who came before them. That is, the Chumash did things right, but the people who came after them didn't always stick to the code. Perhaps population growth resulted in too many groups competing for resources? Perhaps a new group came into the region, lacking the knowledge of those who came before? Whatever happened, the resulting disappearance of predators and other links in the food chain set off a chain of events that have indirectly contributed to modern-day ecological instability, such as toxic algae blooms, dead zones, and outbreaks of disease.[347]

Another study suggests that Native Americans contributed more to greenhouse gases than previously thought (yet, of course, nothing like the levels produced today). They did so through land use practices like burning trees to actively manage the forests and to yield the nut and fruit trees which were large parts of their diets.[348] And we have already read about the same processes occuring in Africa, with the most recent desertification of the Sahara being caused, at least in part, by the clearing of West African forests for game. Even then, we had techniques to "get the best part" from whatever conditions we encountered. For example, the Garamantes of the Sahara were still able to develop a thriving civilization, even in the desert, because they knew how to mine for groundwater. They created underground qanats (or tunnels) to bring groundwater to their settlement. They effectively converted a desert site to an oasis, "turning back the clock" to a time when the entire Sahara was a lush, fertile region. Unfortunately, they too, overexploited their resources and eventually ran out of steam, so to speak. But all of this happened within the past 20,000 years. Considering we've been here for millions of years, that's a pretty good track record, compared with those who have nearly destroyed the entire planet in less then two thousand seasons.

Regarding the "failures" of indigenous people within this time period, Keefer adds:

> It's true, as some hasten to point out, that not every indigenous society has acted sustainably. While much of the "evidence" used to support this point, like the "Pleistocene Overkill" hypothesis, is nonsense, and often leads cynics to go too far in condemning indigenous groups, care must still be taken to not exaggerate the truth or make an equally egregious mistake by claiming that indigenous societies were universally "perfect" in any respect. The fact that very few or no non-indigenous societies have ever definitively demonstrated their own sustainability, and that for the most part the only comprehensive, effective sustainability promoting worldviews are indigenous ones, however, causes this specific advice [to learn from indigenous examples] to remain sound.[349]

STRINGS ON OUR FINGERS

SUPREME UNDERSTANDING

Back in the old days, people would tie a red string around their finger before leaving their homes, to serve as a reminder for later in the day. Today, we put these reminders in our phones, but since we tend to ignore our phones many alerts and alarms, perhaps a string isn't such a bad idea after all. Turns out, when we wrote this universe in the language of mathematics, we put quite a few strings on our fingers to remind us of its authorship.

This is a concept I touched on in *How to Hustle and Win, Part Two,* where I wrote:

"O Lord, our Lord, how majestic is Your name in all the earth, who have displayed Your splendor above the heavens!" – Psalms 8:1

> The spiral found in the human fingerprint is the same spiral found everywhere throughout nature, from a common shell to the shape of our galaxy. The human brain cell is structured almost exactly like the structure of our universe. The Earth's surface is 75% water, the human body is 75% water, and the human brain is 75% water. And just as the most powerful natural forces on the Earth (and in the universe) involve electricity and magnetism, your body has its own magnetic field, and your brain operates using electricity…Melanin, the chemical substance that gives you your skin color is present in the brain as neuromelanin, and throughout the blackness of space. Carbon, a black element, is fundamental to all life…and your melanin.
>
> Religious people often look at the birth process as proof of a miracle-working unseen God. God IS in the details, but is – as the Qur'an describes – "nearer to you than your jugular vein." In a woman's body, the egg can be found in only one fallopian tube at a time. A man's sperm, upon reaching the woman's uterus, consciously decides which way to turn, and picks the correct path 90% of the time. This sperm has almost nothing in it, except DNA. That's your mind still at work. Why do you think you get tired when she's the one pregnant? A human fetus then progresses through several stages, which resemble the different points in human evolution, once again reminding us of the brilliance of our design.

"As above, so it is below" - Ancient Egyptian maxim

Some of these connections, like the analogous structure of the universe and the brain cell, are "true correspondences." They derive from the fractal nature of the mathematical language of this universe. That is, patterns are replicated across the various scales of existence. Some are more generalized correspondences, like the parallels between the "impregnation" of the Earth with the first seeds of life and the way this same process occurs within the human body. That is, everything involved is not an exact one-to-one match, but the fundamentals involved are more or less reproduced at both scales (and probably at many other scales as well, if we should look). Others are more specific, but the numbers involved aren't "easy" numbers like the 3/4s ratio of water to "land" in the Earth and the human body. These correspondences involve numbers scientists call universal constants. There are several such "magic numbers" identified in John Gribbin's book *Cosmic Coincidences: Dark Matter, Mankind, and Anthropic Cosmology*. They represent the same principle in more specific terms, that of our universe being the same at nearly scales (until we look beneath the surface of it all, at the quantum level). And at the center of these scales, as Primack and Abrams point out in *The View from the Center of the Universe*, is us.

MNEMONIC DEVICES VS. TRUE CORRESPONDENCES

Some correspondences aren't as mathematical. They may be linguistic or historical patterns which can be used as mnemonic devices to draw connections we wouldn't otherwise see. When I first heard a Five Percenter breaking down ALLAH as the "Arm-Leg-Leg-Arm-Head," it drew me into this culture almost immediately. It just made sense. About five years in, I encountered a common objection, that Allah was an Arabic word and had no relation to the English words for arms and legs and whatnot. But that was never the intention behind this breakdown. Consider the way the word "Allah" traveled from Arabia into Islamic West Africa via the Moors, and then was carried into the Americas by kidnapped Africans...only to survive as residual elements of Islam well into the 1920s...when Islam was revived by the MSTA, the NOI, and various other groups...until the Five Percenters emerged in the 1960s to bring it all back to self...by explaining Allah as the "Arm-Leg-Leg-Arm-Head."

There are similar "re-emergences" of historical patterns and language throughout history. Consider how long (and how far) we carried the artistic elements we find in prehistoric cave art...only to find those same elements reproduced in modern graffiti. The cultural continuum is just one part of this puzzle. The other part is the fact we inserted these elements into the design of things for future discovery. Some seem overly simplistic, while some are too much of a stretch, but at least some of them exist simply so that we could later rediscover these strings on our fingers.

The Signs we Left Behind

SUPREME UNDERSTANDING

Shakespeare – in translating the Bible – worked his name into Psalm 46. The builders of Khufu's Great Pyramid inscribed their names in the walls after finishing this monumental job (in an early instance of graffiti, but not the world's first, as you'll soon see). And proud authors and artists don't just work themselves into their creations; they sometimes add elements to their creations that represent them on a smaller scale. That's why the creators of many video games will go beyond putting their name in the credits and actually put themselves in the game somewhere. We did the same thing…with the Earth. When we built, we built according to the design principles we employed in constructing the Earth itself, and added elements to remind ourselves that man and his creation were inseparable.

I'll explain. By 4,000 BC, we were employing serious city planning and design principles to accommodate the many citizens of our urban centers. This allowed us to incorporate complex symbolic designs into our city layouts (from the fractal design of many African villages to the astronomical alignment of ancient Egypt) as well as our buildings themselves. Some of our designs were abstract, representing the laws and principles themselves. Other designs were representational (representing us, the law-makers). This is why Solomon's temple was built in the shape of a man.

And this is why huge renderings of a man were carved into the hillside of Nazca in Peru. It's not a spaceman. It's just us. And the hundreds of lines (the abstract) scattered throughout Nazca were crafted on ground too soft for an actual "runway," but they certainly follow the pattern of Original people "shaping" the Earth with lines and circles, which often served very specific purposes. Just as our earliest writing was composed of lines and circles and symbolized greater meanings,

A Quick Note on Indigenous Accounts of Alien Encounters

Within the past 50 years, Western writers have attributed "spaceship" myths to indigenous cultures throughout the world, yet none of these myths are found in anthropological accounts from the 200 years preceding. Some of them are obvious misinterpretations of the available evidence (like describing a throne as a jetpack), while others may have derived from missionary influence or some other Europeans. Considering how much Europeans have influenced indigenous conceptions of God, it is not surprising that the "cosmic egg" of the Aborigines (which symbolized the singularity from which the cosmos sprung) has now become a flying saucer. One problem with oral tradition is that it's quite difficult to tell how old a tradition is, or when it was influenced or altered. This is compounded by the fact that Europeans actually DID crashland flying vehicles (cargo planes) throughout the early 1900s, leaving many indigenous people with ideas of "white men coming down from heaven." At the end of the day, most alien mythology really serves the interests of white-dominated ideology. That is, EVERYTHING we learned either came from white people or space people (who are assumed to be more like white people than us). Thus the following "Aborigine" account is not surprising: "Long long ago, far back in the Dream Time, a great red coloured egg (spaceship) came down from the skies. It tried to land safely on the ground but broke (crash landed). Out of it emerged white-skinned culture-heroes (gods) and their children." This is the kind of "evidence" used to "prove" aliens have been everywhere and gave everyone whatever knowledge they have today.

DID YOU KNOW?
Geoglyphs have been found throughout the world, including the following places: the Blythe Intaglios in California; the Pintados in the Atacama Desert of Chile; Big Horn Medicine Wheel in Wyoming; the Uffington Horse in England; the Cerne Abbas Giant in England; the Gummingurru Arrangement in Australia; Nazca Lines in Peru; the Quebrada de Santo Domingo in Peru; at Abd Al Kuri Island near Northeast Africa; and at Malabo Island near Cameroon, West Africa.

the same forms were incorporated into the larger landscape and supplied meaning there as well. Some lines were markers for the paths of heavenly bodies, while some holes in the ground were made for mining or storage. Everything in our world has meaning and function.

Oh and the representational drawings of people and animals don't point to outer space visitors. Even though they are only visible from the skies, they still point back to us! How? The evidence suggests they were constructed by humans who actually used hot air balloons (at least 1500 years ago!) to survey (and perhaps enjoy) their creations. (Yes, our history is THAT rich…WITHOUT magic beings to help move the story along.)

Did I mention that there are carvings that look just like the Nazca pictures in Australia? These images are known as geoglyphs (large drawings in the ground) and petroglyphs (smaller drawings on rock surfaces). They're another form of "rock art," and you can't go anywhere Original people have been (everywhere), and not find at least a hundred examples.

We'll return to evidence like this in Volume Two. For now, let's explore the evidence that we actually "made" this Earth the way it is. To do so, we'll have to travel waaaay back before the Nazca lines or Solomon's Temple. We'll have to look at man's role in changing the face of the Earth since it all began. But first, let's clear something up.

Aliens Didn't Build It

C'BS ALIFE ALLAH

"Extraordinary claims require extraordinary evidence." – Carl Sagan

When people use aliens as a "*deus ex machina*"* to explain "jumps" in civilization or scientific knowledge, where advanced technology or culture appears to "come out of nowhere," they're typically telling half-truths (omitting some of the story), or unaware of the actual chronol-

* A plot device where an "unsolvable" problem becomes answered by inserting some new person, thing, etc. From ancient plays where when such a problem arose they lowered a character playing a god into the piece.

ogy of how things came to be. Often this is because of religious bias or racist bias.

As a stand-in for religion, aliens are inserted to explain the "creation of man." So instead of having God "create man," some alien race comes to earth and "creates man" (though the problem of who created the aliens is never tackled). As a stand in for racism, aliens are utilized to explain away the acomplishments of the Original Man throughout history (architecture, medicine, technology in general). Instead of looking for the traces of the developmental steps leading up to those accomplishments by Original People, by searching in the vicinity of a particular find, or in the records of their history, any "extraordinary" feat is attributed to aliens.

For example, the Great Pyramids at Giza didn't "just pop up." It seems that people who talk like that don't realize that the infamous "Bent Pyramid" of Snefru came before the Pyramid of Khufu. It's called the Bent Pyramid because it was an early attempt to build a true pyramid, and the architects realized it wouldn't work at the angle they were building it at, so they changed the angles halfway through construction. Sounds pretty human to me. Before that, there were step-pyramids, which provided the structural basis for later "true pyramid" design. And let's not forget that burial mounds predated all such structures.

And as you can see in the *National Geographic* documentary on the "Real" Scorpion King, long before there were any stone temples, the

A QUICK NOTE ON STUDYING ALIEN CIVILIZATIONS

If you're interested in life on other planets, first realize there are some serious limitations to how much you can find out, even with the wealth of information (and misinformation) that's currently out there. By comparison, how much do you know about other countries and cultures on this planet? "Alien" just means "foreign" so there may be a lot of human cultures that are entirely alien to you. And there's no reason you should be able to tell us more about Planet X than Mexico! Matter fact, how well do you know your neighbors?

And, in one sense, man himself is part extraterrestrial if you consider the story of the origins of life. That carbon-rich meteorite that impregnated the Earth with life about four billion years ago is the closest we can get to saying that "we came here from outer space." It might not have been a voyage from the planet Krypton, but the seeds of life (and us) did, in part, start in the heavens above. Well, perhaps instead of "heavens" we should say a black interstellar cloud nebula, because otherwise people might think we're talking about something religious.

Yet, speaking of religion, there is a notable verse in Genesis that describes the sons of God mating with the daughters of men. If Genesis is a book of genetics (as the name implies), this could be interpreted as the meeting between the celestial and terrestrial elements that eventually "begat" all life on Earth.

first Pharaoh, known as King Scorpion (circa 3450 BC), had a wooden temple. The Egyptians built in wood before they built in stone, just as their cousins in England built a Wooden Henge before raising up Stonehenge. In fact, there were probably several transition stages.

We know the Sumerians had a transition phase where they went from wood to brick to stone. In Egypt, the evidence suggests it went from wooden huts, to wooden temples, to small stone or brick buildings, to large stone temples and burial mounds, to step pyramids, to the Bent and Red Pyramids, and then finally to the Pyramids at Giza. An editiorial published in Architectural Science Review explains why the ancient Egyptians knew how to build using arches, but did not use this technology in their later temples:

> So why didn't the Egyptians use arches for their stone structures to avoid the use of a veritable forest of columns in the interior? Temples were originally of timber, which was considered a building material superior to mudbrick. One can see that from the decoration of the stone columns in both Ancient Egypt and Ancient Greece, and even more clearly from some early stone structures, whose ceiling soffit is carved to look like round logs of timber; the processional corridor of Pharaoh Sozer's Step Pyramid at Sakhara is an example of a very early Egyptian stone structure (about 2800 BC), whose stone ceiling is carved to imitate round timber logs. So the early stone construction was based on the existing timber technology, and not on the more appropriate technology used for mudbrick.[350]

And unless you think that space aliens started with wood, it looks like it was US all along.

But what about people who see a UFO? As the science of astronomy developed, ancient man had to discern between starts, planets, and comets. The average observer can't make such a distiniction. With the plethora of manmade objects and visual effects in the modern sky, one has to be grounded in basic astronomy before declaring that everything that one sees in the night (or day) sky is a flying saucer. In modern day lingo, UFO has become synomymous with alien craft. It is important to note that the true meaning of UFO is "unidentified flying object" not flying saucer. Making a jump from something that you haven't seen before and can't fully identify to an alien flying machine is just dumb and you're lying to yourself. It's called confirmation bias.[351] There are many shows addressing objects that can and have been misidentified as "flying saucers,"* as well as books which deal with how to

* For example, the show *Fact or Faked* on the SyFy channel features an investigation team of analysts who break down various sightings and offer a host of alternative theories simpler than flying saucers. This brings us back to Occam's Razor, which

> DID YOU KNOW?
> Widespread Western bias against any notion of indigenous scientific knowledge is what led Marcel Griaule and Germaine Dieterlin to "speak up" for the science of the Dogon people of West Africa in their book *The Pale Fox*. Griaule and Dieterlin's methodology wasn't strong (he relies entirely on one Dogon elder and one translator), but there's still good reason to think the Dogon knew of Sirius B, the companion star to Sirius. There is not good reason, however, to support the conclusions of the book that attempted to reinterpret Griaule's work, Robert Temple's *The Sirius Mystery*, which uprooted this knowledge from its African origins and gave credit to aliens from outer space...even though the Dogon have no such tradition.

get your observation game up.[352]

But theoretically, it is possible that life exists somewhere else in the universe. With that said, that doesn't mean that every unidentified flying object should be classified as a flying saucer and that every episode of X-Files is now revealing some hidden truth. Just like anything in life, there's a scientific method which you can utilize to look for, identify, and qualify alien life.

If you walk a few yards in any direction you will have passed hundreds of different forms of life from bacteria to plants to insects. It will probably be a minute before you cross paths with another human. Traveling through our universe would be similar. Even if you found life, intelligent life would be the exception, not the rule. And if you did find life, it will most like not look anything like you. All those science fiction aliens with arms, mouths, eyes, fingers, etc. – that ain't happening. The conditions which formed the human form and structure happened within the specific environment of Earth. The probability of alien life in general is very small. The probability of intelligent alien life is even smaller. The probability for intelligent alien life that looks like us…think about that…really?

So why so all the depictions of aliens look so much like people? (from hairy man to shiny suit man to bug-eyed "greys," etc.). That's because – whatever type of alien form was depicted in a specific decade, that was the type that was popular in the science fiction media of that time. Since the 1970s, however, the "grey" form has persisted with the strongest impact.

Why? One of the first reported cases of "alien abduction" was in New Hampshire in 1961, known as the Betty and Barney Hill abduction. Themes of alien abduction had been going strong in the movies and television shows of the 50's and 60's. Those who study science fiction know that those themes were strong in science fiction during those times because of the fear of the threat of Communism. It is notable that rises in abduction stories also coincided with higher rates of

---

states that one should begin first with more simpler theories to a problem

xenophobia (fear of others) in the United States.

But the Betty and Barney Hill story is significant because it really hit home with the American public when a movie version of the story, staring James Earl Jones, aired on TVs across the country in 1975. This story quickly became the prototype for most modern day stories of abductions (missing time, grey aliens, bright lights, etc.).* Though held as gospel by many in the alien community, you might want to examine the skeptic's report of the incident. In fact, any time you encounter an audacious claim, check out the skeptic's report too.

If you're still interested in intelligent alien life you should check out the current research on the intelligence of other forms of life on earth, from squid to slime mold. This will help you become familiar with a more "universal" framework of intelligence and consciousness, and get an idea of how "alien" an alien intelligence would be. It's not going to be like E.T., to say the least.

If you're really interested in the science of this stuff (not just the fun stories and fantasies), you should also study the lifeforms that exist in the Earth's most extreme climates. These extremophiles are the closest thing we have to aliens on Earth, because they're so unlike anything else. For example, there's GFAJ-1, a micro-organism found in Mono Lake in California, which replaces one of the six essential elements (carbon, hydrogen, oxygen, nitrogen, sulfur and phosphorus) with arsenic. There are bacteria that are able to live in the Antarctic ice in a sort of suspended animation where they live on and off for millions of years. There are even strange micro-organisms living in our city's sewers who consume methane as if it's a nutrient. Basically, the whole biology of an alien species is going to be alien from physical appearance down to their genetic code.

Still interested in the search for alien life? Well guess what, there's already a scientific framework set up for that type of research. The scientific discipline is called Astrobiology. It's an interdisciplinal study which involves chemistry, physics, astronomy, biology, geology and general planetary science. It oversees the search for viable planetary environments where types of life have thrived in the past, are thriving now or can thrive in the future. It also develops the format by which humans can detect said life.

Get familiar with the research done by SETI (Search for Extraterres-

* In fact, that story itself might trace back to a specific episode of the television show, *The Outer Limits*, as well as the 1950's movie, *Invaders from Mars*.

trial Intelligence), as well as the Drake Equation,* the Fermi Paradox,† the Mediocrity Principle,‡ the Rare Earth Hypothesis,§ the Inverse Gambler's Fallacy,** the Goldilocks Principle,†† and other explanations that can help you sift through the relevant data. And if you're not interested in doing any of that, please don't go around making people ignorant by spreading ideas you haven't looked into.

"You can't convince a believer of anything; for their belief is not based on evidence, it's based on a deep-seated need to believe." – Carl Sagan

Much of the modern alien hype is just a substitute for old time religion. Many of the themes of "alien abduction" match perfectly against ancient stories of "ascending" and "meeting the gods." Modern day adherents just say that these stories are cases of ancient alien abductions, instead of simply seeing a replay of the same religious, mythical themes in a modern context and setting.[353] People are starting to replace savior gods and demonic forces with savior spacemen and evil aliens. It's not the evidence that's swaying us. It's our own psychological dispositions!

And now that the History Channel has sold its soul for steady viewers by airing shows like Ancient Aliens (where the "experts" tell lies than you can easily look up as false), we're seeing the spread of what's known as the ancient astronaut theory. This theory revolves around the notion that several times in the past, aliens visited Earth, created man, directed his cultures, inspired his religions, and gave him technology. Proponents of this theory identify most, if not all of the ancient deities with aliens. Though alien narratives became famous in the early 1900s, the ancient astronaut theory didn't become famous until the end of the 60's and the beginning of the 70's. This parallels the rise of Black studies and history in American academia and in Black conscious circles in general. Before this, the idea of a "lost white tribe"

---

* The equation used to estimate the number of detectable extraterrestrial civilizations in the Milky Way galaxy.

† The contradiction between high estimates of the existence of extraterrestrial civilizations and the lack of evidence for, or contact with, such civilizations.

‡ The notion in the philosophy of science that there is nothing very unusual about the evolution of our solar system, the Earth, any one nation, or humans.

§ Argues that the emergence of complex multicellular life (metazoa) on Earth required an improbable combination of astrophysical and geological events and circumstances.

** The fallacy of concluding, on the basis of an unlikely outcome of a random process, that the process is likely to have occurred many times before.

†† That something must fall within certain margins, as opposed to reaching extremes.

was used to explain any impressive feats of architecture or technology that Europeans encountered when they came in contact with ancient Black and brown civilization and cultures. They didn't want to attribute these feats to the indigenous people, so they introduced a white tribe, as a deus ex machine, who supposedly went all over the world spreading knowledge and then all of a sudden disappeared. But this theme went out of vogue because there was no evidence to support this mythical white race. In their art, these ancient civilizations attributed their accomplishments to irrefutably dark-skinned people who resembled them. Thus the idea of the white tribe gave way to the ancient astronaut theory, which again, stole credit from Original people by making those Black and brown icons out to be spacemen.[354] In some cases, they describe traditional African hairstyles as "spacehelmets"!

In many cases, whole areas of the ancient astronaut theory have long been debunked due by modern evidence or critical analysis of the claims, yet this may not be well known among the theory's adherents. For example, the notion that the Annunaki are some alien creators of humans is derived from one source. It doesn't matter if you are a white New Age adherent or within the new Black conscious circles, the source is Zecharia Sitchin, whose translations of the Sumerian cuneiform have long been debunked.* So people are making "arguments" based on wrong translations! While Sitchin spread these ideas (and his faulty translations) far and wide, the pioneer of this premise was Erik Von Daniken, who also has been thoroughly debunked. You can read *The Space Gods Revealed* by Richard Story for details on *that* story.

But seriously, how much information do you really need to see through some of that ancient astronaut stuff? Critical thinking alone should have you raising questions like:

- They can fly all the way across space yet need to carve giant directions in the desert rocks?
- They need to come to earth to mine gold yet travel across the universe using various means of nuclear technology (which even humans know how to use to make gold)?
- There are no skeletons (oh wait, they've ALL been hidden by the government) and there are no pieces of their technology (ships, clothes, etc.) ever found (oh wait, they've ALL been hidden by the government), yet the government CAN'T hide the 5,000 bestselling books on the subject? That don't make no sense.

* Check out the work of Michael S. Heiser (www.SitchinIsWrong.com), who does an in-depth presentation of ancient languages and how they show Sitchin's many errors.

When you consider an extraordinary claim, the elimination of easy logical alternatives (Occam's Razor) is the first thing that you're supposed to do. If I come back home and my door is open, my first thought isn't that the ghost from Christmas Past opened it. Did I leave it open? Is there a robber in the house? That's the type of thought rational people have. The person who goes straight to things like ghost and mystical forces would be seen as psychologically "off."

And honestly, going straight to the alien thing without even examining other data is off also. It's rooted in religious beliefs and racist assumptions, yet most of us who claim to be anti-religion and anti-racism don't see this. We don't need aliens to explain the presence and progress of man on Earth. The Original Man is needed to explain the presence and progress of the Original Man on earth.

WHAT MAKES RAIN, HAIL, SNOW AND EARTHQUAKES?

SUPREME UNDERSTANDING

Now, let me say this again. Everything in our world has meaning and function. So all this carving, digging, and shaping wasn't just for aesthetic value. It was both actual and solar (or literal and symbolic). The literal meaning was its functionality. Much of that was intentional, but some of it had unintended results. So when we dug up a massive underground mine and piled all the excavated Earth beside it, we were effectively creating mountains thousands of years ago. Realizing this, manmade hills and mountains were used to protect herds during times of flooding (in a technique used until recent years in the Netherlands), to create highlands for groups of people trying to avoid predators or rival clans, and even to artificially affect the local weather patterns. We did that. And we did all that so long ago that there's no way to tell when we first started doing it. At the same time, these artificial constructions often created problems due to failures (on our part) to properly calculate potential outcomes. Even then, it's entirely possible that these "mistakes" were expected by the wisest among us, but allowed, so that other predicted events could come to pass. Looking at things that way, it's difficult to peg any development as a true "mistake," since all things lead to other things, and everything follows the cycle.

So when we needed water supplies, we created rivers by chanelling water using dikes and inlets. We used the same technology to hold back the sea from consuming a coastline we'd made our home. But doing so would cause the water to go elsewhere, consuming someone else's strip of the coast. And that's how people start beefing, by the way. But beyond that, even a small change produced by a local group of people can have very significant and widespread effects. For example, changing the distribution of the Earth's mass (even fractionally) by transporting or displacing significant amounts of water, earth and stone can actually cause earthquakes in other parts of the globe. And changing the local fauna (animals) in an ecosystem (through hunting or displacement) can cause a change in the local flora (plants), and vice versa. Changes like this can have even more significant consequences when they begin affecting local weather patterns, which over the course of thousands of years, can also contribute to the shaping of the Earth. There are currently technologies like HAARP which can affect weather patterns through the use of sonar and other frequencies (but please don't go crazy and believe that HAARP is the cause of ALL weather events on the planet!).

Before that, farmers knew that dropping dust into clouds would bring about rain. Well, even before that, we did all that and more. Don't think that we did all this earth-shaping cluelessly. We've always been scientists, so even when its trial and error, there's still a process being used to make our decisions. We put the pyramids in the center of the Earth's landmass for a reason. We helped redirect the Nile River at least 4 times for a reason. We even came up with rain dances (in the Americas and Africa) and other uses of sound (sonar!) to call down water because we knew that sound could play a role in natural phenomena. Even in our infancy on this planet, we could observe an echo or a loud noise causing an avalanche and come to the conclusion that sound travels in waves that have physical effects. Of course, we'd already tracked the weather to know the water was coming, so the traditional rain dance wasn't really about begging for water from nowhere. In fact, Native American accounts describe how the Windigokan and other groups in government-suppressed areas simply called their solar ceremony a "rain dance" instead of a "sun dance" to avoid federal prosecution, since all Native American solar traditions had been outlawed by the Europeans. Other accounts describe how Native American meteorologists (sometimes called "weather-trackers") would observe patterns in nature to see when rain was coming, and then tell white settlers they would do a rain dance and bring rain…in exchange for goods. Of course, rain would come, and the settlers would keep coming back for more "help." Sounds like a hell of a hustle, but it's the way we "made" the weather. That process was more like thousands of years in the making, not the result of a twenty-minute dance. So there's a lot of layers to our story. We certainly did shape the Earth, but not in a magical, mysterious way either.

With all that said, I hope it's plain to see that man's role – even 100,000 years ago – was much more significant than people currently think. I understand that it might be tough to visualize how the grassy landscape of modern Europe (which was once mostly trees) was probably created by widespread fires started by humans, just as Native Americans created many of America's grasslands and prairies through burning.[355]

This practice – used in conjunction with scientific planting and hunting practices – was intended to improve, without overburdening, the local ecology. On the other hand, we have the so-called Dust Bowl of the midwestern U.S., which was caused by white folks ignoring the advice of the local Indians, and creating a single crop economy that killed the fertile soil. This caused the ground to become like dust, which – no longer held in place by plant life – would sweep through neighboring states in massive dust storms that shut down everything and made life unbearable. White settlers also abused the regions further west with massive herds of cattle, which they simply relocated instead of taking measures to preserve the land, thus turning another fertile area to desert! So a lot of the "barren" regions in the U.S. were actually created by white people in the past 400 years! Now just imagine what WE did to the Earth over the course of, say, 4,000,000 years!

WE MADE MOUNTAINS, RIVERS, AND EVEN THE DESERT

Andrew Goudie's book, *The Human Impact on the Natural Environment*, details how even the earliest hominids were responsible for major developments in the natural world. For example, many savannahs, once thought to have been the product of climate, appear to have been produced by man, often through the controlled use of fire. Geographer I. G. Simmons' book *Changing the Face of the Earth*, a tribute to W.L Thomas' 1956 edited collection, *Man's Role in Changing the Face of the Earth*, illustrated – like its predescessor – that man has affected the development of this planet in ways that may today seem unfathomable.

I'll give you one of the strongest examples of OUR work. Would you believe that we made the Sahara Desert? As we discussed earlier, the Sahara region was once covered in forest and supported many human settlements, some of them hunter-gatherers, some of them pastoral (herding) communities. But if the settlement at Nabta Playa is any indication, these people were definitely not "primitive." In fact, archaeological finds at Nabta reveal that these prehistoric peoples were at a higher level of organization than their contemporaries in the Nile Valley.*

But the Atlantic heat conveyor stopped transporting heat from the region, dried up lakes and turned the forest to desert. People living throughout this region either migrated north into Europe, further south in Africa, or eastward towards what would become Egypt, which was lush and fertile. In effect, the most recent periods of Saharan desertification (beginning around 6,000 BC but reaching their peak about 4,000 BC) are what forced human settlements to merge together and consolidate into large, heavily-populated urban complexes like ancient Egypt (and the other Near Eastern civilizations, since much of the Fertile Crescent was becoming desert around the same time).[356] The Nile Valley region – now full of Black astronomers, medicine men, and city planners from all over the region – soon became something Africa didn't typically need – a highly-organized "inner city" with centralized leadership. While the people of the Nile Valley had already been building massive structures like the Sphinx (and possibly some of the pyramids) long before this influx, it wasn't until the Nile Valley became "flooded" with people (hence the name KMT, meaning the Black Land), that we start seeing the massive urban complexes and huge temples (and tombs) we identify with dynastic Egypt.

So why did the Sahara become a desert? The most likely reason is because of us. That is, human activity in the area, particularly the slash-and-burn farming techniques still used throughout Africa today and the clearing of forest for hunting wild game, compromised the protective features built into the land.[357] In effect, through unchecked deforestation within the last 10,000 years, Africans set in motion a course of events which changed the climate, which, in turn, further changed the environment. Up until about 6,000 years ago, the Sahara may have fluctuated, but it always recovered. After this last period, there's never been more than a partial recovery. This continues to be relevant today, as widespread deforestation, along with other factors, are continuing to change the course of Atlantic thermohaline circulation, better known as the Atlantic heat conveyor belt. New research indicates there is a 45% chance that the thermohaline circulation in the North Atlantic Ocean could shut down by 2100, suggesting that the scenario proposed in the film *The Day After Tomorrow* is more realistic than we'd like to think. A weakening of the system could cause a cool-

* See "Nabta Playa" in Volume Two.

ing in northwest Europe and worsen droughts in equatorial Africa.[358] As it is, the Sahara Desert continues to grow by 4 inches every year.

CREATING UNDERGROUND RIVERS

We never stopped "remaking" the Earth. While scientists are currently working on new ways to bring water to the Sahara and building man-made islands in Dubai, we were doing this kinda stuff thousands of years ago. For example, the reason why the "desert country" of Iran is somehow also a "farming country," growing its own food, cotton, and so on, is because we built underground rivers there over 3,000 years ago. To this day, Iran depends on an ingenious, indigenous system for tapping underground water, called *qanat* (meaning "to dig"). The ancient qanat system was once so popular that it spread to many other regions of the Middle East and around the Mediterranean. In fact, archaeologists once said the underground water supply systems of the ancient world started in Rome, but now it's clear that these methods were developed by Original people. Spreading from a Persian homeland, qanats have been found in Pakistan, Turkistan, southern Russia, Iraq, Syria, Arabia, Yemen, North Africa, Spain, Sicily, and several oasis settlements in the Sahara region. Whereas the Roman aqueducts now are only a historical curiosity, the system in Iran is still in use. There are 22,000 qanat units in Iran, comprising more than 170,000 miles of underground channels, supplying 75% of the country's water supply.[359] We'll dig deeper (no pun intended) into archaeology and ancient technology in Volume Two.

FINAL REVIEW

Throughout this book we've explored how matter comes into being, how being is born into life, and how life evolves into us. Throughout this process, we've seen that this process is both non-linear and has multiple dimensions. At every level, every scale, and every frequency, however, you will find us. We are the living mathematics that turn Mind into matter, word into flesh, and keep life moving forward. Throughout this book, we've shown you the mathematical structure by which these things happen, and how we can apply this same understanding of life's dynamics into our own lives (at least we hope you caught those parts!). This book is about more than just science and history. It is a handbook to existence. Thus, we've laid out the principles of predictive planning, adaptive strategy, resource management, mutual aid, and dozens of other principles (along with countless examples, across nearly 14 billion years of history).

As we close out this volume, we leave you with the knowledge of how the Original Man and Woman came to be. Yet, then what happened? In Volume Two, you'll learn what happened as the human population grew, eventually exceeding our capacity to simply "live off the land," leading to developments like agriculture, animal domestication, urbanization, and property ownership. We'll talk about the earliest cultural developments in Africa, and we're talking about 2,000,000 BC, not 2,000 BC. We'll talk about the earliest sites of Black civilization outside Africa, as well as what was happening IN Africa before the rise of ancient Egypt. You'll see how we used the sciences discussed in this volume to accomplish the incredible, everywhere from ancient Mexico to China. We've laid the foundation by showing you how all this came to be, and why the Original Man and Woman are the archetypes for everything in existence. In Volume Two, we'll show you what we built throughout the world. After all, the ancient Black civilizations of Africa, Asia, Europe, Australia, and the Americas are not simply a testament to who we were, but to who we are today.

APPENDIX

ALL THE STUFF THAT DIDN'T FIT

RECOMMENDED READING

A Beginner's Guide to Constructing the Universe by Michael Schneider

A Brief History of Everything by Ken Wilbur

Africa Counts by Claudia Zaslavsky

African Fractals by Ron Eglash

African Philosophy by English and Kalumba

Ancient Future by Wayne Chandler

Blackfoot Physics by F. David Peat

Civilization or Barbarism by Cheikh Anta Diop

Cosmos by Carl Sagan

Hidden Dimensions: The Unification of Physics and Consciousness by B. Alan Wallace

Holographic Universe by M. Talbot

Knowledge of Self by Supreme Understanding, C'BS Alife Allah, and Sunez Allah

Man, God, and Civilization by John G. Jackson

Mind and Matter by Erwin Schrödinger

Mind and Nature by Hermann Weyl

Mind, Matter and Quantum Mechanics by Henry Stapp

My View of the World by E Schrodinger

Native Science by Gregory Cajete

Parallel Worlds by Michio Kaku

Quantum Activist by Amit Goswami

Quantum Aspects of Life by Abbott, Davies, and Pati

Red Earth White Lies by Vine DeLoria

Sacred Geometry by Robert Lawlor

Science and Mysticism by R. H. Jones

Tao Te Ching translated by M. Kwok

The African Abroad by William H Ferris

The Dancing Wu Li Masters by Gary Zukav

The Elegant Universe by Brian Greene

The Fabric of the Cosmos by Brian Greene

The Field by Lynne McTaggart

The Hood Health Handbook by Supreme Understanding and C'BS Alife Allah

The Inner Life of Numbers by Andrew Hodges

The Loom of God by Clifford Pickover

The Pale Fox by Marcel Griaule

The Science of Melanin by T. Owens Moore

The Space Gods Revealed by R. Story

The Star of Deep Beginnings by Charles S. Finch

The Tao of Physics by Fritjof Capra

The Tao of Wu by the RZA

The Universe in a Single Atom by Tenzin Gyatso, the Dalai Lama

The Web of Life by Fritjof Capra

The World of Quantum Culture by Manuel Caro and John Murphy

Upanisads trans. by Patrick Olivelle

View from the Center of the Universe by Joel Primack and Nancy Abrams

Wholeness and the Implicate Order by David Bohm

About the Authors

About Dr. Supreme Understanding

Supreme Understanding is a community activist, educator, and researcher dedicated to the cultural, socio-economic and psycho-logical struggles of oppressed people.

As a teenager, he became involved with the Nation of Gods and Earths, which helped transform him – then a juvenile delinquent who had recently been expelled from high school – into a world class scholar and youth worker. Supreme's experience with the NGE taught him a deep-rooted appreciation for Original people (Black and brown people, or indigenous people, throughout the world), a love of ancient history, and a strong interest in scientific thinking.

Supreme received his bachelor's degree in world history from Morehouse College, followed by a master's degree in urban education from Georgia State University. At 26, he received his doctorate in education from Argosy University, where he focused his research on the benefits of non-formal education with at-risk youth and disadvantaged populations. His dissertation was a study of education methods used by the Nation of Gods and Earths.

Supreme Understanding's research has brought him to Ghana, Mexico, India, the Dominican Republic, Japan, Thailand, the Czech Republic, Austria, Germany, England, the Caribbean, and dozens of cities throughout the United States. The author's travels have provided him in-depth opportunities to learn about the history struggles of oppressed people throughout the world, and their similarities with the struggles of people of color in the U.S.

Through Supreme Design Publishing, the publishing company he founded in 2008, Supreme reaches urban audiences with nonfiction literature focused on teaching practical, research-based solutions to community problems. Although praxis forms the crux of his work, Supreme's most recent contributions to the field have focused on promoting themes of critical thinking, scientific inquiry, self-knowledge, historical awareness, and being able to draw lessons from the past (and from under the microscope) to produce solutions that work for us today. Through collaboration with visionary C'BS Alife Allah, The Science of Self series represents the culmination of this work. It represents the completion of a project that has taken over a decade to finish.

ABOUT C'BS ALIFE ALLAH

C'BS ALife Allah is the son of a minister and a teacher. Both instilled within him a drive to experience his environment and to imbue it with meaning. The academic bar was always set high in his household and he met the challenge becoming accepted to several academic programs and camps in his youth. One of those programs was a summer program at the Yale Peabody Museum which enhanced his already deep love of science.

During high school, he worked at a Waldenbooks bookstore, which started an almost lifelong relationship with books and knowledge. While attending Wesleyan college, he worked in the library, where he would hide in the stacks, consuming volumes of books. While working as the Special Orders Supervisor for Barnes and Noble, he would spend almost more than he earned on crates of books. As a resident of New Haven, Connecticut, he has found his way into every library at Yale University, where he sets up camp, reading and taking notes until closing time.

In high school, he was also introduced to the Nation of Gods and Earths (also known as the Five Percenters). The pro-scientific focus of this group further enhanced his views on science and its place in the Original community. In the early 90's he was one of the first Gods online through IRC chat. He eventually had a hand in every manifestation of the Gods and Earths' online presence, from listservs, to chat groups, to NGE webpages, to having one of the first Nation-centric blogs (*Sol Power Network* at allahsfivepercent.blogspot.com). However, this "nation building" was never limited to online activity, as he has become a well known builder in his immediate area and within the Nation at large. He has been instrumental in promoting and ensuring that there is a science fair at every annual Educational and Scientific Show and Prove in Harlem.

C'BS has traveled abroad to England, Kenya, Zimbabwe and Tanzania. His forays into Africa allowed him to see indigenous problem-solving methodologies applied to pressing issues. It was in many of these observations that he noticed that the adoption and/or forcing of Western views on science and technology exacerbated the problems at hand. He also noticed that there wasn't a full appreciation for indigenous science. When he returned to the states, he realized that the lack of scientific awareness in the Original community had much to do with the veiling of the indigenous roots of western science, the veiling of the Original man's contributions to modern day science, and a lack of the connection between theory and application in addressing the issues of Original people.

C'BS ALife Allah's full platform is science literacy for Original people worldwide. Through this work, along with Supreme Understanding, he hopes to develop a template whereby Original people can regain their place at the center of scientific discourse and whereby research-based theory can become dynamically linked with community-controlled problem-solving initiatives.

KNOWLEDGE OF SELF

C'BS ALIFE ALLAH & SUPREME UNDERSTANDING

As we explore in this book, the 5% are a timeless element of human society. That is, the 5% have always been here. In different periods of history, they have developed different paradigms and occupied different roles in moving history along.

"A new scientific truth does not triumph by convincing its opponents and making them see the light, but rather because its opponents eventually die and a new generation grows up that is familiar with the ideas from the beginning." – Max Planck, 1918 Nobel Prize in Physics

Between 1964 and 1969, a paradigm shift took place among the 5%, when a man named Clarence 13X declared himself ALLAH. Hailing from Danville, Virginia, he moved to NYC as a youth and did a tour of duty in the Korean War. Following his return, he joined the Nation of Islam under the Honorable Elijah Muhammad. During his time in the temple, he came to a unique understanding of the "science" embedded in the theology of the Nation of Islam. He chose to leave the temple and began teaching street youth a curriculum in which he emphasized and promoted the non-theistic elements. Those who came after him, the modern Nation of Gods and Earths, became the heirs to a non-theistic system of thought rooted in the scientific method. Allah told his students, "If you can't science it up, don't deal with it!"[360] This system of thought fully blossomed into a way of life which is the culture of the Gods and Earths.

In 2000, the need was recognized for re-dedication to the scientific legacy of the Gods and Earths. Due to the heritage of the Gods and Earths transferring through the temples of the Nation of Islam, there was still a strong theistic current present within some circle of the Gods and Earths. To counter this, Supreme Understanding Allah secured writings from many Gods and Earths across the Nation. The end result was a free text entitled *The Black God: An Anthology of Truth*, released in 2006. This book took the focus from "God as man" to "Man as God."

A Quick Note on Self Definition

The worldwide hegemony of White Supremacy, through its vehicles of colonization, slavery, and genocide executed a massive attack against the knowledge-base of Original People worldwide. One element of this knowledge-base that sustained heavy damage was the plethora of languages utilized by Original People. European imperialism promoted mono-linguistic societies. Malcolm X said one thing stolen from Original people during slavery was their language. Yet ironically, many of our indigenous languages inluenced the colonizer's languages.[361] By examining some of the remaining indigenous languages, you can see they contain many concepts that don't even exist in English or the various Romance languages.[362] As linguist K. David Harrison explained while reporting the discovery of Koro, a previously unknown language in a remote part of northeast India: "It contains very sosphicated knowledge that these people possess about this valley, the ecosystem, the animals, the plants, how they survive, how they adapted. So if they switch over to another language, a lot of that knowledge will simply be lost."[363]

As a result, we lost the ability to self-define, using our own linguistic parameters. So we're stuck with English, but we'll make it work through self-definition. Self-definition simply means deciding what you mean for yourself. Linguistic self-definition involves creating a new lexicon or glossary. Europeans did it plenty. When a group of people utilize a certain sound and ascribe to it a meaning, it becomes a word. That's really all it takes. All words are first rooted in the mind and then become manifested in speech. To be specific, these concepts are actually words even when they're only in mind. There are countless cultures whose languages are entirely oral. They don't have written records and they definitely don't have dictionaries. According to a layman's perception, they don't have a "real" language because none of their words are in a dictionary. You'll run across people who think like this, who don't use certain words UNTIL they're validated by "Webster."

Yet, since Webster's dictionary is a generic name that can be used by any publisher, any word can become "real" (for people who think like this) as soon as someone publishes it as a dictionary. As a result, these people will argue that "tricknology" is not a word (it's recognized by linguists as a portmanteau), and neither are "interorientation" or "devilishment" (both are archaic words, no longer published in modern dictionaries)...but won't question the words the *Oxford Dictionary* "added" in 2011, which include LOL, OMG, sexting, bloggable, and the text-message symbol for heart, <3. Yes, the symbol for heart. That's a word now. What's our point? That language is defined by the user, and even moreso by the community in which that language is in use. So if we ascribe a definition to a word, and use it consistently in that context, we are reclaiming words for our own purposes. In reclaiming our identities from those constructed by the dominant society (which are often distorted and stripped-down versions of our Original selves), we must first and foremost exercise our right of self-definition. It reminds me of mind a joke I once heard: A white man said to a Black man, "Negro, Colored, Afro-American, African-American, Black...you all change your name so much…what am I supposed to call you?" The Black man replied, "Whatever we tell you to call us." In other words, self-definition is a major part of self-determination.

In 2008, a contingent of the Gods and Earths who had been involved in the production of *The Black God: An Anthology of Truth* recognized that not only did the scientific frame work need to be emphasized within the confines of the Nation, it needed to be expressed to the world at large. As a result of this process, Supreme Understanding recruited C'BS ALife Allah and Sunez Allah to compile another anthology. This anthology provided an accurate portrait of what the Five Percent study and teach as well as once again cementing the scientific trajectory of the Nation. The end result was the paperback bestseller *Knowledge of Self: A Collection of Wisdom on the Science of Everything in Life.*

With both of these projects receiving very positive feedback, Supreme Understanding and C'BS ALife Allah decided to revisit a project that they had discussed between the years of 1998-1999. It was an ambitious project. It was to be the grand "story of man" from the "beginning of time" at the "big bang," through protozoa, across evolution, weaving in and out of ancient civilization, all the way up into the modern era! Within that comprehensive linear narrative would be clusters of cycles that would point out patterns and sequences. All of this would be tied together with the thread of science, demonstrating how – through a deeper understanding of mathematics – one could see the connectedness of all points in space and time. The book was initiated in 1998, but never finished.

Now, with the success of the above-named anthologies, plus a combined 28 years worth of additional research by both C'BS ALife Allah and Supreme Understanding Allah, the two decided that they were ready to tackle this "ambitious project" with the quality such a task deserves. This is how the *Science of Self* series came into existence. It is part of the broader mission of the Gods and Earths to infuse mathematical thinking and self-knowledge into communities throughout the world. The mission of the 5% (as a collective) was never to amass wealth or followers, but to simply "Teach on!" and thus influence the community-at-large. It is our intention that *The Science of Self* contributes heavily to a new era of thinking on street corners and college campuses across the globe, ultimately blooming into the Hood Renaissance that is yet to come. That is the mission of this series of books. And this is just the first volume.

A FINAL NOTE

Throughout this volume, we've explored several key concepts that will be familiar to those who have read the precursor to this book, *Knowl-*

edge of Self. What we've done in this book is explained the science behind some of the esoteric ideas taught and discussed within the culture of the Nation of Gods and Earths. These concepts, as you've seen, can be found in indigenous/Original traditions from around the world, but the NGE is unique because of the when, where, why, and how of what it teaches. The 5% is a timeless phenomenon and the NGE is one of the most comprehensive vehicles for mobilization of the 5% within the present-day American context. Because the NGE encourages critical thinking rather than indoctrination, we were able to use the teachings as a inspiration for our research, rather than a set of strict rules by which to confine our studies. What we've found, over the past ten years of research, has been that the most cutting-edge findings in physics, anthropology, archaeology, and so on, are only recently discovering what we've known for thousands of years.

Most Five Percenters learn this quite early in their studies. All we've done is explain the data to substantiate the understandings. When you immerse yourself in the culture of the NGE/5%, you'll learn the same ideas, only using a different form of self-definition or linguistic license. For example, consider this interpretation of every word used in the first degree of the Student Enrollment, written by Azmar Blackseed Allah, and circulated in the 90s as a "plus lesson":

> The Original Man is the root of civilization. Asia [as in "Asiatic"] is the largest of the Earth's continents. Atic [means] the Mind, a storage place for the knowledge. Black is the darkest achromatic visual value, and is dominant. Man is a mind. Therefore the Blackman has the most dominant mind in the universe. The original man is the maker, for he formulates the mind of all. The original man is the owner, for confers and possesses the knowledge of all things in existence. The original man is the Cream of the Planet Earth, for he is the best part of the planet, which is a celestial nonluminous body illuminated by the light of the star (Sun), from which it revolves. Earth is the third planet from the sun. Its distance is 93,000,000 miles. The original man is the Father of Civilization, the original ancestor of his nation. The father's duty is to civilize the babies (the uncivilized) by speaking wisdom and teaching knowledge (the truth). The original man is God of the Universe, for he is the sole controller of his mind which is his God! The Universe is the body and is, in fact, controlled by the mind.

A GUIDE TO CRITICAL THINKING[364]

"Critical thinking is the careful and deliberate evaluation of ideas or information for the purpose of making a judgment about their worth or value: It is the ability to construct and evaluate arguments."
– K.T. McWhorter, Study and Thinking Skills in College

CRITICALLY EVALUATING THE LOGIC AND VALIDITY OF INFORMATION

Many articles and essays are not written to present information clearly and directly; instead they may be written to persuade you to accept a particular viewpoint, to offer an, opinion, to argue for one side of a controversial issue. Consequently, one must recognize and separate factual information from subjective content.

Subjective content is any material that involves judgment, feeling, opinion, intuition, or emotion rather than factual information. Recognizing and evaluating subjective content involves distinguishing between facts and opinions, identifying generalizations, evaluating viewpoints, understanding theories and hypotheses, weighing data and evidence, and being alert to bias.

EVALUATING VARIOUS TYPES OF STATEMENTS

DISTINGUISHING BETWEEN FACTS AND OPINIONS:

Facts are statements that can be verified or proven to be true or false. Factual statements from reliable sources can be accepted and used in drawing conclusions, building arguments, and supporting ideas.

Opinions are statements that express feelings, attitudes, or beliefs and are neither true nor false. Opinions must be considered as one person's point of view that you are free to accept or reject. With the exception of informed ones, opinions have little use as supporting evidence, but they are useful in shaping and evaluating your own thinking. "Informed opinion or testimony" means the opinion of an expert or authority on the subject.

RECOGNIZING GENERALIZATIONS:

A generalization is a statement made about a large group or class of items based on observation or experience with a portion of that group or class. It is a reasoned statement about an entire group based on known information about part of the group. It involves a leap from observed evidence to a conclusion which is logical, but unproven. Because writers do not always have the space to describe all available evidence on a topic, they often draw the evidence together themselves and make a general statement of what it shows. But generalizations need to be followed by evidence that supports their accuracy, otherwise the generalization is unsupported and unusable. A generalization is usable when these two conditions exist:

- Your experiences are sufficient in number to merit a generalization.

- ❐ You have sampled or experienced enough different situations to draw a generalization.

TESTING HYPOTHESES:

A hypothesis is a statement that is based on available evidence which explains an event or set of circumstances. Hypotheses are simply plausible explanations. They are always open to dispute or refutation, usually by the addition of further information. Or, their plausibility may be enhanced by the addition of further information. Critical thinking and reading requires one to assess the plausibility of each hypothesis. This is a two-part process. First, one must evaluate the evidence provided. Then one must search for information, reasons, or evidence that suggests the truth or falsity of the hypothesis. Ask questions such as:

- ❐ Does the hypothesis account for all known information about the situation?
- ❐ Is it realistic, within the realm of possibility and probability?
- ❐ Is it simple, or less complicated than its alternatives? (Usually, unless a complex hypothesis can account for information not accounted for by a simple hypothesis, the simple one has greater likelihood of being correct.)
- ❐ What assumptions were made? Are they valid?
- ❐ Weighing the Adequacy of Data and Evidence:

Many writers who express their ideas use evidence or data to support their ideas. One must weigh and evaluate the quality of this evidence; one must look behind the available evidence and assess its type and adequacy. Types of evidence include:

- ❐ Personal experience or observation
- ❐ Statistical data
- ❐ Examples, particular events, or situations that illustrate
- ❐ Analogies (comparisons with similar situations)
- ❐ Informed opinion (the opinions of experts and authorities)
- ❐ Historical documentation
- ❐ Experimental evidence

Each type of evidence must be weighed in relation to the statement it supports. Evidence should directly, clearly, and indisputable support the case or issue in question. One does not need all the types of evidence listed above to support one's ideas. However, it is good practice to use several types of evidence at once.

EVALUATING PERSUASIVE MATERIAL

While the main purpose of textbooks is to explain and present information that can be accepted as reliable, other sources may have very different purposes. Some materials are intended to convince or persuade rather than to inform, and these sources need to be carefully and critically evaluated. Persuasive writers use both language and logical argument to exert influence.

RECOGNIZING PERSUASIVE LANGUAGE

A writer's or speaker's choice of facts and the language used to convey them may influence the reader's or listener's response. Careful choice of details to describe an event shapes a reader's perception of the incident. Selective reporting of details is known as slanted writing. Careful choice of words allows one to hint, insinuate, or suggest ideas without directly stating them. Through deliberate choice of words one can create positive or negative responses. This is often accomplished through manipulation of the connotative meanings.

IDENTIFYING BIASED AND SLANTED WRITING:

Bias is when a statement reflects a partiality, preference, or prejudice for or against a person, object, or idea. Much of what you read and hear expresses a bias. As you read biased material keep two questions in mind:

- ❒ What facts has the author omitted?
- ❒ What additional information is necessary?

Slanting is when a writer or speaker uses a selection of facts, choice of words, and the quality and tone of description, to convey a particular feeling or attitude. Its purpose is to convey a certain attitude or point of view toward the subject without expressing it explicitly. As you read or listen to slanted materials, keep the following questions in mind:

- ❒ What facts were omitted? What additional facts are needed?
- ❒ What words create positive or negative impressions?
- ❒ What impression would I have if different words had been used?

EVALUATING ARGUMENTS

An argument is a logical arrangement and presentation of ideas. It is reasoned analysis, a tightly developed line of reasoning that leads to the establishment of an end result or conclusion. Arguments are usually developed to persuade one to accept a position or point of view. An argument gives reasons that lead to a conclusion. Analyzing arguments is a complex and detailed process. The following guidelines are useful:

- ❒ Analyze the argument by simplifying it and reducing it to a list of statements.

- ❒ Are the terms used clearly defined and consistently applied?
- ❒ Is the thesis (the point to be made) clearly and directly stated?
- ❒ Are facts provided as evidence? If so, are they verifiable?
- ❒ Is the reasoning sound? (Does one point follow from another?)
- ❒ Are counterarguments recognized and refuted or addressed?
- ❒ Does the body of the essay reflect what the author was trying to convey?
- ❒ What persuasive devices or propaganda techniques does the author use (examples: appeal to emotions, name-calling, appeal to authority)?

ASKING CRITICAL QUESTIONS

What is the source of the material? Some sources are much more reliable and trustworthy than others; knowledge of the source will help you judge the accuracy, correctness, and soundness of the material. Articles from professional or scholarly journals are often more useful and reliable than articles in newsstand periodicals. To evaluate a source consider:

- ❒ its reputation
- ❒ the audience for whom the source is intended
- ❒ whether references or documentation are provided

What are the Author's Credentials? You must assess whether the material you are reading is written by an expert in the field who can knowledgeably and accurately discuss the topic.

Why was the Material Written? Identify an author's primary purpose. If the author's purpose is to persuade or convince you to accept a particular viewpoint then you will need to evaluate the reasoning and evidence presented.

Is the Author Biased? Does the author display partiality, preference, or prejudice for or against a person, object, or idea?

Does the Author Make Assumptions? An assumption is an idea or principle the writer accepts as true and makes no effort to prove or substantiate.

Does the Author Present an Argument? An argument is a logical arrangement and presentation of ideas. It is reasoned analysis, a tightly developed line of reasoning that leads to the establishment of an end result or conclusion.

RECOGNIZING PROPAGANDA AND FAULTY LOGIC

PROPAGANDA TECHNIQUES

What are Propaganda Techniques? They are the methods and approaches used to spread ideas that further a cause – a political, commercial, religious, or civil cause. These techniques can also be used to hurt another cause.

Why are they used? To manipulate the readers' or viewers' reason and emotions; to persuade you to believe in something or someone, buy an item, or vote a certain way. Not all forms of propaganda are bad. Anti-smoking ads are a form of propaganda that help promote a healthier life style.

What are the most commonly used propaganda techniques? See which of the ten most common techniques you already know.

TYPES:

Name calling: This technique consists of attaching a negative label to a person or a thing. People engage in this type of behavior when they are trying to avoid supporting their own opinion with facts. Rather than explain what they believe in, they prefer to try to tear their opponent down.

Glittering Generalities: This technique uses important-sounding "glad words" that have little or no real meaning. These words are used in general statements that cannot be proved or disproved. Words like "good," "honest," "fair," and "best" are examples of "glad" words.

Transfer: In this technique, an attempt is made to transfer the prestige of a positive symbol to a person or an idea. For example, using the American flag as a backdrop for a political event makes the implication that the event is patriotic and in the best interest of the U.S.

False Analogy: In this technique, two things that may or may not really be similar are portrayed as being similar. When examining the comparison, you must ask yourself how similar the items are. In most false analogies, there is simply not enough evidence available to support the comparison.

Testimonial: This technique is easy to understand. It is when "big name" personalities are used to endorse a product. Whenever you see someone famous endorsing a product, ask yourself how much that person knows about the product, and what he or she stands to gain by promoting it.

Plain Folks: This technique uses a folksy approach to convince us to support someone or something. These ads depict people with ordinary looks doing ordinary activities.

Card Stacking: This term comes from stacking a deck of cards in your favor. Card stacking is used to slant a message. Key words or unfavorable statistics may be omitted in an ad or commercial, leading to a series of half-truths. Keep in mind that an advertiser is under no obligation "to give the truth, the whole truth, and nothing but the truth."

Bandwagon: The "bandwagon" approach encourages you to think that because everyone else is doing something, you should do it too, or you'll be left out. The technique embodies a "keeping up with the Joneses" philosophy.

Either/or fallacy: This technique is also called "black-and-white thinking" because only two choices are given. You are either for something or against it; with no middle ground or shades of gray. It is used to polarize issues, and negates all attempts to find a common ground.

Faulty Cause and Effect: This technique suggests that because B follows A, A must cause B. Remember, just because two events or two sets of data are related does not necessarily mean that one caused the other to happen. It is important to evaluate data carefully before jumping to a wrong conclusion.

ERRORS OF FAULTY LOGIC

Contradiction: Information is presented that is in direct opposition to other information within the same argument. (Example: If someone stated that schools were overstaffed, then later argued for the necessity of more counselors, that person would be guilty of contradiction.)

Accident: Someone fails to recognize (or conceals the fact) that an argument is based on an exception to the rule. (Example: By using selected scholar-athletes as the norm, one could argue that larger sports programs in schools were vital to improving academic performance of all students.)

False Cause: A temporal order of events is confused with causality; or, someone oversimplifies a complex causal network. It is easy to get causation and correlation confused. In causation there is an immediate and visible cause and effect. While a correlation is something that is connected to an effect but is not necessarily the causative factor.

Begging the Question: A person makes a claim then argues for it by advancing grounds whose meaning is simply equivalent to that of the original claim. This is also called "circular reasoning."

Evading the Issue: Sidestepping an issue by changing the topic.

Arguing from Ignorance: Someone argues that a claim is justified simply because its opposite cannot be proven.

Composition and Division: Composition involves an assertion about a whole that is true of its parts. Division is the opposite: an assertion about all of the parts that is true about the whole.

ERRORS OF ATTACK

Poisoning the Well: A person is so committed to a position that he/she explains away absolutely everything others offer in opposition.

Ad Hominem: A person rejects a claim on the basis of derogatory facts (real or alleged) about the person making the claim.

Appealing to Force: The use of threats or fear to establish a claim.

ERRORS OF WEAK REFERENCE

Appeal to Authority: Authority is evoked as the last word on an issue. (Example: Someone uses the Bible as the basis for his arguments against specific school reform issues.)

Appeal to the People: Someone attempts to justify a claim on the basis of popularity. (Example: Opponents of year-round school claim that students would hate it.)

Appeal to Emotion: An emotion-laden "sob" story is used as proof for a claim. (Example: A politician uses a sad story of a child being killed in a drive-by shooting to gain support for a year-round school measure.) (For a comprehensive list of over 80 logical fallacies, see the Appendix to Volume Two.)

HOW TO READ A BOOK

MORTIMER ADLER & CHARLES VAN DOREN

How to Read a Book was first written in 1940 by Mortimer Adler. He co-authored a heavily revised edition in 1972 with Charles Van Doren, which gives guidelines for critically reading great books of any tradition. The chapters are grouped into four parts.

PART I: THE ACTIVITY OF READING

Author Adler explains for whom the book is intended, defines different classes of reading, and tells which classes will be addressed. He also makes a brief argument favouring the "Great Books,"* and explains his reasons for writing *How to Read a Book*.

* The "Great Books" – like the Western Canon – refers primarily to a group of books that tradition, and various institutions and authorities, have regarded as constituting or best expressing the foundations of Western culture. Expectedly, all of the "Great Books" in Adler and Van Doren's list were written by whites. But curiously, Franz Boas (who we talk about elsewhere in this book) was one of only four authors omitted from the revised list of "Great Books" published in 1972.

There are three types of knowledge: practical, informational, and comprehensive. He discusses the methods of acquiring knowledge, concluding that practical knowledge, though teachable, cannot be truly mastered without experience; that only informational knowledge can be gained by one whose understanding equals the author's; that comprehension (insight) is best learned from who first achieved said understanding – an "original communication."

The idea of communication directly from those who first discovered an idea as the best way of gaining understanding is Adler's argument for reading the Great Books; that any book that does not represent original communication, is inferior, as a source, to the original, and, that any teacher, save those who discovered the subject he or she teaches, is inferior to the Great Books as a source of comprehension.

Adler asserts that very few people can read a book for understanding, but that he believes that most are capable of it, given the right instruction and the will to do so. It is his intent to provide that instruction. He takes time to tell the reader about how he believes that the educational system has failed to teach students the arts of reading well, up to and including undergraduate university-level institutions. He concludes that, due to these shortcomings in formal education, it falls upon the individuals to cultivate these abilities in themselves. Throughout this section, he relates anecdotes and summaries of his experience in education as support for these assertions.

PART II: THE THREE LEVELS OF READING

Here, Adler sets forth his method for reading a non-fiction book in order to gain understanding. He claims that three distinct approaches, or readings, must all be made in order to get the most possible out of a book, but that performing these three levels of readings does not necessarily mean reading the book three times, as the experienced reader will be able to do all three in the course of reading the book just once. Adler names the readings: "structural," "interpretative," and "critical," in that order.

STRUCTURAL READING

The first stage of the third level of reading is concerned with understanding the structure and purpose of the book. It begins with determining the basic topic and type of the book being read, so as to better anticipate the contents and comprehend the book from the very beginning. Adler says that the reader must distinguish between practical and theoretical books, as well as determining the field of study that the book addresses. Further, Adler says that the reader must note any divisions in the book, and that these are not restricted to the divisions laid out in the table of contents. Lastly, the reader must find out what problems the author is trying to solve.

INTERPRETATIVE READING

The second stage of the third level of reading involves constructing the author's arguments. This first requires the reader to note and understand any special phrases and terms that the author uses. Once that is done, Adler says that the reader should find and work to understand each proposition that the author advances, as well as the author's support for those propositions.

CRITICAL READING

In the third stage of the third level of reading, Adler directs the reader to criticize the book. He claims that now that the reader understands the author's propositions and arguments, the reader has been elevated to the level of understanding of the book's author, and is now able (and obligated) to judge the book's merit and accuracy. Adler advocates judging books based on the soundness of their arguments. Adler says that one may not disagree with an argument unless one can find fault in its reasoning, facts, or premises, though one is free to dislike it in any case. This method is sometimes called the *Structure-Proposition-Evaluation (SPE)* method, though this term is not used in the book.

PART III: APPROACHES TO DIFFERENT GENRES

It briefly discusses approaches to reading fiction and poetry, and suggests other books that address it. He suggests one read the books that influenced a given author prior to reading works by that author and gives several examples of that method.

PART IV: THE ULTIMATE GOALS OF READING

The last part of the book covers the fourth level of reading; syntopical reading. At this stage the reader uses several books to inform himself/herself about a subject such as love, war, particle physics, etc. In the final pages of this part the authors expound on the philosophical benefits of reading, such as "growth of the mind" and fuller experience as a conscious being.

A Mathematical Narrative of the Quantum Universe

This narrative is the other side of the previous narrative detailing the Visible Universe's physical formation. As we've learned about quantum mechanics, one of the most important elements to understand is the concept of duality. There is a dual nature to everything. Much of what is "reality" is determined by OUR observation of it. So there's another way to look at the Growth, Order, and Direction of the Universe. We're going to use the lens of quantum mechanics and information theory to consider what was happening behind the scenes while all those classical (large matter) interactions were happening on a visible level. Let's start at the beginning.

1. In the beginning, there is information. Information literally means the act of giving form or shape to something. When form is given to something, other forms are ruled out. As a consequence, information theory defines information as the reduction of possibilities or uncertainty. (Think back to what we learned about quantum mechanics.) In Information Theory, the degree of information is directly proportional to how unlikely that sequence of data would be to occur by random chance. Information doesn't provide meaning, however. Information is not our perception of the data but the data itself and how it is organized and reduced to a constant (in place or time) or simplicity (thus the singularity of information). Thus information is the origin and the foundation of reality, and according to this principle all things are ordered and unified. However, this information doesn't gain context or meaning without (a) a consciouness to "know" the information, and (b) a medium to act as a context for this information. The consciousness is the field in which all this happens, and thus it – and the pattern it provides – predated the beginning itself.
2. The medium, or context, is the physical universe. This medium, however, requires an observer to make it "real." Otherwise, as the Heisenberg Uncertainty Principle shows us, "no-thing" is actually "real" until YOU reduce it to knowledge by observation and perception. But the binary nature of reality allows for it to exist, yet at the same time, the very nature of "existence" results in instability.
3. This instability causes asymmetry, which – when symmetry is broken – gives birth to the multitude. Think of the many wavelengths of the electromagnetic spectrum and how – although they are ac-

tually one unified field – they are perceived differently. Another example is the three-dimensional space we occupy, giving us perception of length, width, and depth. It is only through these many dimensions that we can have perception of reality.

4. And it is within this field of perception that systems and patterns of real phenomena can arise, based on the defining factors that preceded them.
5. These complex systems (which are fractal in nature because they reproduce patterns that are infinitely reducible) naturally grow at different rates because of the minor fluctuations in the blueprint.
6. Some find homeostasis or balance. The energy exchange and other process are, or become, even and homogenous. This sort of homogeneity goes back to the origin, when all was one.
7. But minor fluctuations can make a big difference in the outcome of the data. Minimal +1 inconsistencies tilt the evolution of the physical world towards the emergence of organized matter, towards organic life, and ultimately towards the Original man. The tendency towards such "perfect imperfections" was also preprogrammed into the blueprint from the beginning.
8. Such processes of growth, order, and development rely on, and continue, to create order within themselves through the dissipation of entropy (or disorder). This build/destroy process, known as negentropy, is not only a maintenance mechanism for life, but a means towards completion of any system' (living or nonliving) reproduction cycle.
9. Ultimately the entropy of the universe itself will increase to the point where everything evens out completely, bringing all reality back to the realm of pure information.

For more on this subject, check out *Information Theory, Evolution and the Origin of Life* by Hubert Yockey, *The Fifth Miracle* by Paul Davies, and *The Anthropic Cosmological Principle* by John Barrow and Frank J. Tipler

ANOTHER LOOK AT EVOLUTION

SUPREME UNDERSTANDING

In a future chapter, we'll explore the way that living organisms (indeed all life and matter) naturally "self-organize" and come together to evolve. This is known as "superorganism theory." But the idea of "group evolution" is a serious challenge to the linear thinking of Darwinism. A growing number of scientists argue that neo-Darwinian evolution doesn't explain the jumps in biological complexity that accompany superorganism-type development. And we're talking about people with strong reputations like Carl Woese, who sees bacteria in terms of networked communities rather than individual cells, and interprets their evolutionary history as driven by the non-linear self-organization that's now being studied at all biological scales:

> The jumps – saltations, in complexity parlance – appear to be non-linear emergent phenomena, the result of networked interactions that produce self-organization at ever higher levels. From this perspective, Darwinian evolution is a mechanism of a higher universal law, perhaps even a variant on the second law of thermodynamics.[365]

Woese suggests that this "higher universal law" may also explain the formation of stars and galaxies. He continues, noting that evolution doesn't occur at the individual unit (cellular) level but at the community (organism/superorganism) level, and adds that these evolutionary jumps are not the gradual changes Darwin proposed:

> Saltations are state changes. The simple example would be something like a magnet heated up to a high temperature where the iron dissolves; the magnetic properties are gone; then when you reach a critical temperature in cooling down, the magneticism reappears in a very short temperature change.*
>
> The property is gone in the individual iron atoms, but when they behave collectively, you see the property of the whole. That's a very simple example.
>
> The microbial world is where I work; [saltational evolution] predicts that there should be properties of the collective thing, that arise as the thing collects...Twentieth century biology was structured according to a linear Newtonian worldview. Linear thinking is not the kind of thinking that's needed to study evolution. It doesn't help you understand the nature of systems. Molecular biologists were so set about linearity that when the gene came along, they took the gene to be the be-all and end-all of basic biology. That comes out of thinking in terms of particles and linear interactions...I see evolution as the quintessential non-linear dynamics problem.[366]

A non-linear dynamics problem, huh? Woese, although he only hints at it, is not alone in describing evolution in terms of directed patterns of purposeful organization. Evolutionary biologist Gro Amdam, a honeybee specialist and colleague of Hoelldobler, cites the superorganismal nature of bees as an example of the flaws in mainstream Darwinian ideas about evolution:

---

* Another scale where state changes can be observed is at the quantum level, when an electron "jumps" from one shell of an atom up to outer shell, never traversing the space in between.

> How bees went from solitary to hive-linked creatures is a mystery, but it appears to involve a phenomena known as pre-adaptation, in which gradually accumulating characteristics suddenly make possible a whole new level of complexity – a level that seems to just...emerge.
>
> Darwin had this really great idea, and people added to it later on...but there are phenomena that don't fit within the basic framework....One very interesting concept that is not fully encompassed by [Darwin's] gradualism is the concept of pre-adaptation.
>
> To boil it down: in nature, solitary honeybees go through multiple reproductive stages, each characterized by a particular physiology. In hives, individual bees are suspended at particular reproductive states – that's how specialization occurs – and their state is determined by interaction with other non-reproducing members of the hive. The neo-Darwinian framework...[is] a linear process. But here it's nonlinear...Synergy is the wrong word; preadaptation is the right word. [It] enables you to make a step forward without this linear trajectory...It's an entirely new way of living, a whole new life history.[367]

Sounds like the nature of mathematics. "Pre-adaptation" implies pre-existing knowledge.[368] This brings us back to the idea of a pre-existing structure or blueprint, which we've described in previous chapters on the mind, man, and the mathematics of nature, which are cyclical rathern than linear. Andam continues, explaining that scientist expected to explain the division of labor among bees by finding a gene in individual bees that determined what job they did. Yet, as they've found with other species (from ants to humans), there was no such factor at the individual level. It was all being determined at the community level. The mind of the community produced what it needed in individuals. Perhaps this would explain why "followers" and "leaders" are not simply natural, but necessary. Andam writes:

> It's the colony that succeeds or fails, and is made up of individuals that contribute to success or failure, but it's selection at the colony level that guides it. When you define the colony as the unit of selection, you have the superorganism at the level of individual, and you can explain its changes through the change of its parts.
>
> That's similar to any gradual process...When it's become a stable entity at the level of population, then the neo-Darwinian framework is perfectly matched to how it's changing through time – but when you go from solitary living to a social being, and you use evolutionary building blocks or old genetic networks to rapidly build a new phenotype in which [genes governing individual processes form the basis of] a social division of labor...that's where the gradualism of neo-Darwinism is not the perfect match.[369]

In other words, Darwin got many of the fundamentals right, but gradual changes can't explain the jump that occurs when communities are formed. It appears the idea of "gradual evolution" (at all levels) isn't

DID YOU KNOW?
It seems man, too, has experienced "jumps" in his development, consisting of physical changes that accompanied accelerated cognitive growth. This has happened along man's evolutionary timeline, just as it does in the lifetime of any male or female who goes through the phases of puberty, adulthood, menopause, and other major changes. On a quantum level, electrons tend to "jump" positions from one shell to another, without even passing the space in between. This all goes back to mathematics. Depending on the process you subject the number 4 to, it could either become 5, 3, 2 or 8. Adding one to make 5 would seem like a gradual increase, while multiplying the 4 to make 8 would seem like a saltation, an unnatural jump. But 4 becoming 5 is also a jump, because you're not travelling through the infinite points in between 4 and 5 (4.1, 4.259, etc.). In life, as in evolution, as in mathematics, the right circumstances (or processes) can promote extraordinary rates of growth or change. And it can go both ways. Some changes that took 6000 years to accomplish can be reversed in one day.

supported by geological history, the fossil record, or genetics. All the available evidence suggests that diversification happened rapidly and meaningfully. And every so often, there would be a mass extinction where this teeming diversity was narrowed down to a few select participants who could move on to the next round.[370]

"In fact, evolution became in a sense a scientific religion; almost all scientists have accepted it, and many are prepared to 'bend' their observations to fit with it." – H.S. Lipson, British Physicist

Yet modern biologists typically adhere – unquestioningly – to the doctrine of neo-Darwinism, which says the reason some organisms survive and others do not is purely a matter of random mutation occasionally producing the best fit for those circumstances. As Anna Gillis remarks in her *BioScience* article, the biological community "got locked into its belief that an organism cannot control its own mutation." Recalling the Sociology of Scientific Knowledge, Gillis notes that "Nobody wants to give the appearance of straying from the neo-Darwinism fold." One of those nobodies was John Cairns, who proposed in a *Nature* article that "bacteria living in an unfavorable environment are able to choose which mutations to produce to adapt to the stressful situation." In other words, organisms could consciously direct their own evolution. It would follow that this ability would increase over the course of evolution, with organisms becoming more and more "conscious" over time.[371]

Some scientists have theorized that consciousness is an "emergent property" of the superorganism, which corresponds to the process in physics where particles can come together as Bose-Einstein condensates (or superparticles) and maintain some degree of coherence (or consciousness). But we argue that consciousness is simply the measurable manifestation of the structure known as the Mind, which underlies all life and matter. In effect, the Mind provides the blueprint for growth, order, and direction. Direction means that not only is life or matter, evolving in a particular way for a purpose, but that – to varying extents – it can carry the information or consciousness of that purpose or direction. Cairns may have been right. Even our earliest ancestors (as single-celled organisms) could have been doing what they did with a goal in mind. That goal was us.

After all, the odds of favorable mutations (by random chance, the type that would "eventually" produce something that works) are quite low. And the probability of the right random genetic variations not just occurring but being 'selected for' within a very short space of time, leading to the appearance of a new species, is even more remote than over a very long period. Not to mention that most genetic mutations are harmful or lethal, while nonlethal mutations tend to be recessive. So what activates the recessive mutated genes and makes them produce the features needed by the next generation? Sheldrake calls it the morphogenetic field, while other scientists have used even more elusive concepts.

There's even evidence to suggest that life is "pushed along" in specific directions by phenoma that occur on a quantum level. According to biologist Dr. Lawrence David, quantum effects (like electron movement) may have dictated major steps in life's early evolution, including photosynthesis and respiration, which led to the Archean Expansion, which populated the Earth with more complex microbial ecosystems.[372] This is a major premise of Johnjoe McFadden's book, *Quantum Evolution: How Physics' Weirdest Theory Explains Life's Biggest Mystery*. According to McFadden, Gabor Vattay, Stuart Kauffman, and Samuli Niiranen, the transitional area between the quantum and classical realms is in energy and information transport in organisms.

But there's more to it. As we've noted before, there's a layer of consciousness below the quantum level. Some scientists call it subquantum, but few have – besides the mystics have attempted to define its nature. We've typically called it the Blueprint or Mathematics. Mathematics, again is the structure or order by which things are organized, aka the Mind, aka Self, aka You. All these are variations of the same general "ultimate reality."

Paul Davies has written that "knowledge" (or information creation) may be a fundamental law by which both life and matter are written into being:

> "[E]mergent laws of complexity offer reasonable hope for a better understanding not only of biogenesis, but of biological evolution too. Such laws might differ from the familiar laws of physics in a fun-

damental and important respect. Whereas the laws of physics merely shuffle information around, a complexity law might actually create information, or at least wrest it from the environment and etch it onto a material structure." – Paul Davies, *The Fifth Miracle*[373]

In a *Physics of Life* review critiquing "spontaneous" (meaning non-directed) self-organization in the origin of life, David Abel and Jack Trevors explain:

> [Life] typically contains large quantities of prescriptive information...Prescription requires choice contingency rather than chance contingency or necessity...The only self that can organize its own activities is a living cognitive agent.[374]

In other words, through the Mathematics that structure the Universe, Man wrote all life and matter in his own language.

CHARLATANS AND PSEUDO SHAMANS

SUPREME UNDERSTANDING

Not to beat a dead horse, but when we're talking about fakes, frauds, and phonies, these horses have nine lives, so there's always a need to keep the discussion going. Since the late 1800s – perhaps starting with Helena P. Blavatsky – Westerners have taken indigenous practices and perverted them into distorted versions of the teachings that satisfy European interests (e.g., Blavatsky's invention of "Indian masters" who taught her that white people were the most advanced race). In recent years, Europeans have co-opted nearly every indigenous culture's most sacred traditions, from assuming the identity of Egyptians (as hermeticists and alchemists), to the "Ancient Arabic Order of the Nobles of the Mystic Shrine," to Indian yogis and swamis (e.g. William Walker Atkinson as Yogi Ramacharaka), to countless Native American impersonators, many who are now finding legal ways to self-identify as indigenous. C'BS Alife Allah calls it "New Age Identity Theft." Most recently, there are now even Europeans becoming Sangomas, or traditional Zulu healers! Yes, Zulu!

But this type of behavior is not limited to Europeans. We have to remember who trained or taught many of these European con-artists (some of whom are more delusional than corrupt, as they sincerely believe their own twisted tales). And since Beverly Paschal Randolph, a Black man and contemporary of Blavatsky (who she hated), assumed the identity of an half-Native American, half-white Rosicrucian, there have been increasing numbers of Black and brown people "faking the funk," representing traditions they have not actually been initiated in. This reached its apex in the 1970s and has been a prevailing presence in our communities since.

This is why I thought the following piece was especially insightful. The "Pseudo Shamans Cherokee Statement" by Richard L. Allen, EdD, Research & Policy Analyst of the Cherokee Nation, says much about the modern day charlatans who need not claim Cherokee, nor Shaman, but are simply unable to verify their place in the traditions they claim:

> Greetings. The Cherokee Nation is overwhelmed with those charlatans who fraudulently claim to be shaman, spiritual leaders or descendents of a Cherokee princess. Such individuals make such claims without ever having lived within the Cherokee communities. They claim to be descended from some nebulous and mysterious ancestor who was from "a reservation in North Carolina" (there is only one) or "a reservation in Oklahoma" (there are none). The ancestor is never just a plain ordinary everyday Cherokee citizen but a "Cherokee Princess," a "Cherokee Shaman," or a "Cherokee Pipe carrier" none of which actually exist or ever have. Those who claim to be "shaman" do not reside within the known boundaries of the Cherokee Nation in Oklahoma.
>
> Cherokee medicine people and spiritual leaders are known to the Cherokee people and do not practice medicine for a fee nor sell "shamanic" lessons to anyone. They do not advertise their services through any form of media and certainly not over the internet. Traditional Cherokee healers and spiritual leaders provide their services to the Cherokee people. A Cherokee medicine person or spiritual leader is fluent in the Cherokee language and would conduct any medical or spiritual practices by using the Cherokee language. Therefore, our medicine people are those who were born of a Cherokee mother and a Cherokee father and would have been reared within a Cherokee community speaking the Cherokee language. Our traditional Cherokee healers and spiritual leaders are humble people and would not present themselves as such nor "hang out a shingle" so to speak. Cherokee medicine people are acknowledged and recognized by members of the Cherokee community as effective healers and leaders. It is the recognition of the Cherokee people that validates these persons as medicine people and healers, not self-proclaimation. We may provide them small gifts, a token amount of money or foodstuffs in payment for their services. They do not charge for their services nor would they withhold their services when asked and they certainly would not prescibe payment by credit card. Cherokee medicine people may provide services to recognized members of other tribes or may provide services to non-Indians who would seek them out for treatment, but certainly would not mix their spirituality or medicine with that of other nations. Cherokee medicine and spiritual practices do not include tarot cards, palmistry, psychic readings or sweatlodge ceremonies.
>
> One may assume that anyone claiming to be a Cherokee "shaman, spiritual healer, or pipe-carrier," is equivalent to a modern day medi-

cine show and snake-oil vendor.

For another look at this topic, see "Modern Day Snake Oil" in *The Hood Health Handbook*, which details the history of "miracle cures" that started out as indigenous healing practices and became exploited by Westerners for profit, with none of the healing benefits of the original formula. Such exploitation continues into the modern age, and this article is written to show you how to tell what's useful and what's a scam.

THE BLACK MAN IS THE 7 IN THE CENTER

SUPREME UNDERSTANDING

Man tends to be the anomaly in any equation. He's like the number 7, in the sense that 7, like every other other digit, is a necessary component of the wholeness of 360. Yet 7 is unique in that it's the only digit that can't be evenly divided into 360. It's an anomaly. It's perfect, yet problematic. Man is perfect in form (mentally) and problematic in function (physically). Again, the 7 represents him well here, with the dual nature of the 3 (the triune, "heavenly" aspect) and 4 (the four-cornered, "earthly" aspect) that compose it. This may be why so many ancient cultures used the number 7 to represent God's presence on the Earth. God is both the problem solver, and the source of the problem. He is the author of what could be a perfect system, and yet the source of its imperfection. That's just the way the blueprint works. Without this constant state of flux between error-production and error-correction, there'd be no such thing as life or evolution of any sort.

Although ancient philosophers may have popularized this symbolism, science agrees that the material aspect of reality is quaternary, or four-parted, as all things exist along two binary axis, one for space-time and one for energy-mass. Everything in the visible universe exists somewhere along these axes. Astronomer and science philosopher, Harlow Shapley has said:

> The basic entities (of physics and astronomy) are commonly recognized to be space, time, matter, and energy; the first two can be linked together as space-time, and the last two as mass-energy. It is difficult to isolate any universal quality that is not a variation on these four. Speed, weight, light, distance, momentum, and the like are all derivatives of the four, or combinations.[375]

Meanwhile, the triune nature of man represent's eternal, immaterial, mental Self. Its three properties are growth, order, and direction. This concept can be represented as the three points of a triangle, but also as a circle (with its center point, diameter, and circumference). This is part of the science behind the Vitruvian Man. You know what I'm talking about – the drawing of the man inside the circle and square, kinda looks like a muscular Steven Tyler? Anyway, this is why Leonardo Da Vinci envisioned his rendering of Vitruvian Man as a "cosmografia del minor mondo" (cosmography of the microcosm). He believed the layout and workings of the human body to be an analogy for the universe.

It should be noted that Da Vinci did not invent Vitruvian Man himself. His drawing was based on the details of Roman architect Vitruvius' analysis of body proportions, originally published in Vitruvius' de Architectura. Da Vinci wasn't the first to put this schematic into a visual form either. It appears he copied the idea from his colleague Giacomo Andrea de Ferrara, whose version looked much less like Steven Tyler than Da Vinci's. And where did Vitruvius get his idea from? We don't know much about him, but we do know that when he wrote de Architectura, he described the building methods of various "foreign tribes" and almost certainly worked for some time in Gaul and North Africa. Could this explain why the proportions of Vitruvian Man don't correspond too well with those of Caucasians? For example, Vitruvius gives the ratio of height to feet as 6.0, which is seen in neither Caucasians (6.4) or Japanese (6.7).[376] Is there anyone with a 6.0 ratio? Unlikely, right. Actually, this ratio IS found in Black Americans, along with several other proportions Vitruvius cites.[377] Collectively, these physical traits (longer feet, longer limbs, longer head, etc.) are known as the tropical body plan because of where they evolved. Apparently, Steven Tyler was not the Vitruvian Man. The Black man is. He is the 7 in the center of the circle and square.

A Quick Overview of Quantum Concepts

C'BS ALIFE ALLAH

Quantum Mechanics are the principles governing everything at the subatomic level. They compile our observations up to this point and suggest the unifying laws for all particles. Yet it must be noted that no one is really sure what exactly these laws are as its gets really weird at the subatomic level. Scientists know what they've observed, and they know these results are always going to occur, but they just haven't agreed on exactly what they mean. These are six things that have been observed in the study of quantum mechanics, and you're bound to read about some (or all) of them in any discussion of the topic:

WAVES ARE PARTICLES AND PARTICLES ARE WAVES

On the subatomic level "particles" sometimes behave like particles (discrete and countable) yet at other times they act like waves (in that they show wave traits, like diffraction and interference). This is why some physicists call them energy "quanta" instead of particles or waves. The most popular experiment involving this "dual nature" had photons firing through a slit (or two slits). Depending on whether they were observed or not, they behaved differently (as a particle or a wave). This implies that observation can change or determine reality at the quantum level.

"Intelligence, as I understand it, is buried in the decoder. So the only place where information is well defined seems to sneak in an intelligent agent somewhere." – Stuart Kauffman

MEASUREMENT DETERMINES REALITY

The Heisenberg Uncertainty Principle says that, up until the moment that a quantum particle is measured, its exact state is indeterminate (unknown). After that measurement is made, it is absolutely determined, meaning all measurements of that particle will give you the exact same outcome. Scientists theorize that the particle occupies multiple "possible" states (which is known as the particle's wave function) up until that point of measurement. After that point it can exist in only one state. This is known as the collapse of the wave function. The Copenhagen Interpretation holds that quantum mechanics does not yield a description of an objective reality but deals only with probabilities, until the act of measurement causes the collapse of the wave function, essentially determining reality.

THE POTENTIAL OF EMPTY SPACE

Most people grow up with the traditional picture of an atom and think that it actually looks like a little solar system. That model is less of a picture and more of a symbol. In actuality, an atom is mostly (we're talking like 99%) empty space. So how can something be both solid and empty? And if that is the case why is there so much stuff around us? Well when you marry the above two points that we made about quantum mechanics with this one you realize that we are dealing with the probability that the empty space will be filled by a particle at a particular moment. So atoms are literally balls of probability.

ALL WE'VE GOT IS PROBABILITY

When physicists use quantum mechanics to predict the outcome of an experiment, only one thing can be detected: the probability of the possible outcomes. We can give the percentile chance of each outcome, yet never predict the definite outcome. Thus, even in an experiment for a single particle, we can never say what will definitely be the outcome, no matter how carefully we prepare.

QUANTUM CORRELATIONS ARE NON-LOCAL

We dig into this when we discuss the non-local aspect of what we are referring to as the Mind. In the world of quantum mechanics, non-locality is referred to as entanglement. It means that when two sub-atomic particles are connected in the right way, they will act as ONE particle no matter how far apart they are in space. Again consider that their states haven't even been determined until measured AND they are apart.

SUB-ATOMIC PARTICLES GET IT IN

When a sub-atomic particle is working to get from point A to point B, it's going to get there by any means necessary. It will take every possible path…at the same time! This includes taking paths that include highly rare events such as electron-positron pairs coming out of nowhere and then disappearing again. Since a particle is effectively nowhere until we consciously determine where it is, it can be anywhere until we collapse its wave function and decide where it is. Speaking of travelling, a related concept is quantum tunneling, where the electrons in atoms are constantly disappearing in one location and reappearing at another location without ever crossing the space in between.

FIND THE MATH IN PRENATAL DEVELOPMENT

SUPREME UNDERSTANDING

We've given you lots of information and insight, but our goal is to empower a new generation of scientists and critical thinkers who can make sense of the universe on their own. So here's your challenge. I've outlined the prenatal development process of a human child. It's ordered chronologically, and I've included only the most relevant details. What I want you to do is find the mathematical parallels in this process that make it like processes in (a) the evolution of life on Earth, (b) the formation of the Earth itself, and (c) the development of the visible Universe. Yes, those are in order of increasingly difficulty. And you'll need to know (a) the story of evolution, (b) the story of the Earth's formation, and (c) the story of the Universe, and know them well, to handle this challenge. Why this challenge? Because we want you to see the math on your own. We also want you to develop a mastery of the information we've presented, and this activity will allow you to do that. So let's begin.

A TIMELINE OF PRENATAL DEVELOPMENT

Weeks 1-2: Day 1 is fertilization, when all human chromosomes are present and a unique human life begins. Day 2 is cleavage (or splitting of the cell, leading to exponential growth). Day 3 is compaction (when the mass of divided cells form a tightly packed sphere). Day 4 is differentiation (when the embryo develops layers). Day 5 is cavitation. By the end of the first week, the embryo begins implantation in the uterus. The embryo then begins differentiation, where each cell starts producing the parent cells of future organs.

Week 3: Rapid early growth cools to a more moderate pace. By the end of this week the child's backbone and nervous system emerge. The liver, kidneys and intestines begin to take shape. By day 22, the heart begins to beat with the child's own blood.

Week 4: Cells are specializing, with the outer layer of the fertilized egg becoming the nervous system, skin, and hair. The inner layer will be the breathing and digestive organs. The middle layer will be the skeleton, muscles, circulatory system, kidneys, and sex organs. The child, who resembles a tadpole, is now ten thousand times larger than the fertilized egg four weeks ago.

Week 5-7: Now eyes, legs, and hands begin to develop. By week 6, brain waves are detectable. Mouth and lips form, followed by eyelids and toes, and then a distinct nose. The baby is kicking and swimming. Between 17 and 56 days the embryo is most susceptible to drugs, disease, and other factors that interfere with normal growth.

Week 8: At this point the embryo is developed enough to call a fetus. All organs and structures found in a full-term newborn are present, bones begin to replace cartilage, and fingerprints begin to form. The baby can begin to hear.

Weeks 9-10: Teeth and fingernails develop. The baby can turn his head, hiccup, and frown. The head comprises nearly half of the fetus' size and the face is well formed. The eyelids close now and will not reopen until about the 28th week.

Weeks 10-11: The baby can "breathe" amniotic fluid and urinate. By week 11, the baby can grasp its umbilical cord and all organ systems are functioning. The baby has a skeletal structure, nerves, and circulation. The genitalia are now clearly male or female.

Week 12: The baby has all of the parts necessary to experience pain, including nerves, spinal cord, and thalamus. Vocal cords are complete.

Month 4: Bone marrow begins to form. The heart is pumping 25 quarts of blood a day and beating at 120-150 beats per minute, while brain waves are clearly detectable. By week 17, the baby can have dream (REM) sleep.

Months 5 and 6: At 20 weeks the baby recognizes its mothers voice. The baby practices breathing amniotic fluid and grasping at the umbilical cord when it feels it. The mother can feel the fetus moving and the fetal heartbeat is strong enough to be heard with a stethoscope. Oil and sweat glands are now functioning.

Months 7 through 9: There are now rhythmic breathing movements, the fetal body temperature is partially self-controlled, and there is increased central nervous system control over body functions. The baby opens and closes his eyes. The baby is using four of the five physical senses (vision, hearing, taste, and touch), knows the difference between waking and sleeping, and can sense the moods of the mother. The mother supplies the fetus with the antibodies it needs to protect it against disease. The baby's skin begins to thicken, and a layer of fat is produced and stored beneath the skin. Antibodies are built up, and the baby's heart begins to pump 300 gallons of blood per day. Approximately one week before birth, the baby stops growing, and "drops" usually head down into the pelvic cavity.

EXCERPT FROM THE AFRICAN ABROAD

BY WILLIAM HENRY FERRIS

REPUBLISHED BY TWO HORIZONS PRESS IN 2012

In *The African Abroad*, William Henry Ferris takes his readers on a whirlwind tour of the universe, exploring science, ancient history, philosophy, political ideology, and the biographies of important Black figures in the U.S. and abroad. What makes Ferris' work especially interesting is that, in 1911, he was a Black man exploring subjects that were incredible ahead of his time, including – but not limited to – the shortcomings of his era's Black leadership, the Black presence in ancient world history, and the idea of his own identity as God.

Ferris begins his discussion of God as man, and man as God, when he envisions a future time that witnesses the fall of the Western empire and the "rise of the black, brown and yellow races" – a time when white people have "passed out of the memory of men and have been swallowed up in oblivion."

"Perhaps in that distant time scholars may write books to prove that the world-renowned and world-conquering Anglo-Saxon race never really existed, except in the imagination of rapt poets, and was only a mythical, legendary race," Ferris writes. He peers further into this future and considers a distant when there must be another catastrophe extinction of life on Earth, as has happened countless times in the past, and asks what this means regarding the nature of man:

> So there will finally, in the course of a few score million years, come a time when every semblance of vegetable or animal life will have disappeared from the planet, and every trace or vestige of man's civilization will have become completely effaced. Annihilation is the ultimate fate and destiny of the human race on this earth of ours.
>
> And then I asked, was this the end of Nature's strivings, was this the final destiny of man's aspiration, was this the consummation of man's hopes, to be swallowed up in the dark midnight of nothingness and oblivion, to fade away forever out of existence? Was human life a dream? Was human history a farce? Were the ideal dreams which have lured on mankind to higher heights of achievements, illusions? Were the heroic ideals to do and dare and strive and achieve, nothing but hallucinations? Were the mighty hopes which made us men but mirages in the desert? In a word, is our striving to realize and embody ethical ideals in our lives and characters a vain struggle, which will finally end in defeat?

Ferris says man instinctively cannot consider such a nihilistic prospect:

> That which is most fundamental and basal in human nature asserts itself, rises in protest and cries out, No! No! No! And yet, if the fate and destiny of this earth on which we live and of the human life it sustains, is to vanish and be blotted out of existence, this would seem to be the nature of man's ideals, and the end of his strivings, unless man were immortal. But if the universe were not the fortuitous play of blind, unthinking atoms and ions, if it were not a chaos but a cosmos, if reason were embedded in the very structure of the universe, if a world drama were being enacted, in which a Master Mind were behind the curtains, shifting the scenes, then it would seem that man is immortal and that his strivings are not in vain. And I asked myself, is there a God, is there a Master Mind behind the mechanism of Nature, who utters his eternal decrees in the immutable laws that regulate the movements of the starry hosts above, and who thunders in trumpet tones in the ideals of man?

Ferris refers back to his studies of nature and astronomy, where he finds "a unity dominated by law and order, so that we have an orderly totality and a cosmos instead of chaos." He observes:

> I reflected that it takes light traveling at the rate of 186,000 miles a second, fifty years to reach us from the Pole Star and four hundred centuries to reach us from the furthest of these dim specks, and then considered that this immense universe hangs together and is governed

> by law and order.

He argues that this order can only come from the Mind of God, who is, on Earth, personified as man himself:

> The only explanation that satisfies the reason of man, for the law and order that makes the universe an orderly totality and a cosmos, is that the universe is the forthputting of an Infinite Mind who manifests his own ideas in the laws of Nature.
>
> But I am a child of Nature. I have been generated in Nature's womb, and I am an offspring of the universe and an integral part of the universe. Hence, I am the manifestation of the Infinite Being, who manifests himself in the universe. The same Being who wells up in grass and Bower, who registers his laws in the movements of the heavenly bodies, also manifests himself in me. Truly has the Apostle said, "In Him we live and move and have our being." And in seeking to understand the Power or Infinite Energy who manifests himself in the universe we must find a being big and brainy enough to beget man…So, then, the Superhuman and Supersensible Cause and Source of the universe of mind and matter must be a universal life, which is as coextensive as the life of the universe, and a universal self-consciousness which is coextensive with the mental and physical changes in the world of finite mind and finite matter, and embraces them in the totality of its own being, manifesting its mind and its ideas in the orderly sequence which we term the laws of Nature and its will in the force of Nature.
>
> The happenings in the universe of finite minds and finite matter are not only phases and aspects and doings and forthputtings of finite selves and things, but they are also movements in the life of the Absolute, facts in his consciousness. For he is the Immanent and Causal Ground of all the psychic and physical changes in the universe, and of the system of things. He is the Absolute Self of whom all finite things and finite selves are but partial and fragmentary manifestations.
>
> To destroy me, the Absolute must destroy a part of himself and destroy his own offspring. The question then arises, Is the Absolute interested in the ideals and strivings of man? Have the ideals and strivings of man an eternal value and significance for the Absolute? Have our personalities and individualities an eternal meaning and value for the Absolute? On the answer to these questions hangs the immortality of man. In developing our manhood, we are not only realizing our latent potentialities and developing the germs of divinity that slumber in our natures, but we are developing a bit of the Absolute. If we share in the life of the Absolute, and are partakers of his divine nature, what is more natural than that we should share in and partake of his eternity and immortality?

Ferris describes the attributes of man as those of a God on Earth, specifically drawing attention to the Black man in America:

> The Being who shot us through the crucible of his own nature, ran us through certain molds of thinking and feeling, stamped us with the

> impress of his own personality, and then launched us forth from the shores of eternity out into the sea of time, endowed us with a reason that can fathom the secrets and mysteries of nature, extract the ores from the earth, harness the forces of nature, water, wind, steam and electricity, to run our mills, carry us over land and water and transmit messages across space,...with a will that enables a man to chisel and carve and hew out his own career and be the architect of his own fate and destiny; with a will that enables a man to tunnel mountains, bridge chasms, brave dangers and defy obstacles and obstructions; with a will that exultingly cries, "There shall be no Alps, I will find a way or make it," – this Being will call us back home some day and then he will longingly – look to see whether we have preserved or effaced the divine impress. The impression of the Almighty is upon us; a divine spark slumbers in us; divinity stirs within us; we are men, not beasts of the field.
>
> That throbbing, divine life, with which the universe pulses and which transforms the world into a fairyland every spring, breaking into expression in leaf and blade and flower and covering the earth with a garment of verdure, wells up in us as the fountain source of the impulses, the instincts, that lift us above the plane of animal life. It manifests itself in the divine discontent and dissatisfaction with our present mode of life. It utters itself in the stirrings within us that prompts us to transform the actual into the likeness of the ideal. It voices itself in the strivings after the higher life, that are the springs of human progress and of the development of man in history. And while the Negro needs to buy all the land, and get as large a bank account as he can, while he needs to branch out into the mercantile world, and go into business, he must remember that this is not the end and goal of our existence. That end and goal is to realize the mighty hopes which make us men.
>
> Every living thing fulfils the laws of its being and realizes the immanent idea that Nature implants in it. The grass grows; the seed buds and blossoms into fruit and leaf and flower; the acorn develops into the wide-spreading oak; the majestic lion stalks the forests, monarch of all he surveys; the eagle soars aloft on his powerful wings and sights his prey from afar, and man develops from a babbling babe into a Godlike being, in whom reason and conscience are inthroned. The plant and animal do this unconsciously, obeying their instincts. Man does this consciously, through the guidance of reason, and by the power of choice.

Ferris warns that the mystery of God would not be apparent to those who could make sense of the language in which he wrote the universe. He writes:

> I then saw that God was the great Geometer, that the universe is crystallized mathematics and that nature is the time vesture of the Eternal, the garments we see Him by, which reveals Him to the wise and hides Him from the foolish...

> I know that this is the age of practical atheism, of agnosticism, the age when men say, "We don't know whether there is any God." But when I reflect that I am living in a universe which is built up out of millions of some seventy odd different kinds of atoms, of millions of minute gyrating and revolving ions, in which law and order reign, when I reflect that certain fundamental laws of reason govern my thinking and the constitutional mode of the operation of my mind, when I reflect that from the depths of my nature, beneath the subsoil of my conscious life, rises the impulse of instincts that make me a moral personality, I cannot believe that this vast universe, and myself a mental and moral being, were formed by the fortuitous concourse of blind and unthinking atoms.

"The universe needs a God back of it to explain it. I need a God back of myself to explain myself to myself," Ferris says. Yet he questions the purpose of man coming to be, and the purpose of the universe in general:

> "For what final purpose was man created? Why did the universal mind never pause in his struggle and striving, in his manifestation until he begot man? For what reason did the Absolute impose the ethical idea upon man, as the supreme law of his nature?"
>
> Unless we believe the universe, with all its vastness and splendor and glory and grandeur and law and order, to be begotten by blind chance, and unless we believe the world ground to be having fun with us, there must be an infinite and eternal meaning and significance to the supremacy of the ethical ideals of man. And I am inclined to believe that the Absolute will preserve whatever is of eternal significance and value in the universe. But the scientists tell us that the immortality of the mind of man is an impossibility because the mind states are epiphonema thrown off by the brain.

But, long before theories of Mind had become well-established in science, Ferris dispute this claim, arguing that the mind and brain "are two separate things, which are causally related to and reciprocally influence each other during the temporal life of man." He continues:

> But the mind of man has a life and nature that is peculiarly its own, that is sui generis, that behaves in ways peculiar to itself alone, and that transcends the functioning of the brain...It may be true, as the late Professor William James of Harvard has said, that the brain is the medium for the transmission of thought as welt as for the production of thought. And it certainly seems clear that the brain, instead of producing the mind, is the occasion for the mind's manifesting its own peculiar nature and activities.

Noting that mind cannot "be conjured up from that which is not mental in its inner structure," Ferris explains that the reason man has a Mind is because he shares the consciousness of the Infinite Mind himself:

> How, then, could two such separate and distinct beings as the brain and mind reciprocally influence each other and be causally related to each other? Only on the hypothesis that they are the manifestations of the same Infinite Mind, who manifests himself in the world of mind and matter and whose self-consciousness is coextensive with and inclusive of the psychical and physical changes in the universe. The mind of man is the manifestation of the mind of God. And if God so wills, the conscious, rational life of man will survive the death of the body and the destruction of the brain. God has the whole universe at his command.
>
> The mind develops in constant and ceaseless dependence upon the protoplasmic molecules of the brain; but in its growth and development it evolves an ego, a unity of personality, a center of self-consciousness that persists and endures during the modifications in the substance of the brain and changes in psychic states. Nerve cells in the brain wear out and are replaced by new nerve cells; psychic states come and go, and succeed each other in the stream of consciousness like the waves of• the sea. But the self, the ego, the unity of personality, the center of self-consciousness, the permanent subject of the psychic states, the "I" who thinks, perceives, imagines, remembers, feels, and wills, remains.

This Mind, Ferris explains, is what makes man immortal:

> And, is it strange that this unity of personality, this center of self-consciousness should survive the destruction of the body, should persist during the physical change of death and be clothed in a new garment and raiment and be attached to a new medium and organ of expression? If God wills it, the rational life of man will survive the death of the body and the destruction of the brain.

Only faintly touching on the Christian themes of his AME upbringing throughout his work, Ferris closes his discussion by subtly critiquing the idea of heaven, and describing death as a reunion with the consciousness that created and ordered the universe:

> No wanderer who has ever set sail on the dreaded sea that laves these terrestrial shores, has ever returned to tell of the sights he saw, the sounds he heard, or what beautiful visions greeted his eye on yonder shore; no one has ever returned to tell of the strange land and countries beyond the sea. But when I must shuffle off this earthly coil…it may be that the noble spirits who have spent their lives doing something to lighten the sins, sufferings, miseries and wretchedness of the world will realize the words of the Apostle when he said, "Beloved, now are we the sons of God, and it doth not yet appear what we shall be; but we know that when he shall appear, we shall be like him, for we shall see him as he is."

What should you do now that you're done reading?
Here are some suggestions:

- ❑ Complete any activities mentioned in this book, especially the discussions. See any of the films mentioned, but with others.
- ❑ Tell somebody about this book and what you've learned. Invite them to come read it. Don't let them steal the book.
- ❑ As another option, let them steal the book. It might help them.
- ❑ Mentor some young people or teach a class using this book as a handbook or reference.
- ❑ Talk about this book online, but don't stay on the Net forever.
- ❑ Join an organization or group that discusses concepts like the ones in this book and get into those discussions.
- ❑ Leave this book away somewhere it will be picked up and read.
- ❑ Identify the people in your community who could use a copy of this book. If they're people would want to buy a book like this, let em read a few pages and see if they can afford to buy a copy.
- ❑ If they're people who don't normally buy books – but you know that givin em a copy could change their life – give em a copy and tell em to come see you when they're ready for another one. This is why you can order copies at wholesale rates at our site.

We hope this helps you keep the knowledge contagious.

ALSO FROM OUR COMPANY

How to Hustle and Win, Part 1: A Survival Guide for the Ghetto

By Supreme Understanding
Foreword by the Real Rick Ross

This is the book that started it all. Now an international bestseller, this book has revolutionized the way people think of "urban literature." It offers a street-based analysis of social problems, plus practical solutions that anyone can put to use.

CLASS	PAGES	RETAIL	RELEASE
I-1	336	$14.95	Jun. 2008

ISBN: 978-0-9816170-0-8

How to Hustle and Win, Part 2: Rap, Race, and Revolution

By Supreme Understanding
Foreword by Stic.man of Dead Prez

Seen here in its original green cover, the controversial follow-up to *How to Hustle and Win* digs even deeper into the problems we face, and how we can solve them. Part One focused on personal change, and Part Two explores the bigger picture of changing the entire hood.

CLASS	PAGES	RETAIL	RELEASE
I-1	384	$14.95	Apr. 2009

ISBN: 978-0-9816170-9-1

Knowledge of Self: A Collection of Wisdom on the Science of Everything in Life

Edited by Supreme Understanding, C'BS Alife Allah, and Sunez Allah, Foreword by Lord Jamar of Brand Nubian

Who are the Five Percent? Why are they here? In this book, over 50 Five Percenters from around the world speak for themselves, providing a comprehensive introduction to the esoteric teachings of the Nation of Gods and Earths.

CLASS	PAGES	RETAIL	RELEASE
I-2	256	$14.95	Jul. 2009

ISBN: 978-0-9816170-2-2

The Hood Health Handbook, Volume One (Physical Health)

Edited by Supreme Understanding and C'BS Alife Allah, Foreword by Dick Gregory

Want to know why Black and brown people are so sick? This book covers the many "unnatural causes" behind our poor health, and offers hundreds of affordable and easy-to-implement solutions.

CLASS	PAGES	RETAIL	RELEASE
PH-1	480	$19.95	Nov. 2010

ISBN: 978-1-935721-32-1

The Hood Health Handbook, Volume Two (Mental Health)

Edited by Supreme Understanding and C'BS Alife Allah

This volume covers mental health, how to keep a healthy home, raising healthy children, environmental issues, and dozens of other issues, all from the same down-to-earth perspective as Volume One.

CLASS	PAGES	RETAIL	RELEASE
MH-1	480_	$19.95	Nov. 2010

ISBN: 978-1-935721-33-8

A Taste of Life: 1,000 Vegetarian Recipes from Around the World

Edited by Supreme Understanding and Patra Afrika

This cookbook makes it easy to become vegetarian. In addition to over 1,000 recipes from everywhere you can think of, plus over 100 drink and smoothie recipes, this book also teaches how to transition your diet, what to shop for, how to cook, as well as a guide to nutrients and vitamins.

CLASS	PAGES	RETAIL	RELEASE
W-1	400	$19.95	Jun. 2011

ISBN: 978-1-935721-10-9

La Brega: Como Sobrevivir En El Barrio

By Supreme Understanding

Thanks to strong demand coming from Spanish-speaking countries, we translated our groundbreaking How to Hustle and Win into Spanish, and added new content specific to Latin America. Because this book's language is easy to follow, it can also be used to brush up on your Spanish.

CLASS	PAGES	RETAIL	RELEASE
0-1	336	$14.95	Jul. 2009

ISBN: 978-0981617-08-4

Locked Up but Not Locked Down: A Guide to Surviving the American Prison System

By Ahmariah Jackson and IAtomic Allah
Foreword by Mumia Abu Jamal

This book covers what it's like on the inside, how to make the most of your time, what to do once you're out, and how to stay out. Features contributions from over 50 insiders, covering city jails, state and federal prisons, women's prisons, juvenile detention, and international prisons.

CLASS	PAGES	RETAIL	RELEASE
J-1	288	$14.95	Jul. 2012

ISBN: 978-1935721-00-0

The Science of Self: Man, God, and the Mathematical Language of Nature

By Supreme Understanding and C'BS Alife Allah

How did the universe begin? Is there a pattern to everything that happens? What's the meaning of life? What does science tell us about the depths of our SELF? Who and what is God? This may be one of the deepest books you can read.

CLASS	PAGES	RETAIL	RELEASE
I-4	360	$19.95	Jun. 2012

ISBN: 978-1935721-67-3

The Science of Self: Man, God, and the Mathematical Language of Nature (Hardcover Edition)

By Supreme Understanding

A beautiful hardcover edition of the bestselling work, *The Science of Self*. Under the full-color dust jacket is an embossed clothbound hard cover. Autographed and numbered as part of a special limited edition series, this book also includes the 16 full-color inserts found in the paperback edition.

CLASS	PAGES	RETAIL	RELEASE
I-4	360	$34.95	Jun. 2012

Only available direct from publisher.

365 Days of Real Black History Calendar (2012 Edition)

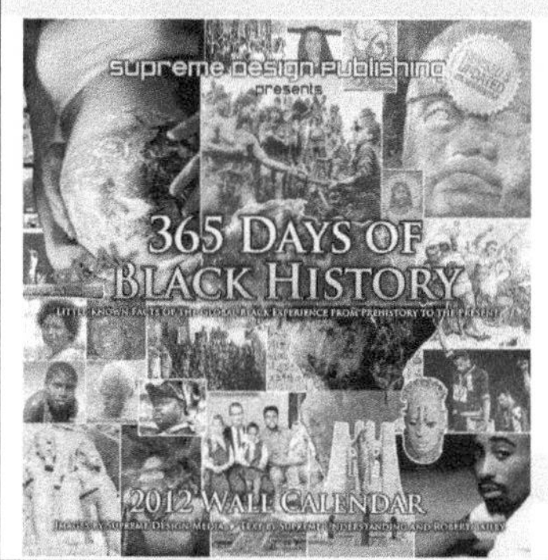

By Supreme Understanding and Robert Bailey

A calendar that'll never be outdated! Over 365 important facts and quotes covering little-known, but important, moments in Black history. Written in brief chunks and easy language for all audiences.

CLASS	PGS	PRICE	RELEASE
I-2	26	$2.95	2011

Only available direct from publisher.

365 Days of Real Black History Calendar (2013 Edition)

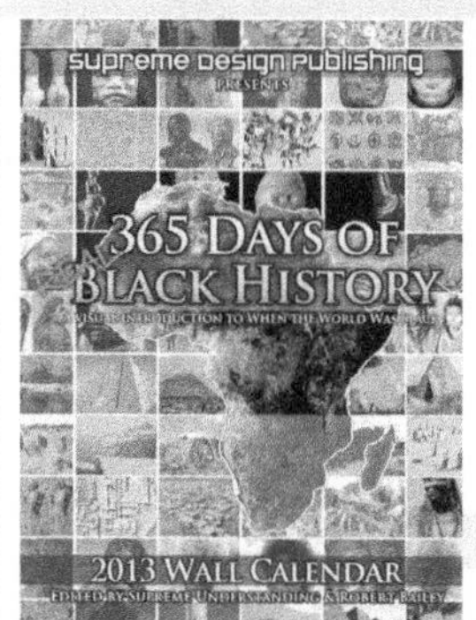

By Supreme Understanding

Our 2013 calendar and planner was also designed to be timeless, as it's a beautifully-designed companion to *When the World was Black*. You'll find dozens of striking full-color images that help tell the stories of global Black history.

CLASS	PAGES	PRICE	RELEASE
I-2	26	$4.95	2012

Only available direct from publisher.

When the World was Black, Part One: Prehistoric Cultures

By Supreme Understanding
Foreword by Runoko Rashid

When does Black history begin? Certainly not with slavery. In two volumes, historian Supreme Understanding explores over 200,000 years of Black history from every corner of the globe. Part One covers the first Black communities to settle the world, establishing its first cultures and traditions. Their stories are remarkable.

CLASS	PAGES	RETAIL	RELEASE
I-3	400	$19.95	Feb. 2013

ISBN: 978-1-935721-04-8

When the World Was Black, Part Two: Ancient Civilizations

By Supreme Understanding

Part Two covers the ancient Black civilizations that gave birth to the modern world. Black people built the first urban civilizations in Africa, Asia, Europe, and the Americas. And every claim in these books is thoroughly documented with reputable sources. Do you want to know the story of your ancestors? You should. We study the past to see what the future will bring.

CLASS	PAGES	RETAIL	RELEASE
I-3	400	$19.95	Feb. 2013

ISBN: 978-1-935721-05-5

When the World was Black, Parts One and Two (Hardcover)

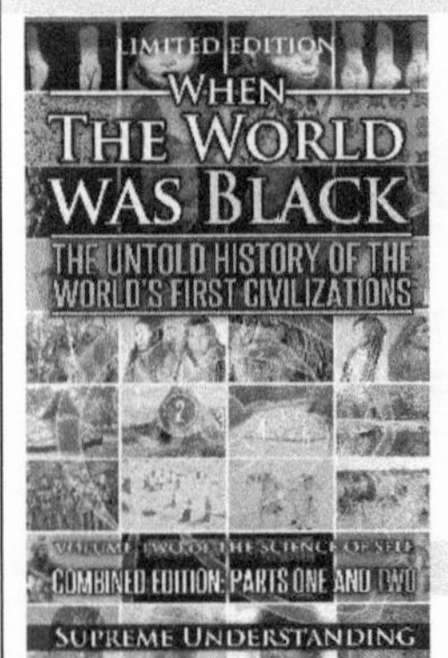

By Supreme Understanding

An incredible limited edition that combines Part One and Part Two into a single book, cased in an embossed clothbound hardcover and dust jacket. Autographed and numbered, this collector's item also includes both sets of full-color inserts.

CLASS	PAGES	RETAIL	RELEASE
I-3	800	$19.95	Dec. 2013

Only available direct from publisher.

Black Rebellion: Eyewitness Accounts of Major Slave Revolts

Edited by Dr. Sujan Dass

Who will tell the stories of those who refused to be slaves? What about those who fought so effectively that they forced their slavers to give up? Black Rebellion is a collection of historical "eyewitness" accounts of dozens of major revolts and uprisings, from the U.S. to the Caribbean, as well as a history of slavery and revolt.

CLASS	PAGES	RETAIL	RELEASE
P-3	272	$14.95	May 2010

ISBN: 978-0-981617-04-6

The Heroic Slave

By Frederick Douglass

Most people don't know that Douglass wrote a novel...or that, in this short novel, he promoted the idea of violent revolt. By this time in his life, the renowned abolitionist was seeing things differently. This important piece of history comes with *David Walker's Appeal*, all in one book.

CLASS	PAGES	RETAIL	RELEASE
P-3	160	$14.95	Apr. 2011

ISBN: 978-1-935721-27-7

David Walker's Appeal

By David Walker

This is one of the most important, and radical, works ever published against slavery. Rather than call for an end by peaceful means, Walker called for outright revolution. His calls may have led to over 100 revolts, including those described in *Black Rebellion*. This important piece of history comes with Douglass' *The Heroic Slave*, which it may have helped inspire.

CLASS	PAGES	RETAIL	RELEASE
P-3	160	$14.95	Apr. 2011

ISBN: 978-1-935721-27-7

Darkwater: Voices from Within the Veil, Annotated Edition

By W.E.B. Du Bois

This book makes Du Bois' previous work, like *Souls of Black Folk*, seem tame by comparison. *Darkwater* is revolutionary, uncompromising, and unconventional in both its content and style, addressing the plight of Black women, the rise of a Black Messiah, a critical analysis of white folks, and the need for outright revolution.

CLASS	PAGES	RETAIL	RELEASE
I-4	240	$14.95	Jun. 2011

ISBN: 978-0-981617-07-7

The African Abroad: The Black Man's Evolution in Western Civilization, Volume One

By William Henry Ferris

Who would think a book written in 1911 could cover so much? Ferris, chairman of the UNIA, speaks up for the Black man's role in Western civilization. He discusses a wealth of history, as well as some revolutionary Black theology, exploring the idea of man as God and God as man.

CLASS	PAGES	RETAIL	RELEASE
I-5	570	$29.95	Oct. 2012

ISBN: 978-1935721-66-6

The African Abroad: Volume Two

By William Henry Ferris

The second volume of Ferris' epic covers important Black biographies of great leaders, ancient and modern. He tells the stories of forty "Black Immortals." He also identifies the African origins of many of the world's civilizations, including ancient Egypt, Akkad, Sumer, India, and Europe.

CLASS	PAGES	RETAIL	RELEASE
I-5	330	$19.95	Oct. 2012

ISBN: 978-1-935721-69-7

From Poverty to Power: The Realization of Prosperity and Peace

By James Allen

Want to transform your life? James Allen, the author of the classic *As a Man Thinketh,* explores how we can turn struggle and adversity into power and prosperity. This inspirational text teaches readers about their innate strength and the immense power of the conscious mind.

CLASS	PAGES	RETAIL	RELEASE
I-3	144	$14.95	May 2010

ISBN: 978-0-981617-05-3

Daily Meditations: A Year of Guidance on the Meaning of Life

By James Allen

Need a guidebook to a productive and healthy year? This is it. James Allen delivers another great work in this book, this time offering 365 days of inspiration and guidance on life's greatest challenges. This book includes sections for daily notes.

CLASS	PAGES	RETAIL	RELEASE
C-3	208	$14.95	Apr. 2013

ISBN: 978-1-935721-08-6

The Kybalion: The Seven Ancient Egyptian Laws _

By the Three Initiates

Thousands of years ago, the ancients figured out a set of principles that govern the universe. In *The Kybalion*, these laws are explored and explained. This edition includes research into the authorship of the book, and where the laws came from.

CLASS	PAGES	RETAIL	RELEASE
C-4	130	$14.95	Oct. 2012

ISBN: 978-1-935721-25-3

Real Life is No Fairy Tale (w/ Companion CD)

By Sujan Dass and Lord Williams

Looking for a children's book that teaches about struggle? Written for school age children, this full-color hardcover book is composed entirely in rhyme, and the images are as real as they get. Includes a CD with an audio book, animated video, review questions, and printable worksheets and activities.

CLASS	PGS	RETAIL	RELEASE
CD-4	36+	$16.95	Jun. 2010

ISBN: 978-0-9816170-2-2

Aesop's Fables: 101 Classic Tales and Timeless Lessons

Edited by Dr. Sujan Dass

What's better to teach our children than life lessons? This easy-to-read collection of classic tales told by an African storyteller uses animals to teach valuable moral lessons. This edition includes dozens of black-and-white images to accompany the timeless fables. Color them in!

CLASS	PAGES	RETAIL	RELEASE
CD-3	112	$14.95	Feb. 2013

ISBN: 978-1-935721-07-9

Heritage Playing Cards (w/ Companion Booklet)

Designed by Sujan Dass

No more European royalty! This beautiful deck of playing cards features 54 full-color characters from around the world and a 16-page educational booklet on international card games and the ethnic backgrounds of the people on the cards.

CLASS	PGS	RETAIL	RELEASE
CD-2	16+	$6.95	May 2010

UPC: 05105-38587

Black God: An Introduction to the World's Religions and their Black Gods

By Supreme Understanding

Have you ever heard that Christ was Black? What about the Buddha? They weren't alone. This book explores the many Black gods of the ancient world, from Africa to Europe, Asia, and Australia, all the way to the Americas. Who were they? Why were they worshipped? And what does this mean for us today?

CLASS	PAGES	RETAIL	RELEASE
C-3	200	$19.95	Jan. 2014

ISBN: 978-1-935721-12-3

Black People Invented Everything

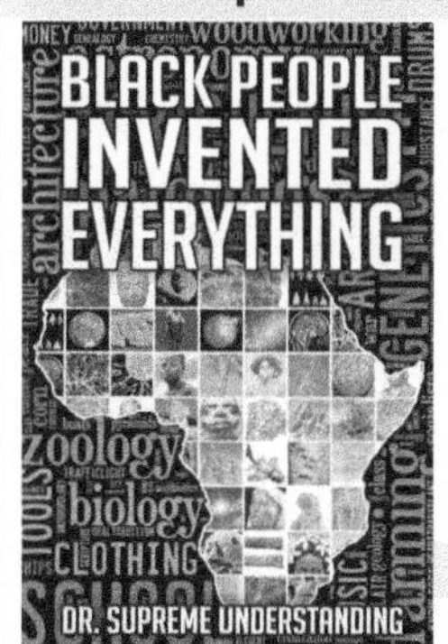

By Supreme Understanding

In *The Science of Self* we began exploring the origins of everything that modern civilization depends on today. In this book, we get into specifics, showing how Black people invented everything from agriculture to zoology, with dozens of pictures and references to prove it!

CLASS	PAGES	RETAIL	RELEASE
I-3	180	$14.95	Feb. 2014

NOT YET PUBLISHED

The Yogi Science of Breath: A Complete Manual of the Ancient Philosophy of the East

By Yogi Ramacharaka

A classic text on the science of breathing, one of the most ignored, yet important, aspects of our physical and emotional health. This book has been used by both martial arts experts and legendary jazz musicians. This edition explores the "secret science" of breath, and where its mysterious author learned such teachings.

CLASS	PAGES	RETAIL	RELEASE
PH-4	112	$14.95	Apr. 2012

ISBN: 978-1-935721-34-5

IT'S A DIRTY GAME...NOW THERE'S A HANDBOOK.
HOW TO HUSTLE AND WIN
A Survival Guide for the Ghetto
FOREWORD by the REAL
SUPREME UNDERSTANDING
YOU CAN'T WIN THIS GAME...YOU HAVE TO CHANGE IT.
HOW TO HUSTLE AND WIN PART 2
Rap, Race and Revolution
FOREWORD by of DEAD PREZ
SUPREME UNDERSTANDING
KNOWLEDGE OF SELF
A COLLECTION OF WISDOM ON THE SCIENCE OF EVERYTHING IN LIFE
DIET & NUTRITION • SICKNESS & DISEASE • HEALING & PREVENTION • HYGIENE
THE HOOD HEALTH HANDBOOK
VOLUME ONE
EDITED BY
C'BS ALIFE ALLAH and SUPREME UNDERSTANDING
EXERCISE & WEIGHT LOSS • MENTAL HEALTH • SURVIVAL & FIRST AID • SEX
HOOD HEALTH
VOLUME TWO
SUPREME UNDERSTANDING and C'BS ALIFE ALLAH
WWW.SUPREMEDESIGNONLINE.COM
The Publishing Company that's "Reinventing the World"!
REAL LIFE IS NO FAIRY TALE
by Sujan Dass
HERITAGE PLAYING CARDS
THE WHOLE WORLD IN THE PALM OF YOUR HAND
54 Full Color Playing Cards and Educational Booklet
FROM POVERTY TO POWER
JAMES ALLEN
INTRODUCTION BY C'BS ALIFE
BLACK REBELLION
EYEWITNESS ACCOUNTS OF MAJOR SLAVE REVOLTS
EDITED BY DR. SUJAN DASS
supreme design publishing presents
365 DAYS OF BLACK HISTORY
2011 WALL CALENDAR

How to Get Our Books

To better serve our readers, we've streamlined the way we handle book orders. Here are some of the ways you can find our books.

In Stores

You can find our books in just about any Black bookstore or independent bookseller. If you don't find our titles on the shelves, just request them by name and publisher. Most bookstores can order our titles directly from us (via our site) or from the distributors listed below. We also provide a listing of retailers who carry our books at www.bestblackbooks.com

Online (Wholesale)

Now, you can visit our sites (like www.supremeunderstanding.com or www.bestblackbooks.com) to order wholesale quantities direct from us, the publisher. From our site, we ship heavily discounted case quantities to distributors, wholesalers, retailers, and local independent resellers (like yourself – just try it!). The discounts are so deep, you can afford to GIVE books away if you're not into making money.

Online (Retail)

If you're interested in single "retail" copies, you can now find them online at Amazon.com, or you can order them via mail order by contacting one of the mail order distributors listed below. You can also find many of our titles as eBooks in the Amazon Kindle, Nook, or Apple iBooks systems. You may also find full-length videobook or audiobook files available, but nothing beats the pass-around potential of a real book!

By Mail Order

Please contact any of the following Black-owned distributors to order our books! For others, visit our site.

Afrikan World Books
2217 Pennsylvania Ave.
Baltimore, MD 21217
(410) 383-2006

Lushena Books
607 Country Club Dr
Bensenville, IL 60106
(800) 785-1545

Special Needs X-Press
3128 Villa Ave
Bronx, NY 10468
(718) 220-3786

ENDNOTES

1 John Edmonds. (2000). Tyrian or Imperial Purple: The Mystery of Imperial Purple Dyes, Historic Dye Series, no. 7. Little Chalfont, Buckinghamshire, England.
2 Arna Sontemps. (1954). The Story of George Washington Carver. Grosset & Dunlap.
3 Shirley Graham and George D. Lipscomb. Dr. George Washington Carver, Scientist. Washington Square Press.
4 W. Mills. (Apr 1999). Tone Magazine. Ottawa, Ontario.
5 Mundaka Upanishad. (1957). Prabhavanada & Manchester, pp. 45-6.
6 Barry Hallen and J.O. Sodipo. (1997). Knowledge, Belief, and Witchcraft:
Analytic Experiments in African Philosophy. Stanford University Press.
7 The Minoans – possibly the same brown-skinned "Mediterranean" population who later became the Phoenicians – may have created this dye much earlier, about 4,000 years ago. Reese, D. S. (1987). Palaikastro Shells and Bronze Age Purple-Dye Production in the Mediterranean Basin, Annual of the British School of Archaeology, 82, 201-6; Stieglitz, R. (1994), "The Minoan Origin of Tyrian Purple," Biblical Archaeologis,t 57, 46-54.
8 Barry Hallen and J.O. Sodipo. (1997). Knowledge, Belief, and Witchcraft:
Analytic Experiments in African Philosophy. Stanford University Press.
9 "Tenskwatawa - Shawnee." http://www.firstpeople.us/FP-Html-Wisdom/Tenskwatawa.html
10 Ibid.
11 Frank Waters. (1993). Brave Are My People. Santa Fe: Clear Light Publishers. p. 64.
12 Eugenia Herbert. (1975). "Smallpox Inoculation in Africa." Journal of African History 16.
13 Grizzard F.E. et al. (1985). George Washington to Major General Horatio Gates, 5–6 February 1777. In The papers of George Washington. University Press of Virginia; Winslow, Ola. (1974). A Destroying Angel: The Conquest of Smallpox in Colonial Boston.
14 Henri Trilles. (1945). L'Ame du Pygmee d'Afrique (The Soul of the African Pygmy). Les Editions du Cerf. p. 95-96.
15 Jeffrey B. Perry, ed. (2001). A Hubert Harrison Reader. Wesleyan University Press. Harrison's ideology can be seen in works like Black Man's Burden (1915), The Negro and the Nation (1917), and When Africa's Awakes (1920), but Perry's A Hubert Harrison Reader is the most comprehensive source on this uncompromising Black man, who I consider a predescessor Harlem's Five Percenters.
16 Keith Devlin. (1996). Mathematics: The Science of Patterns: The Search for Order in Life, Mind and the Universe. Holt Paperbacks. p. 3.
17 Ubiratan d'Ambrosio. (1985). "Ethnomathematics and its Place in the History and Pedagogy of Mathematics," For the Learning of Mathematics 5, No.1. p 44-48
18 Jim Barta and Tod Shockey. (2006). "The Mathematical Ways of an Aboriginal People: The Northern Ute," The Journal of Mathematics and Culture 1, no.1.
19 Ecker, Michael. (Decemeber 1992). "Caution: Black Holes at Work," New Scientist, p. 38.
20 Sean Henahan. "Art Prehistory," http://www.accessexcellence.org/WN/SU/caveart.php
21 Tapper, Richard. (March 1989). "Changing Messages in the Genes," New Scientist, p. 53.
22 Davies, Paul. (March 1992). "Is Nature Mathematical?" New Scientist, p. 25.
23 Aboriginal kids count without numbers. (August 2008). Cosmos Online. http://www.cosmosmagazine.com/news/2142/aboriginal-kids-count-without-numbers
24 Izard Veronique; Pica Pierre; Spelke Elizabeth S.; et al. Flexible intuitions of Euclidean geometry in an Amazonian indigene group. PANS. pnas.1016686108
25 Lea, Hilda. (1990). Informal Mathematics in Botswana: Mathematics in the Central Kalahari, Faculty of Education, University of Botswana, p. 1-7.
26 Baiter, Michael. (2001) "What Makes the Mind Dance and Count?" Science, 292: 1635.
27 Nicholas E. Humphries et al. (June 2010). Environmental context explains Lévy and Brownian movement patterns of marine predators. Nature, 465, 1066–1069. http://www.nature.com/nature/journal/v465/n7301/full/nature09116.html
28 Esther Inglis-Arkell. (Mar 23, 2012). Why do cicadas know prime numbers? io9.com

http://io9.com/5895840/why-do-cicadas-know-prime-numbers?tag=math
29 J. D. Murray (1989). Mathematical Biology. Springer-Verlag.
30 I. Prigogine and R. Lefevre. (1968). Journal of Chemical Physics (48) p.1695.
31 A.M. Turing. (1992). A Diffusion Reaction Theory of Morphogenesis in Plants (with C. W. Wardlaw). In the Collected Works of A.M. Turing, Elsevier, Amsterdam,
32 Horgan, John. (May 1996) "Plotting the Next Move." Scientific American, 274:16.
33 Roger Penrose & Stuart Hameroff (2011). "Consciousness in the Universe: Neuroscience, Quantum Space-Time Geometry and Orch OR Theory". Journal of Cosmology 14. Stuart Hameroff later added to Penrose's theory, contributing possible mechanisms by which quantum processing could occur in the brain, and suggesting that such activities could explain gamma wave phenomena in the brain.
34 Alvarez, L., et al. (2012). The rate of change in Ca2 concentration controls sperm chemotaxis. Journal of Cellular Biology doi: 10.1083/jcb.201106096
35 Denise Martin, Ph.D. (March 2008). Pan African Metaphysical Epistemology: A Pentagonal Introduction. The Journal of Pan African Studies, vol.2, no.3. p. 209-227.
36 John Fleming and Hugh Honour. (1977). Diction of the Decorative Arts.
37 Mohr, Peter J, et al. (2008). "CODATA Recommended Values of the Fundamental Physical Constants: 2006." Rev. Mod. Phys. 80 (2): 633-730. doi:10.1103/RevModPhys.80.633.
38 Barrow, John D. (2002) The Constants of Nature; From Alpha to Omega - The Numbers that Encode the Deepest Secrets of the Universe, Pantheon Books.
39 Finch, Steven (2003). Mathematical constants. Cambridge University Press. p. 453.
40 Trudi H. Garland. (1987). Fascinating Fibonaccis: Mystery of Magic in Numbers. Dale Seymour Publications,
41 Woorama. (Jun 2006). Linear Vs. Circular Logic: Conflict Between Indigenous And Non-Indigenous Logic Systems. Aboriginal Rights @ Suite101. http://woorama.suite101.com/linear-vs-circular-logic-a2993
42 P. English and K. M. Kalumba. (eds.) (1996). African Philosophy. Prentice Hall.
43 Wheeler, John Archibald. (1981). "Bohr, Einstein, and the Strange Lesson of the Quantum," in Mind in Nature. Edited by Richard Q. Elvee, p.18.
44 Greenstein, George. (1988). The Symbiotic Universe: Life and Mind in the Cosmos. William Morrow, p 223.
45 Roger Penrose. (2010). Cycles of Time: An Extraordinary New View of the Universe. Bodley Head.
46 Adam McDowell. (Dec 2010). What does the Penrose Big Bang theory mean for religions? National Post. http://news.nationalpost.com/2010/12/05/what-does-the-penrose-big-bang-theory-mean-for-religions/
47 Mandelbrot, B. B. (1982). The Fractal Geometry of Nature. W. H. Freeman and Company.
48 Glanz, James. (1996). "Precocious Structures Found," Science, 272:1590.
49 Maddox, John. (1987). "The Universe as a Fractal Structure," Nature, 329:195.
50 Ron Eglash. (1999). African Fractals: Modern Computing and Indigenous Design. Rutgers University Press.
51 Maddox, John. (1985). "No Pattern Yet for Snowflakes," Nature, 313:93.
52 Hayes, Brian. (1995). "SpaceTime on a Seashell," American Scientist, 83:214.
53 Dietrich, R.V., and Chamberlain, Steven C. (September/October 1989). "Are Cultured Pearls Mineral?" Rocks and Minerals, 64:386. Cr. R. Calais. A vestibular system is the sensory system that contributes to balance in most mammals and to the sense of spatial orientation. It also contributes to our physical symmetry.
54 Nisbet, E.G., et al. (1994). "Can Diamonds Be Dead Bacteria?" Nature, 367:694.
55 James R. Hein, Hsueh-Wen Yeh and John A. Barron. (March 1990). Eocene diatom chert from Adak Island, Alaska. Journal of Sedimentary Research. v. 60 no. 2 p. 250-257.
56 Jürgen Schieber, Dave Krinsley & Lee Riciputi. (August 2000). Diagenetic origin of quartz silt in mudstones and implications for silica cycling. Nature 406, 981-985.
57 Bruce H. Lipton, Steve Bhaerman. (2009). Spontaneous Evolution: Our Positive Future (and a Way to Get There from Here). New York: Hay House Publishing.
58 Clifford A. Pickover. (2009). The Loom of God: Tapestries of Mathematics and Mysticism.

Sterling Publishing Company, Inc.
59 C. Sagan. (1979) Broca's Brain: Reflections on the Romance of Science. Ballantine Books.
60 Brooks, A.S. and Smith, C.C. (1987): "Ishango revisited: new age determinations and cultural interpretations," The African Archaeological Review, 5: 65-78.
61 de Heinzelin, Jean. (June 1962) "Ishango," Scientific American, 206:6, p. 105-116.
62 Marshack, Alexander. (1991). The Roots of Civilization, Colonial Hill, Mount Kisco, NY.
63 Zaslavsky, Claudia. (1992) "Women as the First Mathematicians," International Study Group on Ethnomathematics Newsletter, Volume 7, Number 1.
64 Peter D. Beaumont. (1973) Border Cave - A Progress Report, S. Afr. J. Science 69.
65 Graham Flegg, Numbers: their history and meaning, Courier Dover Publications, 2002
66 AW Howitt, (1998) "Notes on Australian Message Sticks and Messengers," Journal of the Anthropological Institute, pp 317-8, London, 1889, reprinted by Ngarak Press.
67 Cyrus L. Day. (Jan 1957). "Knots and Knot Lore: Quipus and Other Mnemonic Knots" Western Folklore, Vol. 16, No. 1, p. 8-26
68 Mann, Charles (2005). "Unraveling Khipu's Secrets." Science 309: 1008-1009.
69 Lu, Wei; Aiken, Max. (Nov 2004). "Origins and evolution of Chinese writing systems and preliminary counting relationships" Accounting History.
70 Donald Smeltzer. (1958). Man and Number. Emerson Books.
71 Radin, Dean I., and Ferrari, Diane G. (1991). "Effects of Consciousness on the Fall of Dice: A Meta-Analysis," Journal of Scientific Exploration, 5:61, no. 1.
72 Marissa A. Ramsier, et al. (Feb 2012). Primate communication in the pure ultrasound. Biology Letters. doi: 10.1098/rsbl.2011.1149
73 Sandved, K. B. (1996). The Butterfly Alphabet. Scholastic. www.butterflyalphabet.com
74 M.J. Denton, J.C. Marshall & M. Legge, (2002) "The Protein Folds as Platonic Forms: New Support for the pre-Darwinian Conception of Evolution by Natural Law," Journal of Theoretical Biology 219 (2002): 325-342.
75 Paul Davies. (1999). The Fifth Miracle, The Search for the Origin of Life. Simon & Schuster, p. 259
76 Bill Bryson. (2003). A Short History of Nearly Everything. Broadway Books. p. 167.
77 Supreme Understanding Allah. (2009). "The Greatness in Man." In Knowledge of Self: A Collection of Wisdom on the Science of Everything in Life. Supreme Design Publishing.
78 L.E. Threlkeld. (1892). An Australian Language as Spoken by the Awabakal: Their Language, Traditions, and Customs. Charles Potter.
79 Soho, Takuan. (1986). The Unfettered Mind. Kodansha International Ltd.
80 Gia-Fu Feng and Jane English (trans.). (1972). Laozi/Dao De Jing. Vintage Books; Komjathy, Louis. (2008) Handbooks for Daoist Practice. 10 vols. Yuen Yuen Institute.
81 David Bohm. (1980). Wholeness and the Implicate Order. Routledge.
82 Michael Talbot. (1991). The Holographic Universe. HarperCollins.
83 Sheldrake, Rupert. (1981). A New Science of Life: the hypothesis of formative causation. J.P. Tarcher; Sheldrake, Rupert. (1988). The Presence of the Past: morphic resonance and the habits of nature. Times Books.
84 Jung, Carl (1972). Synchronicity - An Acausal Connecting Principle. Routledge.
85 Wilson, Robert Anton. (1992) Prometheus Rising, New Falcon Publications.
86 McFadden, Johnjoe. (2000) Quantum Evolution. HarperCollins.
87 Dalai Lama. (2005). The Universe in a Single Atom: The Convergence of Science and Spirituality. Morgan Road Books. p 130.
88 J. Horgan, (1999). The Undiscovered Mind: How the Human Brain Defies Replication, Medication, and Explanation. Touchstone Books, 228.
89 Valerie Ross. (Oct 2010) Unknowns, From Matter to DNA to Brain Cells. Discover. http://discovermagazine.com/2010/oct/15-numbers-unknowns-matter-dna-brain-cells
90 Supreme Understanding Allah. (2009). "The Greatness in Man." In Knowledge of Self: A Collection of Wisdom on the Science of Everything in Life. Supreme Design Publishing.
91 Wilson, Robert Anton. (1992) Prometheus Rising, New Falcon Publications.
92 C. G. Jung. (1996) The Archetypes and the Collective Unconscious.
93 Leary, Timothy. (1989) Info-Psychology, New Falcon Publications.

94 Leary, Timothy. (1977). Exo-Psychology; Leary, Timothy. (1998). The Politics of Ecstasy, Ronin Publishing (Original Ed. 1970); Wilson, Robert Anton. (1990) Quantum Psychology.
95 Edward Bruce Bynum, (March 1999). The African Unconscious: Roots of Ancient Mysticism and Modern Psychology. Teachers College Press.
96 Caplan, Mariana (2009). Eyes Wide Open: Cultivating Discernment on the Spiritual Path. Sounds True Publishing.
97 Sala S. Shabazz. (2002). The best of the little known Black history facts. Lushena Books.
98 Rein, G., and McCraty, R. (1994). "Structural Changes in Water and DNA Associated with New Physiologically Measurable States," Journal of Scientific Exploration, 8:438.
99 Ramana Maharshi, (1977), The Sage of Arunacala. Mahadevan, Allen & Unwin, 90-91.
100 Chandogya Upanishad, In S. Prabhavananda & F. Manchester (eds.), (1957), The Upanishads: Breath of the Eternal. New American Library, 45-46.
101 Weisfeld, Matt (2009). The Object-Oriented Thought Process, Third Edition. Addison-Wesley; Abadi, Martin, Luca Cardelli (1998). A Theory of Objects. Springer Verlag; Armstrong, Deborah J. (February 2006). "The Quarks of Object-Oriented Development." Communications of the ACM 49 (2): 123–128. doi:10.1145/1113034.1113040
102 This translation comes from Swami Nirmalanda Giri's A Commentary on the Upanishads. Another translation reads: "In the beginning this world was only Brahman, and it knew only itself (atman), thinking: 'I am Brahman.' As a result, it became the Whole. Among the gods, likewise, whosoever realized this, only they became the Whole. It was the same also among the seers and among humans. Upon seeing this very point, the seer Vamadeva proclaimed: 'I was Manu [the first/Original man], and I was the Sun.' This is true even now. If a man knows 'I am Brahman' in this way, he becomes this whole world. Not even the gods are able to prevent it, for he becomes their very self (atman). So when a man venerates another deity, thinking, 'He is one, and I am another', he does not understand. As livestock is for men, so is he for the gods." Patrick Olivelle (trans.) (1996). Upanisads. Oxford University Press. p. 15
103 Placide Tempels. (1959). Bantu Philosophy. Presence Africaine.
104 Denise Martin, Ph.D. (March 2008). Pan African Metaphysical Epistemology: A Pentagonal Introduction. The Journal of Pan African Studies, vol.2, no.3. p. 209-227.
105 P. English and K. M. Kalumba. (eds.) (1996). African Philosophy. Prentice Hall.
106 Age continued: "There's a lot of metaphysical stuff relating to African spirituality. The main problem you're going to have is reliable sources, particularly online. Plus only some Africans know about precolonial culture in good detail (my uncle being one of them – he's a chief still involved in traditionalist affairs)...Although most Ndi Igbo (Igbo people) are Christian... because of the similarities with Judaic religion (down to circumcision and many other rituals), many ideas/beliefs have been preserved (yet not documented, and often corrupted) among Igbo families, who may believe that their traditions are somehow related to the Bible when the root is actually a lot deeper."
107 Placide Tempels. (1959). Bantu Philosophy. Presence Africaine.
108 Christopher Holmes. (2010). The Heart Doctrine: Mystical Views of the Origin and Nature of Human Consciousness. Zero Point Publications.
109 Paul Pearsall (1998) The Heart's Code, p.55
110 Science Digest, August 1984, p.90.
111 Mbiti, John S. (1990) African Religions and Philosophy, 2d ed. Heinemann; Mbiti, John S. (1991) Introduction to African Religion, Heinemann.
112 Fu-Kiau, Kimbwandende Kia Bunseki. (2001) African Cosmology of the Bantu-Kongo Tying the Spiritual Knot Principles of Life and Living. Athelia Henrietta Press.
113 Thompson, Robert Farris. (1984) Flash of the Spirit: African and Afro-American Art and Philosophy. Vintage.
114 Denise Martin, Ph.D. (March 2008). Pan African Metaphysical Epistemology: A Pentagonal Introduction. The Journal of Pan African Studies, vol.2, no.3. p. 209-227.
115 Ivan Van Sertima. (1994). Egypt: Child of Africa. Transaction Publishers. p. 370-371.
116 Jean-Pierre Hallet and Alex Pelle. (1973). Pygmy Kitabu. Random House. p. 342
117 Mádhava Áchárya (1996) [first edition 1882]. The Sarva-darsana-samgraha: or Review of

the Different Systems of Hindu Philosophy. Motilal Banarsidass.
118 Isaac Asimov, "The Subtlest Difference" In Abell, G. & Singer, B., eds., (1981). Science and the Paranormal. Scribners Sons, 158.
119 Opoku, K.A. (1978) West African Traditional Religion, FEP Int. Pvt. Ltd., Accra.
120 Paul Davies. (1999). The Fifth Miracle. Simon & Schuster, p. 259
121 Karl R. Popper; J. C. Eccles, (1977). The Self and Its Brain. Springer-Verlag Intl, p. 120
122 T. Dobzhansky, (1967). The Biology of Ultimate Concern. New American Library.
123 John C. Eccles, (1989). Evolution of the Brain: Creation of the Self. Routledge. p. 71
124 C.D.L. Wynne, "The Soul of the Ape", American Scientist, 89 (2001), 120-122;
125 Masataka Fukugita and P. J. E. Peebles, (December 2004) "The Cosmic Energy Inventory," Astrophysical Journal, 616, 643-68.
126 Mary Baker Eddy. (1957) Science and health: with key to the Scriptures. p. 243.
127 Mermin, N.D. (1981) "Bringing Home the Atomic World: Quantum Mysteries for Anybody," American Journal of Physics, 49:940.
128 John Matson. (May 2010). Fermilab Finds New Mechanism for Matter's Dominance over Antimatter. Scientific American. scientificamerican.com/article.cfm?id=muons-mesons
129 Joel R. Primack; Nancy Ellen Abrams. (2006). The View from the Center of the Universe: Discovering Our Extraordinary Place in the Cosmos. Riverhead Hardcover. p. 188
130 J. Robinson (Ed.), (1981). Nag Hammadi Library. Harper & Row, pp. 121-3, 127
131 Allama Abu-al-Mazzafar Syed Amir Hassan Shah. (Year Unknown) The Book of Ilm-ul-Jáffár. Original published in Urdu by Maulana Mullazim Hussain Asghar in Pakistan. Translated and made available online by Syed Nadeem Bokhari at www.momineen.com
132 J. Primack, N. Abrams. (2006). The View from the Center of the Universe. Riverhead.
133 Ibid. p. 175
134 Ibid.
135 Frank J. Tipler; John D. Barrow (1986). The Anthropic Cosmological Principle. Oxford University Press.
136 Chandler, Wayne, (1995) "The Principle of Polarity," In Runoko Rashidi and Ivan Van Sertima (eds.). African Presence in Early Asia, Journal of African Civilizations; Carus, Paul, (1907). Chinese Thought, Open Court Press.
137 Corbett, Greville G. (1991) Gender, Cambridge University Press. (A comprehensive study; looks at 200 languages.)
138 Diego Salazar et al, (2011). "Early Evidence (ca. 12,000 BP) for Iron Oxide Mining on the Pacific Coast of South America." Current Anthropology 52, no. 3. p 468-475
139 C. D. Ollier, D. P. Drover and M. Godelier, (1971). "Soil Knowledge amongst the Baruya of Wonenara, New Guinea." Oceania 42, no. 1. p 33-41
140 Moskowitz, Clara (March 2012). "Life's Building Blocks May Have Formed in Dust Around Young Sun." Space.com. http://www.space.com/15089-life-building-blocks-young-sun-dust.html
141 Kokubo, Eiichiro; Ida, Shigeru (2002). "Formation of protoplanet systems and diversity of planetary systems." The Astrophysical Journal 581 (1): 666–680. doi:10.1086/344105.
142 John Emsley. (2011). Nature's Building Blocks: Everything You Need to Know About the Elements. Oxford University Press.
143 Crok, M. (Aug. 2008) Birth of the Moon: a runaway nuclear reaction? Cosmos. Issue 22.
144 Cutler, Winnifred B., et al. (1987). "Lunar Influences on the Reproductive Cycles in Women." Human Biology, 59:959.
145 McClintock, Martha K. (1971) Menstrual Synchrony and Suppression. Nature, 229:244.
146 Kiltie, Richard A. (1982). On the Significance of Menstrual Synchrony in Closely Associated Women. American Naturalist, 119:414.
147 Jonathan Wells, "Do Centrioles Generate a Polar Ejection Force?," Rivista di Biologia/Biology Forum 98 (2005): 37-62.
148 Song, Xiadong, and Richards, Paul G.; (1996) "Seismological Evidence for Differential Rotation of the Earth's Inner Core," Nature, 382:221.
149 Staff Writer. What Drives Earth's Magnetic Field? NOVA. http://www.pbs.org/wgbh/nova/magnetic/reve-drives.html

150 Ibid.
151 When Compasses Point South. NOVA. www.pbs.org/wgbh/nova/magnetic/timeline.html
152 Muller, Richard A., and Morris, Donald E. (1986) "Geomagnetic Reversals from Impacts on the Earth," Geophysical Research Letters, 13:1177.
153 Proco, C.C., and Danielson, G.E. (1982). "The Periodic Variation of Spokes in Saturn's Rings," Eos, 63:156.
154 "Solar Activity and Terrestrial Thunderstorms," (1979). New Scientist, 81:256.
155 Singh, Surendra; (1978) "Geomagnetic Activity and Microearthquakes," Seismological Society of America, Bulletin, 68:1533.
156 Nikola Tesla. (1905). The Transmission of Electrical Energy Without Wires As A Means Of Furthering World Peace. Electrical World and Engineer January 7: 21–24.
157 Klimley, A. Peter. (October 1995). "Hammerhead City," Natural History, 104:33.
158 Braud, William G., and Dennis, Stephen P.; (1989) "Geophysical Variables and Behavior; LVIII. Autonomic Activity, Hemolysis, and Biological Psychokinesis: Possible Relationships with Geomagnetic Field Activity," Perceptual and Motor Skill, 68:1243.
159 Walcott, Charles, et al; (1979). "Pigeons Have Magnets," Science, 205:1027.
160 Liane Young, et al. (March 2010). Disruption of the right temporoparietal junction with transcranial magnetic stimulation reduces the role of beliefs in moral judgments. PNAS. http://www.pnas.org/content/early/2010/03/11/0914826107
161 Persinger, M.A. (1995) on the possibility of directly accessing every human brain by electromagnetic induction of fundamental algorithms. Perceptual and Motor Skills, 80:791.
162 Staff Writer. (2007) Encounter South Africa Magazine, http://www.encounter.co.za/cgi-bin/magazine/ds.cgi?id=7
163 Michaud, Louise Y. and Persinger, Michael A. (1985). Geophysical Variables and Behavior: XXV. Alterations in Memory for a Narrative Following Application of Theta Frequency Electromagnetic Fields. Perceptual and Motor Skills. 60:416.
164 Walmsley, D.J.; Epps, W.R. (1987) "Direction-Finding in Humans: Ability of Individuals to Orient towards Their Place of Residence," Perceptual and Motor Skills, 64:744.
165 Simon, C. (1983). "Tiniest Fossils May Record Magnetic Field," Science News, 124:308.
166 Semm, P. et al. (1980). Effects of an Earth-Strength Magnetic Field on Electrical Activity of Pineal Cells. Nature 288:607.
167 Harlow Shapley. (1926). Starlight. George H. Doran Company., p. 140, 142.
168 Carl Zimmer. (Nov 2007) "The Decline and Fall of the Animal Kingdom," Wired. www.wired.com/science/planetearth/commentary/dissection/2007/11/dissection_1127B
169 Katherin Harmon. (May 2010). The Proof Is in the Proteins: Test Supports Universal Common Ancestor for All Life. Scientific American. http://www.scientificamerican.com/article.cfm?id=universal-common-ancestor
170 Ann Gibbons. (April 2007). European Skin Turned Pale Only Recently, Gene Suggests. Science 20, Vol. 316 no. 5823 p. 364 . DOI: 10.1126/science.316.5823.364a
171 Petr Kropotkin. (2005). Mutual Aid: A Factor of Evolution; Thomas H. Huxley (ed.) The Struggle for Existence. Porter Sargent Publishers
172 Lovelock, James (2000). Gaia: A New Look at Life on Earth. Oxford University Press.
173 Rupert Sheldrake. (1988). The Presence of the Past. Times Books.
174 Bateson, G. (1979). Mind and Nature: A Necessary Unity (Advances in Systems Theory, Complexity, and the Human Sciences). Hampton Press.
175 H. Bergson (1998). Creative Evolution (L'Evolution créatrice) Dover Publications.
176 John A. Davison, "A Prescribed Evolutionary Hypothesis," Rivista di Biologia/Biology Forum 98 (2005): 155-166.
177 Pierre Teilhard de Chardin. The Phenomenon of Man (1959), Harper Perennial.
178 Pierre Teilhard de Chardin. 1920. The Future of Mankind. Unpublished. Available online. http://www.users.globalnet.co.uk/~alfar2/cosmos/TeilhardFutureOfMan.pdf
179 Pierre Teilhard de Chardin. The Phenomenon of Man (1959), Harper Perennial. 250-75
180 Barrow, J.; Tipler, F. (1986). The Anthropic Cosmological Principle. Oxford. p. 677-682.
181 "Oldest Life-Forms May Have Made Own Food," (1980). Eos, 61:578.
182 Bjorn Carey, (May 2005). "Early Earth Not So Hellish, New Study Suggests," Live

Science. http://www.livescience.com/241-early-earth-hellish-study-suggests.html
183 Hoyle, Fred, and Wickramasinghe, Chandra. (1981). Where Microbes Boldly Went, New Scientist, 91:412.
184 Moore, T. Owens. (2002). The Science and the Myth of Melanin. A&B Publishers Group.
185 Bruno J. R. Nicolaus, Rodolfo A. Nicolaus, Marco Olivieri, (2006). "Speculation on the Chemistry of Interstellar Black Matter," For this paper, and several more from the authors of the study, see http://www.tightrope.it/nicolaus/index.htm
186 Pascale Ehrenfreund, Jan Cami, "Cosmic Carbon Chemistry: From the Interstellar Medium to the Earth," Cold Spring Harbor Perspectives in Biology 2, no. 12 (2010)
187 F.E Barr, J.S Saloma, M.J Buchele (2004, April). Melanin: The organizing molecule. Medical Hypotheses, Volume 11, Issue 1, May 1983, Pages 1–139. http://dx.doi.org/10.1016/0306-9877(83)90122-6.
188 Martins, Zita et al. (20 March 2008). "Extraterrestrial nucleobases in the Murchison meteorite." Earth and Planetary Science Letters.
189 P. Schmitt-Kopplin et al. (2010). "High Molecular Diversity of Extraterrestrial Organic Matter in Murchison Meteorite Revealed 40 years After its Fall," PNAS 107(7): 2763-2768
190 Doreen Walton, (February 16, 2010) "Space Rock Contains Organic Molecular Feast," BBC News, http://news.bbc.co.uk/2/hi/science/nature/8516319.stm
191 Haynes, Raymond et al. (1996). Explorers of the Southern Sky: A History of Australian Astronomy, Cambridge University Press.
192 Michael D. Lemonick, (June 2010). "A New Theory on How Comets Are Born," Time Magazine, http://www.time.com/time/health/article/0,8599,1995914,00.html
193 Francis Crick (1982). Life Itself: Its Origin and Nature, Futura.
194 Paul Davies. (1999). The Fifth Miracle. Simon & Schuster, p. 247-249
195 Michael Schirber, (November 2004). "The Volcanic Origin of Life," Live Science, http://www.livescience.com/24-volcanic-origin-life.html
196 Ura Y, Beierle JM, Leman LJ, Orgel LE, Ghadiri MR. (June 2009). Self-assembling sequence-adaptive peptide nucleic acids. Science. 2009 Jul 3;325 (5936):73-77.
197 Abby Vogel Robinson, (June 14, 2010) "Study: Adding UV Light Helps Form "Missing G" of RNA Building Blocks," www.gatech.edu/newsroom/release.html?nid=58047
198 Stuart A. Kauffman, (August 1996) "Self-Replication: Even peptides do it," Nature 382.
199 News Staff. (April 2008). The Mathematical Probability Of Life On Other Earth-Like Planets. Science 2.0. www.science20.com/news_releases/the_mathematical_probability_of_life_on_other_earth_like_planets
200 Stuart A. Kauffman, (August 1996) "Self-Replication: Even peptides do it," Nature 382.
201 Ker Than, (June 2006) "Odd Rock Structures Could Be Earliest Signs of Life," Live Science, http://www.livescience.com/10524-odd-rock-structures-earliest-signs-life.html
202 Myles McLeod. (Mar. 2006). And life created continents. New Scientist. 189(2544) p. 12.
203 Robert Roy Britt, (August 1, 2005). "How Lowly Bacteria Froze Earth Solid, Live Science, http://www.livescience.com/351-lowly-bacteria-froze-earth-solid.html
204 Mak A. Saito. (2009) "Biogeochemistry: Less Nickel for More Oxygen," Nature 458.
205 Corey Binns, (June 2006) "Early Life Survived 'Snowball Earth.'" Live Science, http://www.livescience.com/10521-early-life-survived-snowball-earth.html
206 James A. Lake. (August 2009) Evidence for an early prokaryotic endosymbiosis." Nature, Vol. 460, No. 7258, http://www.wired.com/wiredscience/2009/08/protoprokaryote/
207 Carl Woese, J. Peter Gogarten, (October 21, 1999) "When Did Eukaryote Cells (cells with nuclei and other internal organelle) First Evolve? What Do We Know About How they Evolved From Earlier Life Forms?" Scientific American. http://www.scientificamerican.com/article.cfm?id=when-did-eukaryotic-cells
208 Bernd Schierwater et al. (Jan. 2009). "Concatenated Molecular and Morphological Analysis Sheds Light on Early Metazoan Evolution and Fuels a Modern 'Urmetazoan' Hypothesis." Public Library of Science Biology.
209 Sagan, D.; Margulis, L. (Mar 1987). "Bacterial Bedfellows," Natural History, 96:26,
210 Abderrazak El Albani et al. (2010). "Large colonial organisms with coordinated growth in oxygenated environments 2.1 Gyr ago," Nature 466, no. 7302: 100-104.

211 Margulis, Lynn, and Sagan, Dorian. (1986). The Origins of Sex. Yale University Press.
212 Michiel Vos. (June 2009). Why do bacteria engage in homologous recombination? Trends in Microbiology. Volume 17, Issue 6, p. 226–232.
213 F. Neuweiler, E.C. Turner, D.J. Burdige, (2009) "Early Neoproterozoic Origin of the Metazoan Clade Recorded in Carbonate Rock Texture," Geology 37, no. 5: 475-478.
214 News Staff, (October 30, 2010) "Snowball Earth to Thank for Evolution of Life on Earth, Says Study," Science 2.0, www.science20.com/news_articles/snowball_earth_thank_evolution_life_earth_says_study
215 P.J. Brenchley, D.A.T. Harper, (1998) Palaeoecology: Ecosystems, Environments and Evolution Psychology Press, p. 359.
216 Alexis Madrigal, (March 5, 2010) "First Microbes Colonized Land By Using Fat for Protection," Wired, http://www.wired.com/wiredscience/2010/03/microbes-colonizing-land
217 "The Worm in Your Brain." (September 2010). Discover Magazine. http://blogs.discovermagazine.com/loom/2010/09/03/the-worm-in-your-brain/
218 News Staff. (Aug 2008). Trichoplax Adhaerens Genome Shows Simplest Animal Has Complex Metabolic Capabilities. Science 2.0. www.science20.com/news_releases/tricho plax_adhaerens_genome_shows_simplest_animal_has_complex_metabolic_capabilities
219 News Staff. (Dec 2010). Hypoxia Explained - And Why It Mattered In The Evolution Of Animal Life. Science 2.0. http://www.science20.com/news_articles/hypoxia_explained_and_why_it_mattered_evolution_animal_life
220 Valerie Ross. (Oct 2010) Unknowns, From Matter to DNA to Brain Cells. Discover. http://discovermagazine.com/2010/oct/15-numbers-unknowns-matter-dna-brain-cells
221 Abi-Rached, L. et al. (March 2011). The Shaping of Modern Human Immune Systems by Multiregional Admixture with Archaic Humans. Science. doi:10.1126/science.1209202.
222 Igor Schneidera, Ivy Aneasb, Andrew R. Gehrkea, Randall D. Dahnc, Marcelo A. Nobregab, and Neil H. Shubina. Appendage expression driven by the Hoxd Global Control Region is an ancient gnathostome feature. PNAS August 2, 2011 vol. 108 no. 31
223 Margulis, Lynn. Sagan, Dorion. (2011). Dazzle Gradually: Reflections on the Nature of Nature. Chelsea Green Publishing
224 Skykes, Brian. The Seven Daughters of Eve. W. W. Norton & Company (May 17, 2002).
225 The Y chromosome is one of the fastest evolving parts of the human genome. It also is susceptible to the most mutations of the human genome. Wade, Nicholas (January 13, 2010). "Male Chromosome May Evolve Fastest". New York Times.
226 Bower, B. (1986). "Wyoming Fossils Shake Up Views of Early Primate Migration," Science News, 129: 71.
227 Smriti Rao, (March 2010). "Is the Mysterious Siberian "X-Woman" a New Hominid Species?" Discover Magazine. http://blogs.discovermagazine.com/80beats/2010/03/25/is-the-mysterious-siberian-x-woman-a-new-hominid-species/
228 Johannes Krause et al. (2010). "The complete mitochondrial DNA genome of an unknown hominin from southern Siberia," Nature 464: 894-897.
229 Carl Zimmer, (March 2010). "The X-Woman's Fingerbone," Discover Magazine, http://blogs.discovermagazine.com/loom/2010/03/24/the-x-womans-fingerbone/
230 Cherfas, Jeremy, and Gribbin, John; (1981). "The Molecular Making of Mankind," and "Descent of Man - Or Ascent of Ape?" New Scientist, 91:518 and 91:592.
231 Eliza Strickland, (October 2009). "A Fossil Named Ardi Shakes Up Humanity's Family Tree," Discover Magazine. http://blogs.discovermagazine.com/80beats/2009/10/01/a-fossil-named-ardi-shakes-up-humanitys-family-tree/
232 Mary-Claire King, (1973) Protein polymorphisms in chimpanzee and human evolution, Doctoral dissertation, University of California, Berkeley.
233 Humans and Chimps: Close But Not That Close. (December 2006). Scientific American.
234 Anzai, T. et al. (2003). Comparative sequencing of human and chimpanzee MHC class I regions unveils insertions/deletions as the major path to genomic divergence. Proceedings of the National Academy of Sciences. 100 (13): 7708-13.
235 Joan U. Pontius et al. (2007). Initial sequence and comparative analysis of the cat genome. Genome Research. 17: 1675-1689. doi: 10.1101/gr.6380007

236 Richard Dawkins. (2004). The Ancestor's Tale. Houghton Mifflin.
237 Filler, A.G. (2007) Homeotic Evolution in the Mammalia: Diversification of Therian Axial Seriation and the Morphogenetic Basis of Human Origins. PLoS ONE 2(10): e1019.
238 Filler, A.G. (2007) The Upright Ape. A New Origin of the Species. New Page Books. See more at www.uprightape.net
239 Kivell TL, Schmitt D. (Aug 2009). "Independent evolution of knuckle-walking in African apes shows that humans did not evolve from a knuckle-walking ancestor." Proc Natl Acad Sci U S A. 106 (34): 14241–6. doi:10.1073/pnas.0901280106.
240 Yvette Deloison, (2000). 'L'homme ne descend pas d'un primate arboricole! Une évidence méconnue', Biométrie Humaine et Anthropologie, v. 17, 1999, pp. 147-50; Yvette Deloison, 'New hypothesis on hominoid bipedalism', American Journal of Physical Anthropology, Supplement 30, p. 137.
241 Ned Potter, (May 2009). "The Link? Primate Fossil Could Be Key Link in Evolution," ABC News, http://abcnews.go.com/Technology/story?id=7603618#.T4EPD5mJeks
242 John Noble Wilford, (May 2009). "German Fossil Found to Be Early Primate," The New York Times, http://www.nytimes.com/2009/05/16/science/16fossil.html
243 News Staff, (Oct 2010) "Instead Of Originating In Africa, Human Ancestors Colonized There From Asia, Says Study," Science 2.0, www.science20.com/news_articles/instead_originating_africa_human_ancestors_colonized_there_asia_says_study
244 Valerie Ross. (Oct 2010) Unknowns, From Matter to DNA to Brain Cells. Discover. http://discovermagazine.com/2010/oct/15-numbers-unknowns-matter-dna-brain-cells
245 Charles Darwin. (1887). Autobiography. http://www.gutenberg.org/ebooks/2010
246 Charles Darwin, (1871). The Descent of Man, vol. II, p. 403.
247 Charles Darwin, The Descent of Man (1871), vol. I, p. 168.
248 Keim, Brandon. (Nov 2009). Early life hedged its bets to survive. Wired. http://www.wired.com/wiredscience/2009/11/bacteria-hedging/#more-13488
249 Johanson D.C. and Edgar B. (1996): From Lucy to language. Simon and Schuster.
250 Tattersal, I. and Schwartz, J. (2000). Extinct Humans. Basic Books.
251 http://www.time.com/time/magazine/article/0,9171,501050606-1066965,00.html
252 Bednarik, R. G. (1998). An experiment in Pleistocene seafaring. International Journal of Nautical Archaeology 27(2); Bednarik R.G. (1999). Determining the maximum capability of Palaeolithic technologies. Praehistoria Thuringica 3: 80-97; Bednarik R.G, Hobman B. & Rogers P. (1999). Nale Tasih 2: journey of a Middle Palaeolithic raft. International Journal of Nautical Archaeology 28(1): 25-33.
253 C. K. Brain and A. Sillen, (December 1998). "Evidence from the Swartkrans cave for the earliest use of fire," Nature, 336, p. 464-465.
254 Archibald, S, Staver, CA, Levin, SA. (December 2011). Evolution of human-driven fire regimes in Africa. PANS. p. 1091-6490
255 Rixiang Zhu, Zhisheng An, R. Pott, K. A. Hoffman (June 2003). "Magnetostratigraphic dating of early humans in China" Earth Science Reviews 61 (3-4): 191–361.
256 Jean-Pierre Hallet and Alex Pelle. (1973). Pygmy Kitabu. Random House. p. 57
257 Wrangham R, Conklin-Brittain N. (September 2003). "Cooking as a biological trait" (PDF). Comp Biochem Physiol a Mol Integr Physiol 136 (1): 35–46.
258 Oliver Renick, (August 22, 2011). "Homo Erectus Processed Food Like Humans, Harvard Scientists Say," Bloomberg, http://www.bloomberg.com/news/2011-08-22/homo-erectus-processed-food-like-humans-harvard-scientists-say.html
259 New York University, (June 29, 2011) "Human ancestor older than previously thought; Finding offers new insights into evolution," ScienceDaily, http://www.sciencedaily.com/releases/2011/06/110629181853.htm
260 Stephen, Stokes. "Chronology, Adaptation and Environment of the Middle Palaeolithic in Northern Africa." Human Evolution, Cambridge University.
261 Helen M. Leach. (2003). "Human Domestication Reconsidered" Current Anthropology, 44:3, 349-368
262 Peter J. Wilson. (1989) The Domestication of the Human Species. Yale University Press.
263 Weckler, J. E. (1954). "The Relationships Between Neanderthal Man and Homo sapiens."

American Anthropologist. 56: 1003-1025.
264 Gerstenblith MR, Goldstein AM, Fargnoli MC, Peris K, Landi MT. (2007). Comprehensive evaluation of allele frequency differences of MC1R variants across populations. Hum Mutation 28:495-505.doi:10.1002/humu.20476; Healy E, Jordon SA, Budd PS, Suffolk R, Rees JL, Jackson IJ. (2001). Functional variation of MC1R alleles from red-haired individuals; Krause J, Lalueza-Fox C. et al. (2007). The derived FoxP2 variant of modern humans was shared with Neandertals. Curr Biol 17:1-5. doi:10.1016
265 Lalueza-Fox C. et al. A melanocortin 1 receptor allele suggests varying pigmentation among Neanderthals. Science doi:10.1126/science.1147417
266 S.O.Y. Keita. (1993). "Studies and Comments on Ancient Egyptian Biological Relationships," History in Africa 20 (1993) 129-54) J Hum Evol. 1997 Jul; 33(1):33-82.
267 Robert Sanders, (June 11, 2003) "160,000-year-old fossilized skulls uncovered in Ethiopia are oldest anatomically modern humans," UC Berkeley. http://berkeley.edu/news/media/releases/2003/06/11_idaltu.shtml
268 Chris Stringer. (2003). Human evolution: Out of Ethiopia. Nature 423, 692–695.
269 Ibid.
270 Hershkovitz, I. et al. (2011), Middle pleistocene dental remains from Qesem Cave (Israel). Am. J. Phys. Anthropol., 144: 575–592. doi: 10.1002/ajpa.21446
271 Randolph E. Schid. (Apr 2008). Humans flirted with extinction 70,000 years ago. http://www.usatoday.com/tech/science/2008-04-24-human-extinction_N.htm
272 Richard Gray, (May 9, 2009) "African Tribe Populated Rest of the World," The Telegraph, http://www.telegraph.co.uk/science/science-news/5299351/African-tribe-populated-rest-of-the-world.html
273 Griffiths AJF, Miller JH, Suzuki DT, et al. (2000). An Introduction to Genetic Analysis. W. H. Freeman. http://www.ncbi.nlm.nih.gov/books/NBK22068/
274 Wen-Hsiung Li. (1997). Molecular Evolution, Sinauer Associates Publishers, p. 379-401.
275 Deyou Zheng and Mark B. Gerstein,(2007). "The Ambiguous Boundary between Genes and Pseudogenes: The Dead Rise Up, or Do They?" Trends in Genetics, 23: 219-24.
276 Valerie Ross. (Oct 2010) Unknowns, From Matter to DNA to Brain Cells. Discover. http://discovermagazine.com/2010/oct/15-numbers-unknowns-matter-dna-brain-cells
277 Richard Dawkins. (1996). The Blind Watchmaker, W.W. Norton & Co. p. 17-18.
278 T. Cavalier-Smith, (1978) "Nuclear Volume Control by Nucleoskeletal DNA, Selection for Cell Volume and Cell Growth Rate, and the Solution of the DNA C-Value Paradox," Journal of Cell Science 34, p. 247-278.
279 Margaret J. Beaton and Thomas Cavalier-Smith, (1999). "Eukaryotic Non-Coding DNA is Functional: Evidence from the Differential Scaling of Cryptomonad Genomes," Proceedings of the Royal Society of London B, 266 , p. 2053-2059.
280 Wolbach, Wendy S., et al; (1985). "Cretaceous Extinctions: Evidence for Wildfires and Search for Meteoric Material," Science, 230:167.
281 Day, Michael. (November 20, 1999) "Hell on Earth," New Scientist, p. 5.
282 Kerr, R. (1999). A Smoking Gun for an Ancient Methane Discharge, Science, 286:1465.
283 Monastersky, R.; "Global Burp Gassed Ancient Earth," Science News, 156:260, 1999.
284 Lewin, Roger; "Pattern and Process in Extinctions," Science, 241:26, 1988.
285 "Poisonous volcanic gas probably caused worst mass extinction says new study," Mongabay, December 02, 2005, http://news.mongabay.com/2005/1202-permian.html
286 "New ancient fungus finding suggests world's forests were wiped out in global catastrophe," (Oct 2009) Science Codex, http://www.sciencecodex.com/new_ancient_fungus_finding_suggests_worlds_forests_were_wiped_out_in_global_catastrophe
287 "Denisovans? Fossil Discovery Is Neither Modern Human Nor Neanderthal," (Dec 2010) Science 2.0. http://www.science20.com/news_articles/denisovans_fossil_discovery_neither_modern_human_nor_neanderthal
288 "Antarctica served as climatic refuge in Earth's greatest extinction event," (Dec 2009). Science Codex. http://www.sciencecodex.com/antarctica_served_as_climatic_refuge_in_earths_greatest_extinction_event
289 "What triggers mass extinctions? Study shows how invasive species stop new life," (Dec

2010). Science Codex, http://www.sciencecodex.com/what_triggers_mass_extinctions_study_shows_how_invasive_species_stop_new_life
290 Dan Vergano, (December 12, 2009). "Did Easter Island get 'ratted' out?" USA Today, http://www.usatoday.com/tech/science/discoveries/2005-12-05-easter-island_x.htm
291 Mark Teaford and Peter Ungar. (Dec 2000). Diet and the evolution of the earliest human ancestors. PNAS. http://www.pnas.org/content/97/25/13506.full
292 Rhett A. Butler, (Dec 2005). "Easter Island's demise caused by rats, Dutch traders says new theory," Mongabay, http://news.mongabay.com/2005/1206-easter_island.html
293 Giovas CM. (2006). No pig atoll: island biogeography and the extirpation of a Polynesian domesticate. Asian Perspectives 45:69-95.
294 Valerie Ross. (Oct 2010) Unknowns, From Matter to DNA to Brain Cells. Discover. http://discovermagazine.com/2010/oct/15-numbers-unknowns-matter-dna-brain-cells
295 William R. Corliss. (Jan-Feb 2001). Superorganisms: From Simplicity to Complexity. http://www.science-frontiers.com/Sf133/Sf133p08.htm
296 Van Wijk, R. (2001). "Bio-Photons and BioCommunication," Journal of Scientific Exploration, 15:183.
297 Downey, Greg. (May 2008). The Human Super-organism. Neuroanthropology.net http://neuroanthropology.net/2008/05/22/the-human-super-organism/
298 "The Great Fungus," (May 21, 1992). Nature.
299 Flannery, Tim. (Feb 29, 2009). The Superior Civilization. The New York Review of Books. http://www.nybooks.com/articles/archives/2009/feb/26/the-superior-civilization/
300 Joan E. Strassmann, Yong Zhu & David C. Queller. (Dec. 21 2000). Altruism and social cheating in the social amoeba Dictyostelium discoideum. Nature. http://www.nature.com/nature/journal/v408/n6815/full/408965a0.html
301 Franks, N. R. (1989) "Army Ants: A Collective Intelligence," American Scientist, 77:139.
302 Ibid.
303 Lee Alan Dugatkin. (2000). Cheating monkeys and citizen bees: the nature of cooperation in animals and humans. Harvard University Press.
304 Joan E. Strassmann, Yong Zhu & David C. Queller. (Dec. 21 2000). Altruism and social cheating in the social amoeba Dictyostelium discoideum. Nature. http://www.nature.com/nature/journal/v408/n6815/full/408965a0.html
305 Bjomerud, M.G. (1997). "Live Universes," Nature, 385:109.
306 White, Michael. (Feb 2008). How Single-Cell Organisms Evolve Into Multicellular Ones. Science 2.0. http://www.science20.com/adaptive_complexity/how_singlecell_organisms_evolve_multicellular_ones
307 White, Michael. (Feb 2008). How Single-Cell Organisms Evolve Into Multicellular Ones. Science 2.0. http://www.science20.com/adaptive_complexity/how_singlecell_organisms_evolve_multicellular_ones
308 Ibid.
309 Smith, John Maynard and Szathmary, Eors. (1995). The Major Transitions in Evolution. Oxford University Press
310 "The Great Fungus," (May 21, 1992). Nature.
311 Bullock, James. (May 30, 1992). "Huge Organisms," New Scientist, p. 54.
312 Russell, Sharman Apt. (April 1, 2002) Talking Plants. Discover Magazine. http://discovermagazine.com/2002/apr/featplants/
313 DeVries, P.J. (1990). "Enhancement of Symbioses between Butterfly Caterpillars and Ants by Vibrational Communication,"Science, 248:1104.
314 Moore, Peter D. (1983) "Animals As Nutrient Carriers," Nature, 305:763.
315 ASU. (June 2007). Competition, Loss of Selfishness Mark Shift to Supersociety. ScienceDaily. http://www.sciencedaily.com/releases/2007/06/070611094002.htm
316 Keim, Brandon. (July 12, 2007). Thoughts on Ants, Altruism and the Future of Humanity. Wired Science. http://www.wired.com/wiredscience/2007/07/thoughts-on-ant/
317 Planes, Trains and Ant Hills: Computer Scientists Simulate Activity of Ants to Reduce Airline Delays. (April 1, 2008). Science Daily. http://www.sciencedaily.com/videos/2008/0406-planes_trains_and_ant_hills.htm

318 Dickemann, Jeffrey M and Flannery, Tim. (May 28, 2009) Are We Becoming a Superorganism? NY Book Reviews. http://www.nybooks.com/articles/archives/2009/may/28/are-we-becoming-a-superorganism/
319 Ibid.
320 DeGeorges, Paul Andre, & Reilly, Brian Kevin (2008). A Critical Evaluation of Conservation and Development in Sub-Saharan Africa, Book I. Edwin Mellen Press.
321 Bert Hölldobler and Edward O. Wilson. (2009) The Superior Civilization: The Beauty, Elegance, and Strangeness of Insect Societies. Norton.
322 Nicholas Wade. (2006) Before the Dawn. Penguin Press.
323 University of Leeds (March 2008). Royal Corruption Is Rife In The Ant World. ScienceDaily. http://www.sciencedaily.com/releases/2008/03/080311103904.htm
324 Brandon Keim, (February 7, 2008) Honeybee Weapon in War on Cancer, Wired. http://www.wired.com/wiredscience/2008/02/honeybee-weapon/
325 Moore, Peter D.; "Animals As Nutrient Carriers," Nature, 305:763, 1983
326 Lovelock, J. E. (1986) Geophysiology. American Meteorological Society Bulletin 67:392.
327 Miller, Barbra; Bernard Wood, Andrew Balansky, Julio Mercader, Melissa Panger (2006). Anthropology. Allyn and Bacon. p. 768; McClellan. (2006). Science and Technology in World History: An Introduction. JHU Press. p. 6–12
328 Jackson J. Spielvogel (2003). Western Civilization (combined vols.). Wadsworth. p. 2–3
329 Jared Diamond. (May 1987). "The Worst Mistake in the History of the Human Race." Discover. p. 64-66. http://anthropology.lbcc.edu/handoutsdocs/mistake.pdf
330 Begg, Ean. (1997). The Cult of the Black Virgin. Penguin.
331 F. Berkes. (1999). Sacred Ecology: Traditional Ecological Knowledge. Routledge, xiii.
332 John Terborgh et al., (2001). "Ecological Meltdown in Predator-Free Forest Fragments," Science 294, 1923-26.
333 John Terborgh. (1988) The big things that run the world – a sequel to E.O. Wilson. Conservation Biology, 2, 402–403.
334 Mark F. Teaford, Peter S. Ungar, (2000). "Diet and the evolution of the earliest human ancestors," PNAS 97, no. 25: 13506-13511
335 Michael Reilly. (Jun 2009). Ancient Humans Knew Sustainable Fishing. Discovery News.
336 Kent Hubbard Redford, Robert E. Godshalk, Kiran Asher. (1995). What About the Wild animals? Wild Animal Species in Community Forestry in the Tropics. Food and Agriculture Organization of the United Nations.
337 The American museum of natural history. (1998). Biodiversity in the next millennium. http://www.amnh.org/museum/press/feature/biofact.html
338 Michael, J.N., (ed). (2001). The biodiversity crisis: an American museum of natural history book. The New Press.
339 Raven H. (2001). What have we lost, what are we losing? In: Michael, J.N., editor. The biodiversity crisis: An American museum of natural history book. The New Press. p.62.
340 William Mark Adams, Martin Mulligan. (2003). Decolonizing nature: Strategies for conservation in a post-colonial era. Earthscan.
341 A. C. Keefer. How Not to Kill Most Life on This Planet: an introduction to Radical Sustainability. http://zinelibrary.info/files/Radical_Sustainability.pdf
342 DeGeorges, Paul Andre, & Reilly, Brian Kevin (2008). A Critical Evaluation of Conservation and Development in Sub-Saharan Africa, Book I. Edwin Mellen Press. p. 237
343 Ibid. p. 240
344 Ibid. p. 247
345 Ibid. p. 350-351
346 K.H. Redford, R.E. Godshalk, K. Asher. (1995). What About the Wild animals? FAO-UN.
347 Jackson. J.B.C. et al. (2001) Historical overfishing and the recent collapse of coastal ecosystems. Science 293, 629-638.
348 "The carbon footprint of early Native Americans much greater than previously thought," (Apr 2010) Science Codex, www.sciencecodex.com/stalagmite_reveals_carbon_footprint_of_early_native_americans
349 A. C. Keefer. How Not to Kill Most Life on This Planet: an introduction to Radical

Sustainability. http://zinelibrary.info/files/Radical_Sustainability.pdf
350 "Innovations in the Design of Long-Span Building Structures." (2003) Architectural Science Review. Volume 46, Issue 2. p. 103-104 DOI: 10.1080/00038628.2003.9696971.
351 Plait, Phillip. (2002). Bad Astronomy: Misconceptions and Misuses Revealed, from Astrology to the Moon Landing "Hoax." Wiley Books.
352 Ridpath, Ian. Astronomical Causes of UFOs. www.ianridpath.com/ufo/astroufo1.htm
353 Jung, Carl. (July 1997) Flying Saucers. Fine Communications.
354 Colavito, Jason (2005). The Cult of Alien Gods: H. P. Lovecraft and Extraterrestrial Pop Culture. Prometheus Books; Grünschloß, A. (Jun 2006). "Ancient Astronaut" Narrations: A Popular Discourse on Our Religious Past." Marburg Journal of Religion 11(1).
355 Erhard Rostlund. (Dec 1957). The Myth of a Natural Prairie Belt in Alabama. Annals of the Association of American Geographers 47(4), p. 392–411.
356 Brooks, Nick (2006). "Cultural responses to aridity in the Middle Holocene and increased social complexity." Quaternary International 151(1): 29–49.
357 Erle Ellis, Robert Pontius, "Land-use and land-cover change," Encyclopedia of Earth, Eds. Cutler J. Cleveland. Environmental Information Coalition, National Council for Science and the Environment. www.eoearth.org/article/Land-use_and_land-cover_change
358 Rhett A. Butler, (Dec 2005) "45% chance Gulf Stream current will collapse by 2100 finds research," Mongabay, http://news.mongabay.com/2005/1207-uiuc.html
359 H. E. Wulff. (April 1968). The Qanats of Iran, Scientific American. p.94 – 105, http://users.bart.nl/~leenders/txt/qanats.html
360 As reported by Universal Shamgaudd Allah, as cited in Knowledge of Self (2009). originally cited in The Sun of Man, a 5% Periodical.
361 See Lipski's A History of Afro-Hispanic Language: Five Centuries, Five Continents; Rickford's Spoken Soul: The Story of Black English; Williams's Ebonics: The True Language of Black Folk; Mar-Molinero's Politics of Language in the Spanish-Speaking World; and Weatherford's Native Roots: How the Indians Enriched America.
362 de Boinod, Adam Jacot. (2006). The Meaning of Tingo: And Other Extraordinary Words from Around the World. Penguin Press.
363 "Hidden Language Recorded." National Geographic. video.nationalgeographic.com/video/player/news/culture-places-news/enduring-voices-koro-vin.html
364 Adapted from the Cuesta College Academic Support Guide to Critical Thinking.
http://academic.cuesta.edu/acasupp/as/400Index.htm
365 Brandon Keim. (February 2008) "Complexity Theory Takes Evolution to Another Level," Wired. http://www.wired.com/wiredscience/2008/02/complexity-theo/
366 Ibid.
367 Ibid.
[368] Sheldrake, Rupert. (1988). The Presence of the Past. Times Books.
369 Brandon Keim. (February 2008) "Complexity Theory Takes Evolution to Another Level," Wired. http://www.wired.com/wiredscience/2008/02/complexity-theo/
370 News Staff. (Nov 2010). Gradual Evolution Not Supported By Geological History, Says Geologist. Science 2.0 http://www.science20.com/news_articles/gradual_evolution_not_supported_geological_history_says_geologist
371 Gillis, A.M. (1991). Can Organisms Direct Their Own Evolution? BioScience, 41:202.
372 News Staff. (Dec 2010). Reconstructing The Evolution Of Ancient Microbes? Science 2.0 http://www.science20.com/news_articles/reconstructing_evolution_ancient_microbes
373 Paul Davies. (1999). The Fifth Miracle. Simon & Schuster, p. 259
374 David L. Abel & Jack T. Trevors, "Self-organization vs. self-ordering events in life-origin models," Physics of Life Reviews, Vol. 3:211–228 (2006).
375 Harlow Shapley, (1967). Beyond the Observatory. Charles Scribner's Sons, p. 111-2
376 Anthropometric Comparison between Japanese and Caucasia American Male University Students. (1999) Applied Human Science-Journal of Phys. Anthropology 18(1): 9–11.
377 Kevin T. Davis. The Foot-Length to Stature Ratio: A Racial of Racial Variance. Unpublished Masters Thesis. Available online http://thinktech.lib.ttu.edu/ttu-ir/bitstream/handle/2346/8468/31295005963201.pdf

CPSIA information can be obtained
at www.ICGtesting.com
Printed in the USA
FSHW021817121119
64030FS